OPTIMIZATION
Algorithms and Applications

Rajesh Kumar Arora

Senior Engineer
Vikram Sarabhai Space Centre
Indian Space Research Organization
Trivandrum, India

CRC Press
Taylor & Francis Group
Boca Raton London New York

CRC Press is an imprint of the
Taylor & Francis Group, an **informa** business

A CHAPMAN & HALL BOOK

OPTIMIZATION
Algorithms and
Applications

MATLAB® is a trademark of The MathWorks, Inc. and is used with permission. The MathWorks does not warrant the accuracy of the text or exercises in this book. This book's use or discussion of MATLAB® software or related products does not constitute endorsement or sponsorship by The MathWorks of a particular pedagogical approach or particular use of the MATLAB® software.

CRC Press
Taylor & Francis Group
6000 Broken Sound Parkway NW, Suite 300
Boca Raton, FL 33487-2742

© 2015 by Taylor & Francis Group, LLC
CRC Press is an imprint of Taylor & Francis Group, an Informa business

No claim to original U.S. Government works

Printed on acid-free paper
Version Date: 20150205

International Standard Book Number-13: 978-1-4987-2112-7 (Hardback)

This book contains information obtained from authentic and highly regarded sources. Reasonable efforts have been made to publish reliable data and information, but the author and publisher cannot assume responsibility for the validity of all materials or the consequences of their use. The authors and publishers have attempted to trace the copyright holders of all material reproduced in this publication and apologize to copyright holders if permission to publish in this form has not been obtained. If any copyright material has not been acknowledged please write and let us know so we may rectify in any future reprint.

Except as permitted under U.S. Copyright Law, no part of this book may be reprinted, reproduced, transmitted, or utilized in any form by any electronic, mechanical, or other means, now known or hereafter invented, including photocopying, microfilming, and recording, or in any information storage or retrieval system, without written permission from the publishers.

For permission to photocopy or use material electronically from this work, please access www.copyright.com (http://www.copyright.com/) or contact the Copyright Clearance Center, Inc. (CCC), 222 Rosewood Drive, Danvers, MA 01923, 978-750-8400. CCC is a not-for-profit organization that provides licenses and registration for a variety of users. For organizations that have been granted a photocopy license by the CCC, a separate system of payment has been arranged.

Trademark Notice: Product or corporate names may be trademarks or registered trademarks, and are used only for identification and explanation without intent to infringe.

Library of Congress Cataloging-in-Publication Data

Arora, Rajesh Kumar (Engineer)
 Optimization : algorithms and applications / Rajesh Kumar Arora.
  pages cm
 "A CRC title."
 Includes bibliographical references and index.
 ISBN 978-1-4987-2112-7 (hardcover : alk. paper) 1. Mathematical optimization. 2. MATLAB. I. Title.

QA402.5.A7625 2015
519.6--dc23   2015004525

Visit the Taylor & Francis Web site at
http://www.taylorandfrancis.com

and the CRC Press Web site at
http://www.crcpress.com

Dedicated to my mother

Contents

Preface xi
Author xv

1. Introduction 1
    1.1   Historical Review 1
    1.2   Optimization Problem 3
    1.3   Modeling of the Optimization Problem 5
    1.4   Solution with the Graphical Method 11
    1.5   Convexity 13
    1.6   Gradient Vector, Directional Derivative, and Hessian Matrix 16
    1.7   Linear and Quadratic Approximations 23
    1.8   Organization of the Book 25
    Chapter Highlights 27
    Formulae Chart 28
    Problems 29

2. 1-D Optimization Algorithms 35
    2.1   Introduction 35
    2.2   Test Problem 37
    2.3   Solution Techniques 38
            2.3.1   Bisection Method 38
            2.3.2   Newton–Raphson Method 40
            2.3.3   Secant Method 42
            2.3.4   Cubic Polynomial Fit 44
            2.3.5   Golden Section Method 46
            2.3.6   Other Methods 47
    2.4   Comparison of Solution Methods 49
    Chapter Highlights 51
    Formulae Chart 52
    Problems 52

3. Unconstrained Optimization 55
    3.1   Introduction 55
    3.2   Unidirectional Search 57
    3.3   Test Problem 59
    3.4   Solution Techniques 60
            3.4.1   Steepest Descent Method 62
            3.4.2   Newton's Method 63
            3.4.3   Modified Newton's Method 66
            3.4.4   Levenberg–Marquardt Method 66

		3.4.5	Fletcher–Reeves Conjugate Gradient Method	68
		3.4.6	DFP Method	70
		3.4.7	BFGS Method	72
		3.4.8	Powell Method	74
		3.4.9	Nelder–Mead Algorithm	75
	3.5	Additional Test Functions		78
		3.5.1	Rosenbrock Function	78
		3.5.2	Quadratic Function	79
		3.5.3	Nonlinear Function	81
		3.5.4	Wood's Function	82
	3.6	Application to Robotics		83
	Chapter Highlights			85
	Formulae Chart			86
	Problems			87

4. Linear Programming … 93

4.1	Introduction	93
4.2	Solution with the Graphical Method	95
4.3	Standard Form of an LPP	98
4.4	Basic Solution	103
4.5	Simplex Method	105
	4.5.1 Multiple Solutions	112
	4.5.2 Degeneracy	114
	4.5.3 Two-Phase Method	116
	4.5.4 Dual Simplex Method	121
4.6	Interior-Point Method	125
4.7	Portfolio Optimization	127
Chapter Highlights		131
Formulae Chart		133
Problems		133

5. Guided Random Search Methods … 139

5.1	Introduction	139
5.2	Genetic Algorithms	140
	5.2.1 Initialize Population	142
	5.2.2 Fitness Evaluation	143
	5.2.3 Reproduction	143
	5.2.4 Crossover and Mutation	147
	5.2.5 Multimodal Test Functions	148
5.3	Simulated Annealing	154
5.4	Particle Swarm Optimization	157
5.5	Other Methods	160
	5.5.1 Ant Colony Optimization	160
	5.5.2 Tabu Search	163
Chapter Highlights		164

Contents ix

    Formulae Chart 165
    Problems... 166

6. Constrained Optimization 169
    6.1   Introduction 169
    6.2   Optimality Conditions 171
    6.3   Solution Techniques.. 175
            6.3.1   Penalty Function Method 176
    6.4   Augmented Lagrange Multiplier Method 182
    6.5   Sequential Quadratic Programming................................... 184
    6.6   Method of Feasible Directions 190
            6.6.1   Zoutendijk's Method 191
            6.6.2   Rosen's Gradient Projection Method.................... 192
    6.7   Application to Structural Design... 195
    Chapter Highlights.. 196
    Formulae Chart 197
    Problems... 199

7. Multiobjective Optimization 203
    7.1   Introduction 203
    7.2   Weighted Sum Approach 205
    7.3   ε-Constraints Method 210
    7.4   Goal Programming.. 212
    7.5   Utility Function Method 214
    7.6   Application 215
    Chapter Highlights.. 220
    Formulae Chart 220
    Problems... 221

8. Geometric Programming 223
    8.1   Introduction 223
    8.2   Unconstrained Problem 224
    8.3   Dual Problem... 229
    8.4   Constrained Optimization 231
    8.5   Application 235
    Chapter Highlights.. 238
    Formulae Chart 238
    Problems... 240

9. Multidisciplinary Design Optimization 243
    9.1   Introduction 243
    9.2   MDO Architecture... 245
            9.2.1   Multidisciplinary Design Feasible 247
            9.2.2   Individual Discipline Feasible 248
            9.2.3   Simultaneous Analysis and Design 249

		9.2.4	Collaborative Optimization...251
		9.2.5	Concurrent Subspace Optimization..................................252
		9.2.6	Bilevel Integrated System Synthesis................................252
	9.3	MDO Framework 253
	9.4	Response Surface Methodology...254
	Chapter Highlights..257
	Formulae Chart 258
	Problems..259

10. Integer Programming 263
    10.1 Introduction..263
    10.2 Integer Linear Programming 264
           10.2.1 Gomory's Cutting Plane Method......................................265
           10.2.2 Zero-One Problems 272
    10.3 Integer Nonlinear Programming..277
           10.3.1 Branch-and-Bound Method..278
           10.3.2 Evolutionary Method 284
    Chapter Highlights..286
    Formulae Chart 286
    Problems..287

11. Dynamic Programming..289
    11.1 Introduction..289
    11.2 Deterministic Dynamic Programming..289
    11.3 Probabilistic Dynamic Programming..294
    Chapter Highlights..296
    Formula Chart 297
    Problems..297

Bibliography..299

Appendix A: Introduction to MATLAB® 309

Appendix B: MATLAB® Code 321

Appendix C: Solutions to Chapter Problems 401

Index 437

Preface

There are numerous books on the subject of optimization, attributable to a number of reasons. First, the subject itself is mathematically rigorous and there are a number of solution methods that need to be examined and understood. No single solution method can be applied to all types of optimization problems. Thus a clear understanding of the problem, as well as solution techniques, is required to obtain a proper and meaningful solution to the optimization problem. With the progression of time, optimization problems have also become complex. It is necessary not only to obtain the global optimum solution, but to find local optima as well. Today's problems are also of the multiobjective type, where conflicting objective functions are to be handled. There is also a need to simultaneously handle objective functions and constraints of different disciplines, resulting in multidisciplinary design optimization (MDO) problems that are handled using different architectures. Gradient-based methods were popular until the 1990s. At present, a large number of complex problems are solved using guided random search methods such as genetic algorithm, simulated annealing, and particle swarm optimization (PSO) techniques. Even hybrid algorithms, that use a combination of gradient-based and stochastic methods, are also very popular. Different authors have addressed these issues separately, resulting in a number of books in this area.

So how does this book differ from the others? The solution techniques are detailed in such a way that more emphasis is given to the concepts and rigorous mathematical details and proofs are avoided. It is observed that a method can be understood better if different parameters in the algorithm are plotted or printed over different iterations while solving a problem. This can be accomplished by writing a software code for the method or the algorithm. It is often difficult for a newcomer to write a software code if the algorithm such as, say, Broyden–Fletcher–Goldfarb–Shanno (BFGS) or PSO is given to him or her. In this book, a step-by-step approach is followed in developing the software code from the algorithm. The codes are then applied to solve some standard functions taken from the literature. This creates understanding and confidence in handling different solution methods. The software codes are then suitably modified to solve some real-world problems. A few books on optimization have also followed this approach. However, the software code in these books is hard to correlate with the corresponding algorithms mentioned in the book and readers are forced to use them as black box optimization tools. The codes presented in this book are user friendly in the sense that they can be easily understood. A number of practical problems are solved using these codes.

The codes are written in the MATLAB® environment and the use of ready-made optimization routines available in MATLAB is avoided. The algorithms are developed right from computing the gradient or Hessian of a function to a complex algorithm such as for solving a constraint optimization problem. MATLAB is a software package for technical computing that performs both computing and visualization with ease. It has a number of built-in functions that can be used by an individual's application. The main advantage of MATLAB is the ease with which readers can translate their ideas into an application.

The book covers both gradient and stochastic methods as solution techniques for unconstrained and constrained optimization problems. A separate chapter (Chapter 5) is devoted to stochastic methods, where genetic algorithm, PSO, simulated annealing, ant colony optimization, and tabu search methods are discussed. With simple modifications of the basic PSO code, one can also solve nonconvex multiobjective optimization problems. This is probably the first optimization book in which MDO architectures are introduced (Chapter 9). Software codes are also developed for the simplex method and affine-scaling interior point method for solving linear programming problems. Gomory's cutting plane method, branch-and-bound method, and Balas' algorithm are also discussed in the chapter on integer programming (Chapter 10). A number of real-world problems are solved using the MATLAB codes given in the book. Some applications that are included in this book are solving a complex trajectory design problem of a robot (Chapter 3), multiobjective shape optimization problem of a reentry body (Chapter 7), portfolio optimization problem (Chapter 4), and so forth.

I thank my organization, Vikram Sarabhai Space Centre (a lead center of Indian Space Research Organisation [ISRO]), for giving permission to publish this book. The book has been reviewed internally by Dr. Mohankumar D., Head, Computer Division. I thank him for his suggestions and corrections. I thank Mr. Pandian, S., Deputy Director, VSSC for supporting the idea to write this book. I am ever grateful to Prof. M Seetharama Bhat from IISc, Bangalore and Dr. Adimurthy, V. for their support during the last ten years. I thank my colleagues Dr. Jayakumar K., Mr. Priyankar, B., Mr. Sajan Daniel and Mr. Amit Sachdeva for many hours of discussions on book-related aspects.

I am grateful to Taylor & Francis Group for agreeing to publish this book and agreeing to most of my suggestions. Much credit should be given to Ms. Aastha Sharma, Editor, for her prompt actions and follow-up with the reviewers. Thanks are also due to three anonymous reviewers for their critical remarks, corrections, and suggestions. I thank Mr. Sarfraz Khan, assistant to Ms. Aastha Sharma, for providing online support in signing the contract. I also thank Mr. David Fausel for coordinating and reviewing the style and art files of the book. My sincere thanks to Mr. Ed Curtis and Ms. Amor Nanas for language corrections, copyediting, and other production related works. The cover page is designed by Mr. Kevin Craig.

I thank the MATLAB book program for supporting the idea of this book on optimization with MATLAB codes. They have also agreed to give wide publicity to this book on their website, for which I am grateful.

I thank my wife, Manju, and children, Abhinav and Aditi, for their patience during the last two years. In fact my whole family—father, brothers, sister, and in-laws—are eagerly waiting for the launch of this book.

This is the first edition of this book. Some errors and omissions are expected. The MATLAB codes are validated with a number of test problems taken from the literature. It is still possible that some pathways in the codes would not have been exercised during this validation. As a result, no claim is made that these codes are bug-free. I request readers to report corrections and suggestions on this book at rk_arora@vssc.gov.in or arora_rajesh@rediffmail.com.

The MATLAB codes mentioned in this book can be downloaded from the weblink http://www.crcpress.com/product/isbn/9781498721127.

Rajesh Kumar Arora, PhD, MBA, FIE

MATLAB® is a registered trademark of The MathWorks, Inc. For product information, please contact:

The MathWorks, Inc.
3 Apple Hill Drive
Natick, MA 01760-2098 USA
Tel: 508 647 7000
Fax: 508-647-7001
E-mail: info@mathworks.com
Web: www.mathworks.com

Author

Rajesh Kumar Arora is a senior engineer with the Indian Space Research Organization (ISRO), where he has been working for more than two decades. He earned his PhD in aerospace engineering from the Indian Institute of Science, Bangalore. He has published a book titled *Fundamentals of Aerospace Engineering*. His area of research includes mission design, simulation of launch vehicle systems, and trajectory optimization. He also has a master's degree in business administration and is a Fellow of Institution of Engineers (India).

1
Introduction

1.1 Historical Review

Optimization means finding the best solution among many feasible solutions that are available to us. Feasible solutions are those that satisfy all the constraints in the optimization problem. The best solution could be minimizing the cost of a process or maximizing the efficiency of a system. Some simple optimization problems that come to mind are machine allocation and diet problems. In the machine allocation problem, one has to find how jobs are to be allocated to different machines of different capacities and with different operating costs so as to meet the production target with minimum cost. In the diet problem, different food types are available with different nutritional contents at different costs. The aim is to estimate different quantities of food so that nutritional requirements are met for an individual at minimum cost.

Though rigorous mathematical analysis of the optimization problems was carried out during the 20th century, the roots can be traced back to about 300 B.C., when the Greek mathematician Euclid evaluated the minimum distance between a point and a line. Another Greek mathematician, Zenedorous, showed in 200 B.C. that a figure bounded by a line that has a maximum area for a given perimeter is a semicircle.

In the 17th century, Pierre de Fermat, a French mathematician, laid the foundation of calculus. He showed that the gradient of a function vanishes at the maximum or minimum point. Moving further in the timeline, Newton and Leibniz laid mathematical details for the calculus of variations. This method deals with maxima or minima of functionals. The foundation for the calculus of variations is credited to Euler and Lagrange (in the 18th century), as they provided rigorous mathematical details on this topic. Subsequently, Gauss and Legendre developed the least squares method, which is extensively used even today. Cauchy used the steepest descent method to solve unconstrained optimization problems.

The first textbook on optimization was authored by Harris Hancock and published in 1917. In 1939, Leonid Kantorovich presented the linear programming (LP) model and an algorithm for solving it. A few years later in 1947, George Dantzig presented a simplex method for solving LP problems. Kantorovich and Dantzig are regarded as pioneers who provided breakthroughs in the development of optimization techniques. The conditions for constrained optimization were brought together by Harold Kuhn and Albert Tucker in 1951 and also earlier by William Karush in 1939. Richard Bellman laid the principles of dynamic programming problems in which a complex problem is broken down into smaller subproblems. Ralph Gomory's contribution to the development of integer programming is worth noting, as in this type of optimization problem, design variables can take integer values such as 0 and 1.

With the advent of computers in the 1980s, subsequently many large-scale problems were solved. Present-day problems in the optimization area are of the multidisciplinary and multiobjective type. The solution techniques that are employed today to solve complex optimization problems are not just gradient-based algorithms, but also include nontraditional methods such as genetic algorithms, ant colony optimization, and particle swarm optimization that mimic natural processes.

Today, optimization methods are required to solve problems from all disciplines, whether economics, sciences, or engineering. As a result of stiff competition in virtually all disciplines, the role of optimization has become still more substantial as one aims to minimize the cost of a product or wants to allocate resources judiciously. A simple example from the subject field of aerospace engineering can prove this point. The cost of putting 1 kilogram of payload in a low Earth orbit is typically about US$15,000. The fuel and structural weight of the different stages of the rocket strongly influence the payload mass, as does the trajectory of the rocket. Of course, one can reduce the structural weight of a stage only to the extent it should not fail because of aerodynamic and other loads. The optimization problem that aims at maximizing the payload mass is highly complex and requires algorithms that run on high-speed computers. Even if the optimization technique results in few extra kilograms in payload, it represents large revenue for the space agency.

In the next section, we introduce to the optimization problem design variables, constraints, and applications of optimization in different domains. Further in the chapter, modeling aspects of a physical problem are explained that convert the verbal problem to a mathematical form. The solution of simple optimization problems with up to two design variables is explained by the graphical method. The importance of convex function in optimization is then explained. The chapter concludes with an introduction to the mathematical preliminaries of the gradient vector, Hessian matrix, directional directive, and linear and quadratic approximation of the function. The road map of this chapter is given in Figure 1.1.

Introduction 3

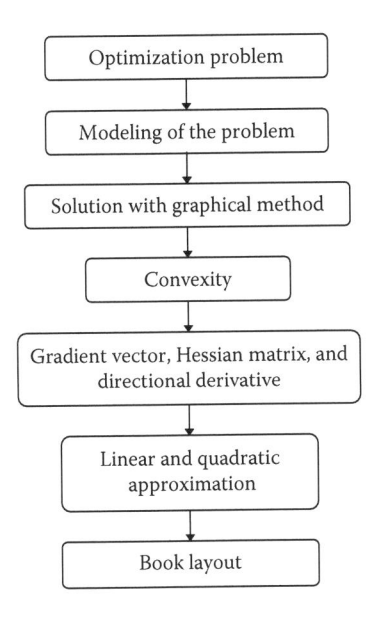

FIGURE 1.1
Road map of Chapter 1.

1.2 Optimization Problem

In an optimization problem, a function is to be maximized or minimized. The function that is being optimized is referred to as the objective function or the performance index. The function is a quantity such as cost, profit, efficiency, size, shape, weight, output, and so on. It goes without saying that cost minimization or profit maximization are prime considerations for most organizations. Certain types of equipment, such as air conditioners or refrigerators, are designed with different optimization criteria to have higher efficiency in terms of reducing energy consumption requirements of the user. However, this higher efficiency evidently comes at a higher cost to the user. Weight minimization is a prime consideration for aerospace applications.

The variables in the objective function are denoted the design variables or decision variables. Typically it could be the dimensions of a structure or its material attributes, for a structure optimization problem. From practical considerations, design variables can take values within a lower and an upper limit only. For instance, the maximum capacity of a machine is limited to a certain value. The design variables can be a real or a discrete number, binary, or integer type. Though a majority of the design variables in the optimization problems are real, some variables can also be discrete. For example, pipe sizes come in standard numbers such as 1, 2, or 5 inches. If pipe size is used as a design variable in an optimization problem, it has to be treated as discrete only. There is no point in selecting it as a real number and getting a solution such as, say, 3.25 inches, a pipe dimension that really does not exist. See Table 1.1 for some typical objective functions and design variables for optimization problems from different disciplines.

The optimization problem can be mathematically expressed as follows.

Minimize

$$f(x) \qquad (1.1)$$

subject to

$$g_i(x) \leq 0 \quad i = 1, 2, \ldots, m < n \qquad (1.2)$$

$$h_j(x) = 0 \quad j = 1, 2, \ldots, r < n \qquad (1.3)$$

$$x_l \leq x \leq x_u$$

where x is a vector of n design variables given by

$$x = \begin{bmatrix} x_1 \\ x_2 \\ \vdots \\ x_n \end{bmatrix}$$

The functions f, g_i, and h_j are all differentiable. The design variables are bounded by x_l and x_u. The constraints g_i are called inequality constraints and

TABLE 1.1
Typical Optimization Problems

Discipline	Design Variables	Objective Function
Manufacturing	Productivity from different machines	Minimize cost
Corporate	Different capitals from projects	Maximize the net present value
Airline	Different aircrafts, different routes	Maximize the profit
Aerospace	Propellant fraction in different stages	Maximize the payload
Agriculture	Different crops	Maximize the yield
Biology	Gene interaction	Network stability
Electronics	Size of the devices	Minimize the power consumption
Portfolio	Investment in stocks/bonds	Maximize the return
Thermal	Dimensions and material properties	Minimize the heat load

Introduction

h_j are called equality constraints. An example in the aerospace industry is to restrict the dimensions of the spacecraft so that in can be accommodated inside the payload fairing of a rocket. These restrictions are the constraints of the optimization problem. The constraints are functions of the design variables. In addition, the number of equality or inequality constraints is lower than the number of design variables (n). If the design variables satisfy all the constraints, they represent a feasible set and any element from the set is called a feasible point. The design variables at which the minimum of $f(x)$ is reached are given by x^*. If the optimization problem does not have any constraints, it is referred to as an unconstrained optimization problem. If the objective function and constraints are linear functions in x then the optimization problem is termed a linear programming problem (LPP).

1.3 Modeling of the Optimization Problem

Modeling refers to expressing observations about a problem in mathematical form using basic building blocks of mathematics such as addition, subtraction, multiplication, division, functions, and numbers with proper units. Observations refer to data obtained for the problem in hand, by varying certain parameters of the problem through experiments. Further, mathematical models provide predictions of the behavior of the problem for different inputs. If the model does not yield expected results, it has to be refined by conducting further experiments. The mathematical model is not unique for different problems, as observed data can be discrete (defined at select intervals) or continuous and can vary in different fashion (say, linear or quadratic) with change in input parameters. Some simple mathematical models of different physical phenomena are presented next.

The pressure (P), volume (V), and temperature (T) relationship of a gas is given by Boyle's law as

$$PV = kT \tag{1.4}$$

where k is a constant. Using this mathematical model, the behavior of a gas can be predicted (say, pressure) for the different input parameters (say, temperature), keeping the volume of the gas constant.

An example from economics could be constructing a mathematical model for the demand–supply problem. The price of a product is to be calculated so as to maximize the profit. It is well known that if the price of the merchandise is kept high, profit per unit will increase but then demand for the product may be low. Likewise, if the price of the product is kept low, profit per unit will decrease, but then demand for the product may be higher. Typically, demand (D) varies with price (P) as

$$D = \frac{c_1}{c_2 + P^2} \tag{1.5}$$

where c_1 and c_2 are constants.

Some problems can be written mathematically in *differential equation* form. A differential equation contains an unknown function and its derivatives. As the derivative represents the rate of change of a function, the differential equation represents the continuously varying quantity and its rate of change. For example, the temperature change (with respect to time) of an object is proportional to the difference between the temperature (T) of the object and that of its surroundings (T_s) and can be represented in differential equation form as

$$\frac{dT}{dt} = -k(T - T_s) \tag{1.6}$$

Equation 1.6 is also referred to as *Newton's law of cooling*. The solution of the differential equation is a function that satisfies the differential equation for all values of the *independent variable* in the domain. As the name suggests, independent variables are changed during an experiment and the *dependent variable* responds to this depending on on the type of the experiment being conducted. A differential equation can have many solutions (referred to as general solution). A particular solution is one such solution. Often, a differential equation has a closed form solution. For example, the solution for the differential equation representing Newton's law of cooling is

$$T(t) = T_s + (T_o - T_s)e^{-kt} \tag{1.7}$$

Not all problems have closed form solutions and such problems have to be numerically simulated to arrive at the solutions.

Therefore, using modeling, one can construct the objective function as well as the constraint functions for the optimization problem. One can then use different optimization techniques for solving such problems. The following examples illustrate how to formulate an optimization problem by constructing the objective and constraint functions.

Example 1.1

In a diet problem, an individual has to meet his daily nutritional requirements from a menu of available foods at a minimum cost. The available food items are milk, juice, fish, fries, and chicken. The nutrient requirements to be met are for proteins, vitamins, calcium, calories, and carbohydrates. Table 1.2 shows the cost in dollars of the food items per serving, nutrient values are shown in rows against their names (such as

Introduction

TABLE 1.2

Data for the Diet Problem

	Milk	Juice	Fish	Fries	Chicken	Required
Cost	1.1	1.2	2.0	1.3	3.0	
Proteins	8	2	15	4	30	60
Vitamins	9	3	3	1	9	100
Calcium	35	3	17	1	16	120
Calories	100	90	350	200	410	2100
Carbohydrates	10	20	40	25	40	400

Note: Construct the objective function and the constraints for this optimization problem.

proteins, vitamins, etc.) per serving, and the last column indicates the minimum daily requirements of the nutrients.

The first step is to select the design variables for the problem. It appears obvious to select quantities of food items such as fish, fries, and so on as the design variables. Let us represent the design variables by $x_1, x_2, x_3, x_4,$ and x_5 for quantities in milk, juice, fish, fries, and chicken respectively. As discussed earlier, the objective function and constraints are a function of these design variables. In this particular problem, the objective function is to minimize the cost of the food items purchased. If x_3 is the quantity of fish ordered and $2 is its unit price, then the total cost of the fish item is $2x_3$. In a similar way, we can evaluate the cost of other items such as milk, juice, and so on. Hence the total cost of the food items is

$$1.1x_1 + 1.2x_2 + 2x_3 + 1.3x_4 + 3x_5$$

Note that the cost function or the objective function is *linear*; that is, x_1 is not dependent on x_2 or any other variable. Having defined the objective function, let us define the constraints for the problem. In the problem it is clearly mentioned that the nutritional needs of the individual have to be met. For example, a minimum protein requirement of 60 units is to be met. Similarly, minimum requirements of other nutrients such as vitamins, calcium, and so forth are also to be met. Now, we can write the first constraint as

$$8x_1 + 2x_2 + 15x_3 + 4x_4 + 30x_5 \geq 60 \qquad (1.8)$$

Note that this constraint is an inequality. In a similar fashion, we can write other constraints. We are now ready to write the objective function and constraints for the diet problem.

Minimize

$$1.1x_1 + 1.2x_2 + 2x_3 + 1.3x_4 + 3x_5 \qquad (1.9)$$

subject to

$$8x_1 + 2x_2 + 15x_3 + 4x_4 + 30x_5 \geq 60 \tag{1.10}$$

$$9x_1 + 3x_2 + 3x_3 + x_4 + 9x_5 \geq 100 \tag{1.11}$$

$$35x_1 + 3x_2 + 17x_3 + x_4 + 16x_5 \geq 120 \tag{1.12}$$

$$100x_1 + 90x_2 + 350x_3 + 200x_4 + 410x_5 \geq 2100 \tag{1.13}$$

$$10x_1 + 20x_2 + 40x_3 + 25x_4 + 40x_5 \geq 400 \tag{1.14}$$

Once the optimization problem is defined, one has to use standard optimization techniques in evaluating the design variables x_1, x_2, x_3, x_4, and x_5. These methods are described in the later chapters. In this chapter, we are focusing on the formulation of the optimization problem.

Example 1.2

A soft drink manufacturer needs to produce a cylindrical can that can hold 330 mL of a soft drink. He wants to make the dimensions of the container such that the amount of material used in its construction is minimized. Formulate the optimization problem by writing down the objective function and the constraint.

The design variables for the optimization problem are the radius and the height of the can. Let these variables be denoted by x_1 and x_2 with units in millimeters (Figure 1.2). The cylindrical can consists of a curved portion and two circular ends. The area of the curved portion is given by $2\pi x_1 x_2$ and the area of two circular lids is given by $2\pi x_1^2$. Hence, the total area that needs to be minimized is $2\pi x_1 x_2 + 2\pi x_1^2$. The volume of the can is given by $\pi x_1^2 x_2$. This volume is to be limited to 330 mL or 330,000 mm³. Now we are ready to formulate the optimization problem.

Minimize

$$2\pi x_1 x_2 + 2\pi x_1^2 \tag{1.15}$$

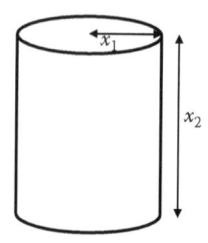

FIGURE 1.2
Cylindrical can.

Introduction

subject to

$$\pi x_1^2 x_2 = 330{,}000 \qquad (1.16)$$

Note that in this optimization problem, the constraint is an equality.

Example 1.3

The shape of a reentry body is a spherical nose, a conical body, and a flared bottom (see Figure 1.3). The design variables through which the configuration of the reentry body can be altered are nose radius (R_n), cone length (l_1), cone angle (θ_1), flare length (l_2), and flare angle (θ_2). By varying the design variables, the area (A) of the reentry capsule is to be minimized. As the reentry capsule has to house electronic packages and other instruments, it must have a certain minimum volume (V), which is specified as 1 m³.

The design variables are bounded between a minimum and maximum value. R_n can take a value between 0.4 and 0.6 m, l_1 and l_2 can take a value between 0.4 and 0.8 m, θ_1 can take a value between 22 and 27 degrees, and θ_2 can take a value between ($\theta_1 + 5$) and ($\theta_1 + 10$) degrees. Formulate the shape optimization problem of the reentry capsule.

The total surface area and volume of the capsule are computed using the equations

$$A = 2\pi R_n^2 (1 - \sin\theta_1) + \pi(R_1 + R_2)\sqrt{(R_2 - R_1)^2 + l_1^2}$$
$$+ \pi(R_2 + R_B)\sqrt{(R_B - R_2)^2 + l_2^2} + \pi R_B^2 \qquad (1.17)$$

$$V = \frac{\pi R_n (1 - \sin\theta_1)}{6}\left(3R_1^2 + R_1^2(1 - \sin\theta_1)^2\right)$$
$$+ \frac{1}{3}\pi l_1\left(R_1^2 + R_2^2 + R_1 R_2\right) + \frac{1}{3}\pi l_2\left(R_B^2 + R_2^2 + R_B R_2\right) \qquad (1.18)$$

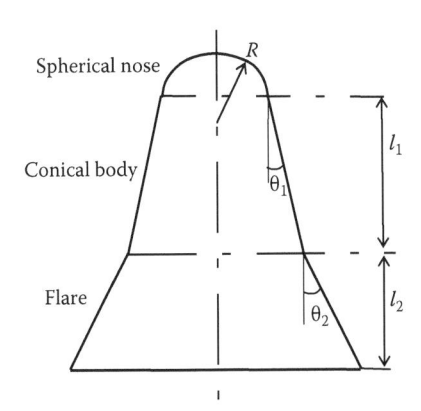

FIGURE 1.3
Reentry capsule.

where

$$R_1 = R_n \cos \theta_1 \quad (1.19)$$

$$R_2 = R_n \cos \theta_1 + l_1 \tan \theta_1 \quad (1.20)$$

$$R_B = R_2 + l_2 \tan \theta_2 \quad (1.21)$$

The optimization problem can now be written as

Minimize

$$2\pi R_n^2 (1 - \sin \theta_1) + \pi (R_1 + R_2)\sqrt{(R_2 - R_1)^2 + l_1^2}$$
$$+ \pi (R_2 + R_B)\sqrt{(R_B - R_2)^2 + l_2^2} + \pi R_B^2 \quad (1.22)$$

subject to

$$\frac{\pi R_n (1 - \sin \theta_1)}{6}\left(3R_1^2 + R_1^2(1-\sin\theta_1)^2\right) + \frac{1}{3}\pi l_1 \left(R_1^2 + R_2^2 + R_1 R_2\right)$$
$$+ \frac{1}{3}\pi l_2 \left(R_B^2 + R_2^2 + R_B R_2\right) \geq 1 \quad (1.23)$$

$$0.4 < R_n < 0.6 \quad (1.24)$$

$$22 < \theta_1 < 27 \quad (1.25)$$

$$\theta_1 + 5 < \theta_2 < \theta_1 + 10 \quad (1.26)$$

$$0.4 < l_1 < 0.8 \quad (1.27)$$

$$0.4 < l_2 < 0.8 \quad (1.28)$$

Example 1.4

It is required to find the optimum diameter (d) of a solid steel shaft whose mass (M) is to be minimized and the first cantilever frequency has to be greater than 20 Hz. Formulate this as an optimization problem by writing down the objective function and the constraint.

If L is the length of the rod (Figure 1.4) and ρ is its density, then the mass of the rod is given by

$$M = \frac{\pi}{4} d^2 L \rho \quad (1.29)$$

Introduction

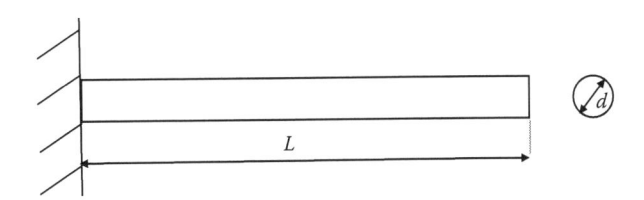

FIGURE 1.4
Cantilever rod.

For this problem, $L = 1$ m and $\rho = 7800$ kg/m³. The first cantilever frequency is given by

$$f_1 = \frac{1}{2\pi} \frac{3.5156}{L^2} \sqrt{\frac{EI}{k}} \qquad (1.30)$$

where E is Young's modulus of steel and its value is 2×10^{11} N/m². The variable k is mass per unit length. The moment of inertia I for the rod is given by

$$I = \frac{\pi}{64} d^4 \qquad (1.31)$$

The optimization problem can be written as follows.

Minimize

$$\frac{\pi}{4} d^2 L \rho \qquad (1.32)$$

subject to

$$\frac{1}{2\pi} \frac{3.5156}{L^2} \sqrt{\frac{EI}{k}} \geq 20 \qquad (1.33)$$

1.4 Solution with the Graphical Method

Having formulated the optimization problems in the previous section, it is tempting for readers to get solutions for these problems. The graphical method is a simple technique for locating the optimal solution for problems with up to two to three design variables. Beyond three variables, the representation of the optimization problem through graphs becomes complex.

The optimization problem mentioned in Example 1.2 requires two design variables, x_1 and x_2, to be evaluated such that it minimizes the total surface area and at the same time satisfying the equality constraint. A MATLAB® code, *graph_examp12.m*, given at the end of this book, is used for drawing the graph for this problem. For a quick introduction to MATLAB, see Appendix A.

The variable x_1 is varied from 1 to 100 mm and the variable x_2 is varied from 1 to 200 mm in the code. The surface area is calculated based on the values x_1 and x_2 and contour of the objective function is plotted (Figure 1.5) for different values of x_1 and x_2. The constraint function is then plotted (marked with *). Because this is an equality constraint optimization problem, the minimum value of the objective function is the contour curve that touches the constraint curve and has the lowest value. The minimum value of the objective function is 26,436 mm² corresponding to design variables x_1 as 37.45 mm and x_2 as 74.9 mm. Note that the length of the can is two times its radius at the minimum point. This can be proved analytically using elementary calculus.

Similarly, the optimization problem mentioned in Example 1.4 has only one design variable, the diameter d of the rod. Again, we can use a graphical method to solve this problem. A MATLAB code, *graph_examp14.m*, is written for this problem.

On executing the code, the output is in the form of a graph or a plot as shown in Figure 1.6. The value of the objective function (along the *y*-axis) decreases with the reduction in the value of the design variable (along the *x*-axis). However, the constraint value (also plotted along the *y*-axis) also decreases with the reduction in the value of the design variable. In the optimization problem, it is given that the constraint should have a value that is equal to or greater than 20 Hz. Hence the optimum solution corresponds to

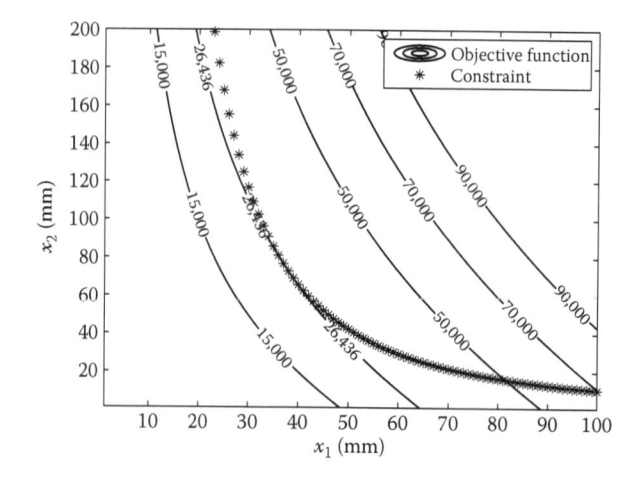

FIGURE 1.5
Function contours for the optimization problem in Example 1.2.

Introduction

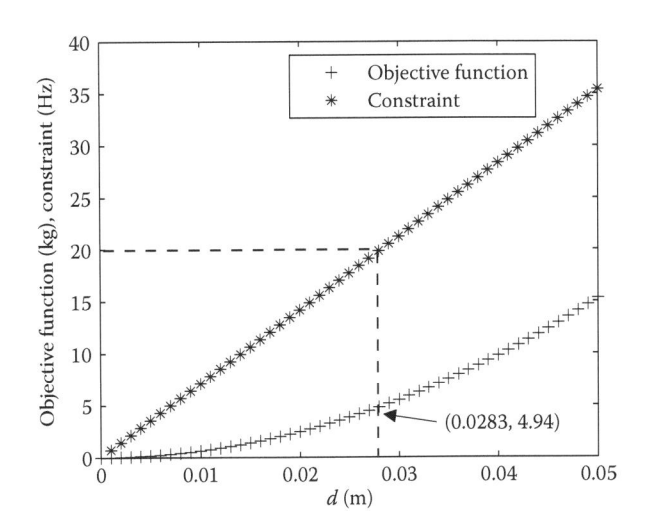

FIGURE 1.6
Objective function and constraint plot for the problem in Example 1.4.

$d = 0.0283$ m, where the value of the objective function mass is 4.94 kg and the constraint value is 20 Hz.

1.5 Convexity

Consider two design points, x_1 and x_2, that belong to a set S. If the line joining these two points is also within the set S, then the set S is a *convex* set. If the line joining the design points x_1 and x_2 does not belong to the set S, then the set S is a *nonconvex* set. See Figure 1.7 for convex and nonconvex sets. In optimization, often we have to check a function for its convexity. Consider a single variable function $f(x)$ as shown in Figure 1.8 and two points x_1 and x_2 at which the value of the function is $f(x_1)$ and $f(x_2)$ respectively. Consider any point $\tilde{x}$ on the line joining x_1 and x_2. If $f(\tilde{x})$ is less than the value of the function at the corresponding point $\hat{x}$ on the line joining $f(x_1)$ and $f(x_2)$ then $f(x)$ is a convex function, that is, for convexity

$$f(\tilde{x}) \leq f(\hat{x}) \tag{1.34}$$

Examples of convex functions are x^2, e^x, etc. If $f(x)$ is a convex function then $e^{f(x)}$ is also a convex function. Hence, e^{x^2} and e^{e^x} are also convex functions. Let us plot (Figure 1.9) these functions in MATLAB (*convexity.m*) to show that these functions are indeed convex.

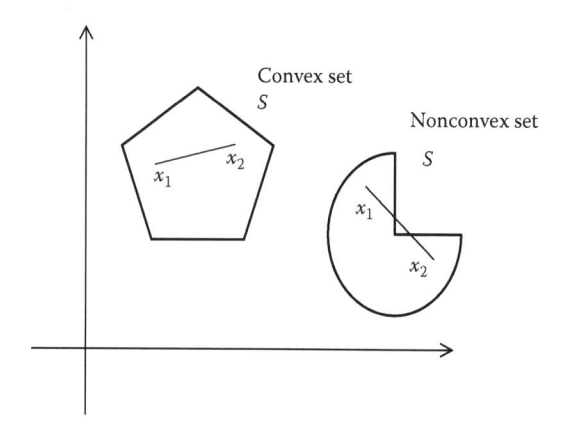

FIGURE 1.7
Convex and nonconvex sets.

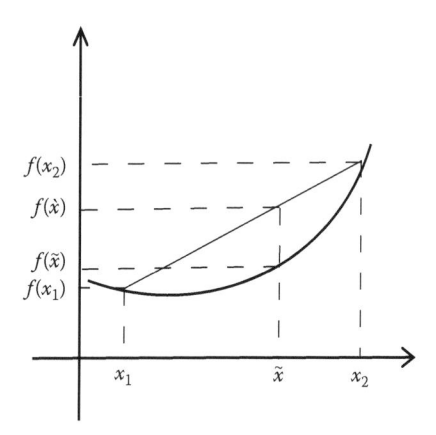

FIGURE 1.8
Convex function.

The concept of convexity is important in declaring that a function has one minimum only. A convex function thus has a global minimum. If a function is nonconvex, the optimum reached might be a local one (see Figure 1.10). Such functions with more than one minimum or maximum are referred to as *multimodal* functions. Traditional gradient-based algorithms have difficulty in locating a global optimum solution. In addition, a designer often has to look for an alternative solution to a global optimum because of the presence of the constraints. For example, at a global optimum solution, the design variables may be such that it might be difficult to manufacture the product or the particular material might be very costly. The task of the designer is thus difficult. He not only has to find a global optimum solution, but also locate local optimum solutions.

Introduction

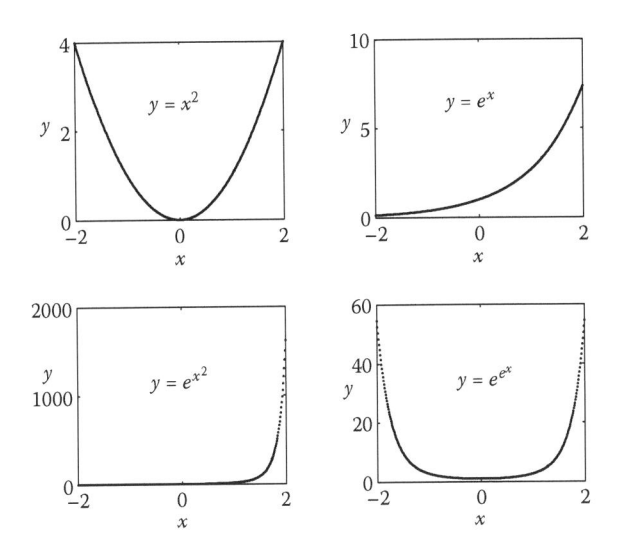

FIGURE 1.9
Examples of convex functions.

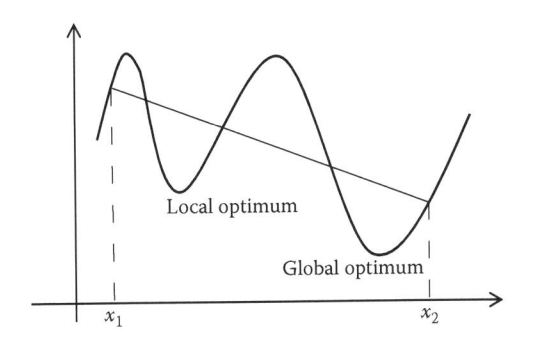

FIGURE 1.10
Local and global optima for a nonconvex function.

If $f(x)$ is a convex function, then $-f(x)$ is a *concave* function. Similarly, if $f(x)$ is a concave function, then $-f(x)$ is a convex function. Figure 1.11 shows both convex and concave functions for $y = e^x$.

Typically, optimization algorithms are developed to minimize the objective function. As discussed earlier, convexity plays an important role for functions where their minima are to be located. However, there can be optimization problems where one needs to maximize the objective function $f(x)$. The maximization problem can be converted to the minimization type by changing the sign of the objective function to $-f(x)$. Mathematically,

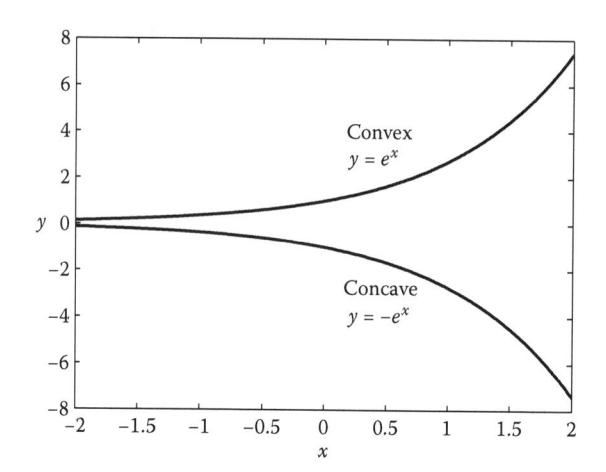

FIGURE 1.11
Concave and convex functions.

Maximize

$$f(x)$$

is the same as
Minimize

$$-f(x)$$

1.6 Gradient Vector, Directional Derivative, and Hessian Matrix

The derivative or gradient of a function $f(x)$ at a point x, generally denoted by $f'(x)$, is the slope of the tangent (see Figure 1.12) at that point. That is,

$$f'(x) = \tan \theta \tag{1.35}$$

where θ is the angle measured by the tangent with respect to the horizontal. Along the gradient direction, there is the maximum change in the value of the function. Thus, gradient information provides the necessary search direction to locate the maximum or minimum of the function.

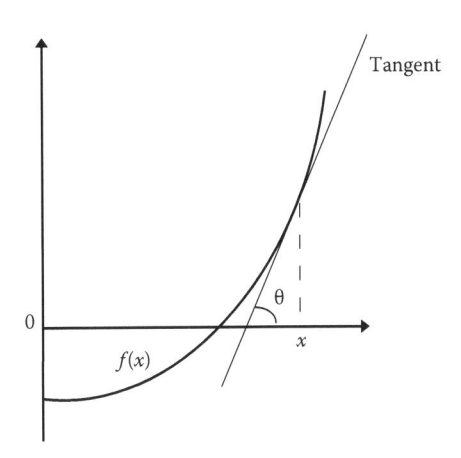

FIGURE 1.12
Concept of derivative.

In most optimization problems, which are generally nonlinear, $f'(x)$ has to be evaluated numerically. We can use *forward difference*, *backward difference*, and *central difference* methods to find the derivative of a function at a point. If the value of a function $f(x)$ is known at a point x, then the value of the function at its neighboring point $x + \Delta x$ can be computed using *Taylor's series* as

$$f(x+\Delta x) = f(x) + \Delta x f'(x) + \frac{\Delta x^2}{2!} f''(x) + \frac{\Delta x^3}{3!} f'''(x) + \cdots \quad (1.36)$$

Rearranging Equation 1.36 gives

$$\frac{f(x+\Delta x) - f(x)}{\Delta x} = f'(x) + \frac{\Delta x}{2!} f''(x) + \frac{\Delta x^2}{3!} f'''(x) + \cdots \quad (1.37)$$

The *forward difference* formula for evaluating the derivative of a function can be written as

$$f'(x) = \frac{f(x+\Delta x) - f(x)}{\Delta x} + O(\Delta x) \quad (1.38)$$

The quantity $O(\Delta x)$ represents that this formula is first-order accurate. In a similar fashion, the *backward difference* formula can be written as

$$f'(x) = \frac{f(x) - f(x-\Delta x)}{\Delta x} + O(\Delta x) \quad (1.39)$$

Using the forward and backward difference formulas, one can derive the *central difference formula* as

$$f'(x) = \frac{f(x+\Delta x) - f(x - \Delta x)}{2\Delta x} + O(\Delta x^2) \qquad (1.40)$$

Because the central difference formula for computing the derivative of a function is of second order, it is more accurate than forward/backward difference method. Again, the second derivative can be evaluated using the equation

$$f''(x) = \frac{f(x+\Delta x) - 2f(x) + f(x - \Delta x)}{\Delta x^2} \qquad (1.41)$$

Let us take a function

$$f(x) = 2 \sin 5x + 3x^3 - 2x^2 + 3x - 5 \qquad (1.42)$$

and compute the first and second derivatives using the central difference formula for x ranging from 0.1 to 1.0 with Δx as 0.01. A MATLAB code, *derivative.m*, is written and the output is plotted in Figure 1.13.

The top plot in the Figure 1.13 is $f(x)$ varying with x. Note that the function has one maximum and one minimum and these points are shown with *. The derivative of the function is plotted in the second plot. Note that $f'(x) = 0$ at the maximum and minimum of the function. From the third plot, observe

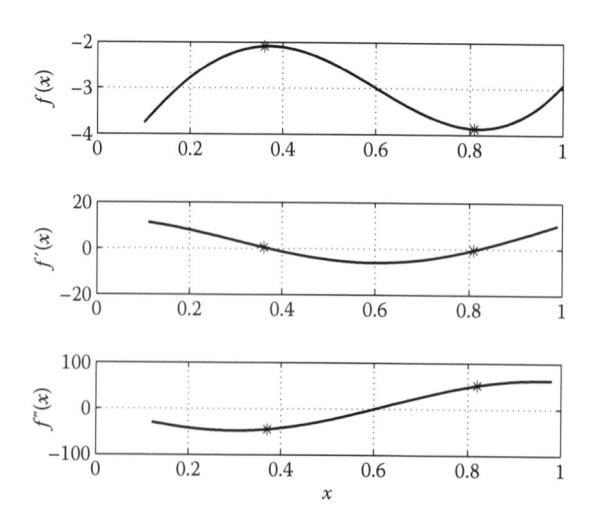

FIGURE 1.13
Plot of a function with its first and second derivative.

Introduction

that $f''(x) \geq 0$ at the minimum and $f''(x) \leq 0$ at the maximum of the function. The second derivative provides curvature information of the function.

For certain functions such as $f(x) = x^3$, both $f'(x) = f''(x) = 0$ at $x^* = 0$. In such instances, one has to look for higher order derivatives. Here $f'''(x) = 6$, which is nonzero. If the first nonzero higher order derivative is denoted by n, then x^* is an *inflection point* (or a *saddle point*) if n is odd and x^* is local optimum if n is even. Therefore, x^* is an inflection point for the function $f(x) = x^3$, as the first nonzero higher order derivative is odd (third derivative). Similarly, it can be shown that the function $f(x) = x^4$ has a local minimum at $x^* = 0$. These two functions are plotted in Figure 1.14.

So far we considered the derivative of a function with one variable only. The gradient is a vector that contains partial derivatives of the function with respect to the design variables $(x_1, x_2, \cdots, x_n)$ and is mathematically written as

$$\nabla f = \begin{bmatrix} \dfrac{\partial f}{\partial x_1} \\ \dfrac{\partial f}{\partial x_2} \\ \vdots \\ \dfrac{\partial f}{\partial x_n} \end{bmatrix} \tag{1.43}$$

Let us plot a tangent and gradient at a given point (x_1, x_2) on the function contours for Example 1.2. For a single-variable case, we observed that the tangent at any point for a function and its gradient are the same (Figure 1.12).

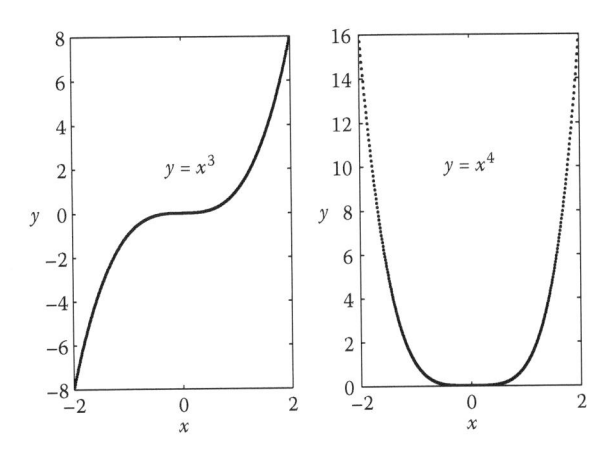

FIGURE 1.14
Saddle point and local minimum functions.

However, for a two-variable case, the tangent for each function contour is different and the value of the function remains the same along the tangent, that is, along a tangent

$$\Delta f = \frac{\partial f}{\partial x_1} \Delta x_1 + \frac{\partial f}{\partial x_2} \Delta x_2 = 0 \tag{1.44}$$

The gradient is normal to the tangent. A MATLAB code, *grad.m*, is written that on execution gives an output shown in Figure 1.15. On the function contour with a value of 15,000, a point (25, 70.493) is located where we desire to plot the tangent and gradient. Using Equation 1.44, we can write

$$\Delta x_2 = -\frac{\frac{\partial f}{\partial x_1}}{\frac{\partial f}{\partial x_2}} \Delta x_1 = -\frac{2x_1 + x_2}{x_1} \Delta x_1 \tag{1.45}$$

Using the incremental Equation 1.45, a tangent can be drawn at the point (25, 70.493). If the slope of the tangent is given by m_t, then the slope of the gradient m_g can be computed from the relation

$$m_g m_t = -1 \tag{1.46}$$

Consider three functions, $f_1(x_1, x_2, x_3)$, $f_2(x_1, x_2, x_3)$, and $f_3(x_1, x_2, x_3)$, which are functions of three variables, x_1, x_2, and x_3. The gradient of these functions

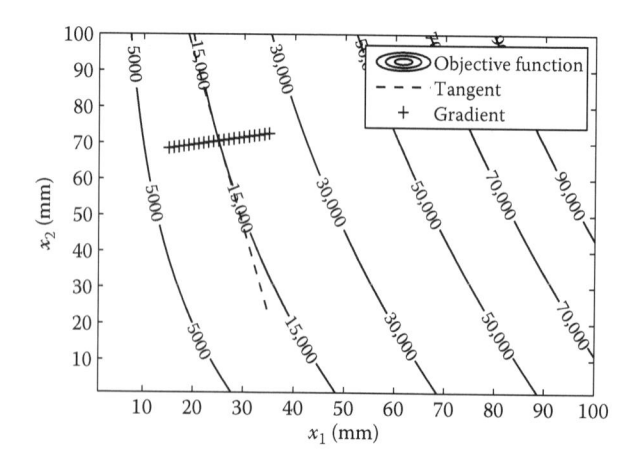

FIGURE 1.15
Tangent and gradient for the objective function given in Example 1.2.

Introduction

can be put in a single matrix referred to as a Jacobian J and is expressed in mathematical form as

$$[J] = \begin{bmatrix} \dfrac{\partial f_1}{\partial x_1} & \dfrac{\partial f_1}{\partial x_2} & \dfrac{\partial f_1}{\partial x_3} \\ \dfrac{\partial f_2}{\partial x_1} & \dfrac{\partial f_2}{\partial x_2} & \dfrac{\partial f_2}{\partial x_3} \\ \dfrac{\partial f_3}{\partial x_1} & \dfrac{\partial f_3}{\partial x_2} & \dfrac{\partial f_3}{\partial x_3} \end{bmatrix} \qquad (1.47)$$

For constrained optimization problems, it is possible that moving in the gradient direction can result in moving into the infeasible region. In such an instance one wishes to move in some other search direction and would like to know the rate of change of function in that direction. The *directional derivative* provides information on the instantaneous rate of change of a function in a particular direction. If u is a unit vector, then the directional derivative of a function $f(x)$ in the direction of u is given by

$$\nabla f(x)^T u$$

The *Hessian* matrix H represents the second derivative of a function with more than one variable. For a function $f(x_1, x_2, x_3)$ with three variables, the Hessian matrix is written as

$$[H] = \begin{bmatrix} \dfrac{\partial^2 f}{\partial x_1^2} & \dfrac{\partial^2 f}{\partial x_1 x_2} & \dfrac{\partial^2 f}{\partial x_1 x_3} \\ \dfrac{\partial^2 f}{\partial x_2 x_1} & \dfrac{\partial^2 f}{\partial x_2^2} & \dfrac{\partial^2 f}{\partial x_2 x_3} \\ \dfrac{\partial^2 f}{\partial x_3 x_1} & \dfrac{\partial^2 f}{\partial x_3 x_2} & \dfrac{\partial^2 f}{\partial x_3^2} \end{bmatrix} \qquad (1.48)$$

The Hessian matrix should be *positive definite* at the minimum of the function. A matrix is positive definite if its eigenvalues are positive. For a square matrix, there exists a nonzero vector such that when multiplied with the square matrix it results in a vector that differs from the original by a multiplicative scalar. The nonzero vector is termed the eigenvector and the multiplicative scalar the eigenvalues. Let us check the eigenvalues for the following matrix by executing a MATLAB code, *positive_definite.m*:

$$H = \begin{bmatrix} 2 & 1 & 1 \\ 1 & 2 & 1 \\ 0 & 2 & 3 \end{bmatrix} \qquad (1.49)$$

The eigenvalues of the matrix are 1, 1.5858, and 4.4142. Because all the eigenvalues are positive, the matrix is positive definite.

Example 1.5

Write a gradient and Hessian matrix for the function

$$f(x) = x_1^2 + 2x_1x_2 + 3x_2^2 + 4x_3^2 - 5x_2x_3$$

Also find the directional derivative of the function at (1, 1, 1) in the direction

$$d = \begin{bmatrix} 1 \\ 2 \\ 3 \end{bmatrix}$$

The gradient is given by

$$\nabla f(x) = \begin{bmatrix} 2x_1 + 2x_2 \\ 2x_1 + 6x_2 - 5x_3 \\ 8x_3 - 5x_2 \end{bmatrix}$$

The Hessian is given by

$$H = \begin{bmatrix} 2 & 2 & 0 \\ 2 & 6 & -5 \\ 0 & -5 & 8 \end{bmatrix}$$

The unit vector in direction d is given by

$$u = \frac{d}{\|d\|} = \frac{1}{\sqrt{1^2 + 2^2 + 3^2}} \begin{bmatrix} 1 \\ 2 \\ 3 \end{bmatrix} = \begin{bmatrix} 1/\sqrt{14} \\ 2/\sqrt{14} \\ 3/\sqrt{14} \end{bmatrix}$$

Now, the directional derivative of the function at (1, 1, 1) in the direction of the unit vector u is given by

$$\nabla f(x)^T u = \begin{bmatrix} 4 & 3 & 3 \end{bmatrix} \begin{bmatrix} 1/\sqrt{14} \\ 2/\sqrt{14} \\ 3/\sqrt{14} \end{bmatrix} = 19/\sqrt{14}$$

Introduction

1.7 Linear and Quadratic Approximations

A quadratic approximation of a function is often desired in optimization, as certain solution methods such as Newton's method show faster convergence for these functions. The Taylor series approximation, as discussed in an earlier section, is used to make linear or quadratic approximations of a function by appropriately considering the number of terms in the series. A MATLAB code, *quadr.m*, is written that demonstrates linear and quadratic approximations (Figure 1.16) of a function e^{-x}.

The Taylor series approximation can be easily extended to a function with n variables and is given by the expression

$$f(x+\Delta x) = f(x) + \nabla f(x)^T \Delta x + \frac{1}{2}\Delta x^T H \Delta x + \cdots \quad (1.50)$$

For a linear approximation of the function, only the gradient term is used and the Hessian term is ignored. For a quadratic approximation of the function, the Hessian term is considered along with the gradient term.

For a function with two variables, as in Example 1.2, a MATLAB code, *quadr_exampl2.m*, is written to make a quadratic approximation of the function. On executing the code, quadratic approximations are plotted (Figure 1.17) along with the function contours. The gradient and Hessian for this function are

$$\nabla f(x) = \begin{bmatrix} 2\pi x_2 + 4\pi x_1 \\ 2\pi x_1 \end{bmatrix} \quad (1.51)$$

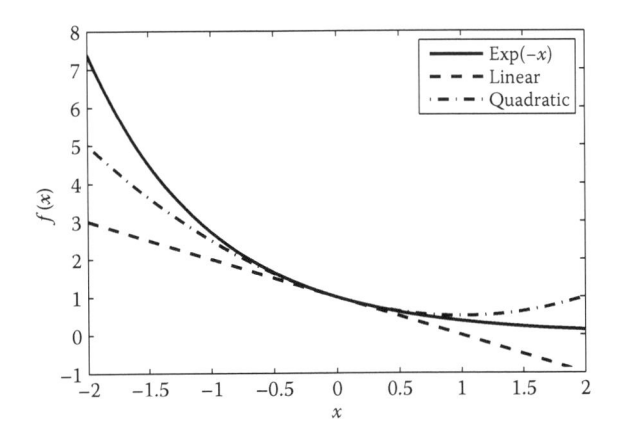

FIGURE 1.16
Linear and quadratic approximation of the function e^{-x}.

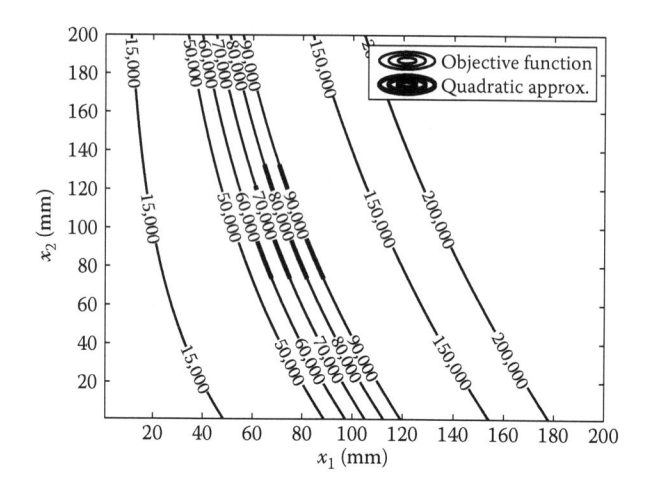

FIGURE 1.17
Quadratic approximation of the objective function in Example 1.2.

$$H = \begin{bmatrix} 4\pi & 2\pi \\ 2\pi & 0 \end{bmatrix} \quad (1.52)$$

Example 1.6

Construct linear and quadratic approximations for the function

$$f(x) = 3x_2 - \frac{x_1}{x_2}$$

at a point (2, 1).

The gradient is given by

$$\nabla f(x) = \begin{bmatrix} -\dfrac{1}{x_2} \\ 3 + \dfrac{x_1}{x_2^2} \end{bmatrix}; \quad \nabla f(x_0) = \begin{bmatrix} -1 \\ 5 \end{bmatrix}$$

The Hessian is given by

$$H = \begin{bmatrix} 0 & \dfrac{1}{x_2^2} \\ \dfrac{1}{x_2^2} & \dfrac{-2x_1}{x_2^3} \end{bmatrix} = \begin{bmatrix} 0 & 1 \\ 1 & -4 \end{bmatrix}$$

The linear approximation of the function is given by

$$l(x) = f(x_0) + \nabla f(x_0)^T (x - x_0)$$

$$= 1 + \begin{bmatrix} -1 & 5 \end{bmatrix} \begin{bmatrix} x_1 - 2 \\ x_2 - 1 \end{bmatrix}$$

$$= -x_1 + 5x_2 - 2$$

The quadratic approximation of the function is given by

$$q(x) = f(x_0) + \nabla f(x_0)^T (x - x_0) + \frac{1}{2}(x - x_0)^T H(x - x_0)$$

$$q(x) = 5x_2 - x_1 - 2 + \left\{ \frac{(x_1 - 2)}{2x_2^2} - \frac{x_1(x_2 - 1)}{x_2^3} \right\} (x_2 - 1) + \frac{(x_1 - 2)(x_2 - 1)}{2x_2^2}$$

1.8 Organization of the Book

The book is organized into 11 chapters. Chapter 2 discusses 1-D algorithms such as the bisection, Newton–Raphson, secant, and golden-section methods. These algorithms form the building blocks for the unconstrained optimization methods such as the steepest descent, Newton, Levenberg–Marquardt, conjugate gradient, Davidon–Fletcher–Powell (DFP), and Broyden–Fletcher–Goldfarb–Shanno (BFGS) methods, which are discussed in Chapter 3. The direct search Powell's method is used to solve a complex robotics problem. Chapter 4 elaborates on linear programming where simplex, dual simplex, and interior-point methods are discussed. A practical portfolio optimization problem is also solved in this chapter. Genetic algorithm, simulated annealing, and particle swarm optimization techniques are elaborated in Chapter 5. Ant colony optimization and the tabu search method are also briefly introduced here. Solution techniques such as penalty function, augmented Lagrangian, sequential quadratic programming, and methods of feasible directions are discussed in Chapter 6 for constrained optimization problems. Multiobjective optimization methods are discussed in Chapter 7. The shape design of a reentry body is optimized and discussed in this chapter. In Chapter 8, both unconstrained and constrained problems are solved using geometric programming techniques. Chapter 9 discusses multidisciplinary design optimization (MDO), where different architectures are considered. The importance of response surface methodology is highlighted for

MDO problems. Gomory's cutting plane method, zero-one problem, Balas' method, branch-and-bound method, and so forth are discussed in Chapter 10 on integer programming. Both deterministic and probabilistic aspects of dynamic programming are discussed in Chapter 11. See Figure 1.18 for a quick glance at the organization of the book.

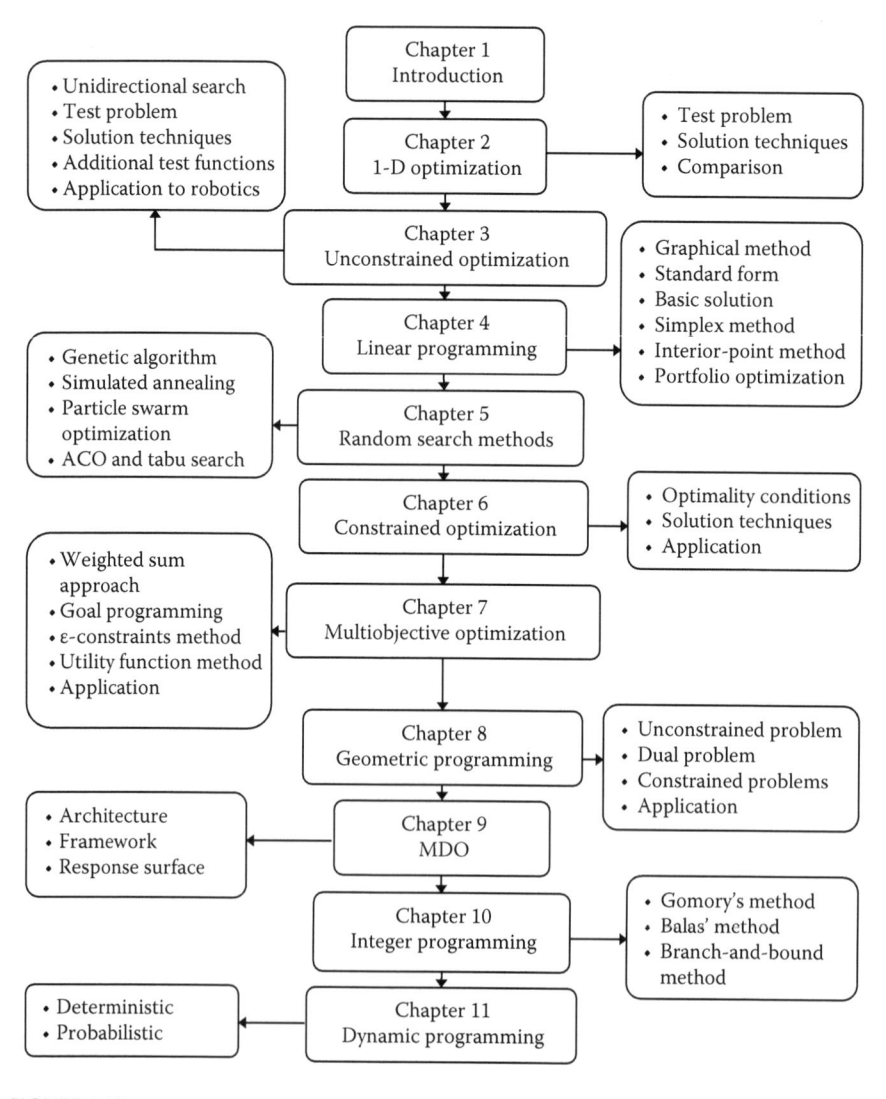

FIGURE 1.18
Organization of the book.

Chapter Highlights

- In an optimization problem, we write the objective function that is to be maximized or minimized along with inequality and equality constraints. The objective function and constraints are a function of the design variables that need to be evaluated by the optimization methods.
- The design variables can be a real number or could be of the discrete, binary, or integer type.
- Modeling refers to writing down the observations of a problem in mathematical form using basic building blocks of mathematics such as addition, subtraction, multiplication, division, functions, and numbers with proper units.
- The gradient at a point is the slope of the tangent at that point.
- If the objective function and constraints are linear functions of the design variables, it is referred to as a linear programming problem. These functions do not contain terms such as $x_1 x_2$ and x_1^2.
- The graphical method can be applied to solve the optimization problem with up to three design variables.
- Functions with more than one minimum or maximum are referred to as multimodal functions.
- The concept of convexity is important in declaring a function to have one minimum only. A convex function thus has a global minimum.
- Typically, optimization algorithms are written to minimize a function. If the objective function is to be maximized, it is negated and then solved as a minimization problem.
- The necessary condition for optimality (either maximum or minimum) is that the gradient vanishes at the point of consideration.
- At the point of optimality, if the second derivative of the objective function is positive, it is a case of the minimum and if the second derivative is negative, it is case of the maximum.
- The derivative of a function can be numerically evaluated using forward, backward, and central difference methods. The central difference method is more accurate than forward or backward difference methods.
- The directional derivative provides information on the instantaneous rate of change of a function in a particular direction.
- The Hessian matrix H represents the second derivative of a function with more than one variable.
- The Hessian matrix should be positive definite at the minimum of the function. A matrix is positive definite if its eigenvalues are positive.

- A quadratic approximation of a function is often desired in optimization, as certain solution methods such as Newton's method show faster convergence for these functions.
- Taylor's series approximation is used to make linear or quadratic approximations of a function by appropriately considering the number of terms in the series.

Formulae Chart

Forward difference:

$$f'(x) = \frac{f(x+\Delta x) - f(x)}{\Delta x}$$

Backward difference:

$$f'(x) = \frac{f(x) - f(x-\Delta x)}{\Delta x}$$

Central difference:

$$f'(x) = \frac{f(x+\Delta x) - f(x-\Delta x)}{2\Delta x}$$

Central difference formula for the second derivative:

$$f''(x) = \frac{f(x+\Delta x) - 2f(x) + f(x-\Delta x)}{\Delta x^2}$$

Jacobian of three functions with three variables:

$$[J] = \begin{bmatrix} \frac{\partial f_1}{\partial x_1} & \frac{\partial f_1}{\partial x_2} & \frac{\partial f_1}{\partial x_3} \\ \frac{\partial f_2}{\partial x_1} & \frac{\partial f_2}{\partial x_2} & \frac{\partial f_2}{\partial x_3} \\ \frac{\partial f_3}{\partial x_1} & \frac{\partial f_3}{\partial x_2} & \frac{\partial f_3}{\partial x_3} \end{bmatrix}$$

Hessian for a three-variable function:

$$[H] = \begin{bmatrix} \dfrac{\partial^2 f}{\partial x_1^2} & \dfrac{\partial^2 f}{\partial x_1 x_2} & \dfrac{\partial^2 f}{\partial x_1 x_3} \\ \dfrac{\partial^2 f}{\partial x_2 x_1} & \dfrac{\partial^2 f}{\partial x_2^2} & \dfrac{\partial^2 f}{\partial x_2 x_3} \\ \dfrac{\partial^2 f}{\partial x_3 x_1} & \dfrac{\partial^2 f}{\partial x_3 x_2} & \dfrac{\partial^2 f}{\partial x_3^2} \end{bmatrix}$$

Quadratic approximation:

$$f(x + \Delta x) = f(x) + \nabla f(x)^T \Delta x + \frac{1}{2} \Delta x^T H \Delta x$$

Problems

1. An airline company in India uses A320 aircraft to fly passengers from New Delhi to Mumbai. Though the maximum seating capacity of the aircraft is 180, the airline observes that on average it flies only 130 passengers per flight. The regular fare between the two cities is Rs. 15,000. From the market survey, the company knows that for every Rs. 300 reduction in fare, it would attract an additional four passengers. The company would like to find a fare policy that would maximize its revenue. Formulate this as an optimization problem.

2. The average yield in a farm is 300 apples per tree, if 50 apple trees are planted per acre. The yield per tree decreases by 3 apples for each additional tree planted per acre. How many additional trees per acre should be planted to maximize the yield? Formulate this as an optimization problem.

3. Determine the area of the largest rectangle that can be inscribed in a circle of radius 5 cm. Formulate this as an optimization problem by writing down the objective function and the constraint. Solve the problem using the graphical method.

4. A field needs to be enclosed with a fence, with a river flowing on one side of the field. We have 300 m of fencing material. Our aim is to use the available fencing material and cover the maximum area of the field. Formulate this as an optimization problem by writing down the objective function and the constraint and clearly stating the design variables.

5. A traveling salesman has to start from city A, cover all other n number of cities, and then come back to city A. The distance between the ith

and jth cities is given by y_{ij}. How could he plan the route so to cover the minimum distance? Formulate this as an optimization problem.

6. A company has initial wealth W and would like to invest this to get maximum returns. It can get higher returns (r_r) if it invests in risky assets, but the return is not guaranteed. A return (r_s) is guaranteed if it invests in safe assets. How much should the company invest in risky assets (R), to maximize its wealth at the end of a stipulated period? Formulate the objective function for the optimization problem.

7. In an experiment, the following observations (see Table 1.3) are made where x is an independent variable and y is a dependent variable. It is desired to fit these data with a straight line

$$\hat{y} = mx + c$$

where m and c are to be determined. The data are to be fitted in the least squares sense, that is, $\sum (y_i - \hat{y})^2$ is to be minimized. Formulate this as an optimization problem.

8. The cost of a solar energy system (King 1975) is given by

$$U = 35A + 208V$$

where A is the surface area of the collector and V is the volume of the storage (Figure 1.19). Owing to energy balance considerations, the following relation between A and V is to be satisfied:

$$A\left(290 - \frac{100}{V}\right) = 5833.3$$

TABLE 1.3

Data Observed from an Experiment

x_i	1	2	3	4	5
y_i	45	55	70	85	105

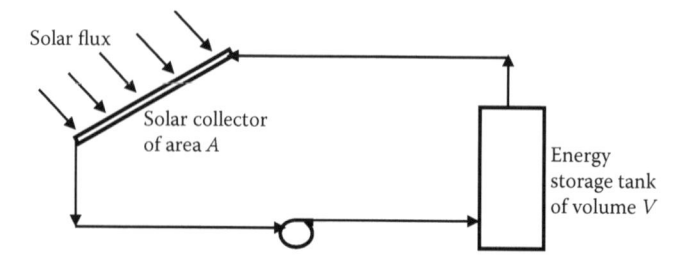

FIGURE 1.19
Solar energy problem.

Introduction

The design variable T is related to V as

$$V = \frac{50}{T-20}$$

The variable T has to be restricted between 40°C and 90°C. The cost U is to be minimized. Formulate this as an optimization problem.

9. Write the gradient and Hessian matrix for the function

$$f(x) = 5x_1 x_2 + \ln\left(2x_1^2 + 3x_2^2\right)$$

10. A company manufactures three products: A, B, and C. Each product requires time for three processes: 1, 2, and 3, and this information is given in Table 1.4.

    The maximum available capacity on each process is given in Table 1.5.

    The profit per unit for the product is given in Table 1.6.

    What quantities of A, B, and C should be produced to maximize profit? Formulate this as an optimization problem.

11. A company has three factories and five warehouses. The warehouse demand, factory capacity, and the cost of shipping are given in Table 1.7.

    Determine the optimal shipment plan to minimize the total cost of transportation. Formulate the optimization problem.

TABLE 1.4
Time Required for Each Process

Product	Time Required (minute)/Unit		
	A	B	C
Process 1	12	25	7
Process 2	11	6	20
Process 3	15	6	5

TABLE 1.5
Maximum Capacity of Each Process

Process	Capacity (minutes)
1	28,000
2	35,000
3	32,000

TABLE 1.6
Profit per Unit of Each Product

Product	Profit/Unit
A	5
B	7
C	4

TABLE 1.7
Cost per Unit of Shipment from Factory to Warehouse

From Factory	A	B	C	D	E	Capacity (No. of Units)
	\multicolumn{5}{c}{Cost per Unit of Shipment}	
P	3	7	4	6	5	150
Q	5	4	2	5	1	110
R	6	3	2	2	4	90
Demand	50	100	70	70	60	

(Warehouse columns: A, B, C, D, E)

12. Plot the function

$$f(x) = (x + 3)(x - 1)(x + 4)$$

and locate minimum and maximum of the function in [−4, 0].

13. An oil refinery company blends four raw gasoline types (A, B, C, and D) to produce two grades of automobile fuel, standard and premium. The cost per barrel of different gasoline types, performance rating and number of barrels available each day is given in Table 1.8.

The premium should have a rating greater than 90 while the standard fuel should have a performance rating in excess of 80. The selling prices of standard and premium fuel are 90 dollars and 100 dollars per barrel respectively. The company should produce at least 6000 barrels of fuel per day. How much quantity of fuel (of each

TABLE 1.8
Cost, Performance Rating, and Production Level of Different Gasoline Types

	Cost/Barrel in Dollars	Performance Rating	Barrels/Day
A	60	75	3000
B	65	85	4000
C	70	90	5000
D	80	95	4000

Introduction

type) should be produced to maximize profit? Formulate this as an optimization problem.

14. Check whether the following functions are convex or not.

   a. $2x^2 - 3x + 5$ $\qquad x \in [-4, 4]$
   b. $x^3 - 2x^2 + 4x - 10$ $\qquad x \in [-3, 1]$
   c. $\dfrac{1}{1-x^2}$ $\qquad x \in [-1.6, -0.8]$
   d. $\sqrt{x^2 + 2x + 5}$ $\qquad x \in [-5, 5]$

15. Write the first three terms of the Taylor series for the function

$$f(x) = \ln(x - 1)$$

   at $x = 3$.

16. Find the linear approximation of the function

$$f(x) = (1 + x)^{50} + (1 - 2x)^{60}$$

   at $x = 1$.

17. Write the Taylor series expansion (up to four terms) for the function e^x centered at $x = 3$.

18. Write the Taylor series expansion (up to three terms) for the function $e^{\cos x}$ centered at $x = \pi$.

19. Find the quadratic approximation of the function

$$f(x) = \ln(1 + \sin x)$$

   at $x = 0$.

20. Find the directional derivative of the function

$$f(x) = x_1^2 x_2 + x_2^2 x_3 - x_1 x_2 x_3^2$$

   at $(1, 1, -1)$ in the direction $\begin{bmatrix} 1 \\ 2 \\ 3 \end{bmatrix}$.

21. Using MATLAB, plot the functions x^4 and $|x|$ and check whether these functions are convex.

22. Solve the following optimization problems using the graphical method.

   i. Maximize $\quad z = 125x_1 + 150x_2$
      subject to $\quad 6x_1 + 11x_2 \leq 66$
      $\quad\quad\quad\quad\quad 8x_1 + 9x_2 \leq 72$
      $\quad\quad\quad\quad\quad x_1, x_2 \geq 0$

   ii. Maximize $\quad z = 3x_1 + 4x_2$
      subject to $\quad 2x_1 + x_2 \leq 30$
      $\quad\quad\quad\quad\quad x_1 + 3x_2 \geq 40$
      $\quad\quad\quad\quad\quad x_1, x_2 \geq 0$

23. Calculate the Jacobian of the following system of equations:

$$\begin{bmatrix} x_1 + 2x_2^2 + 3x_3^3 \\ x_1^2 x_2 x_3^2 \\ 3x_1 x_2 - 2x_1 x_3 + 4x_2 x_3 \end{bmatrix}$$

2

1-D Optimization Algorithms

2.1 Introduction

The one-dimensional (1-D) optimization problem refers to an objective function with one variable. In practice, optimization problems with many variables are complex, and rarely does one find a problem with a single variable. However, 1-D optimization algorithms form the basic building blocks for multivariable algorithms. As these algorithms form a subproblem of multivariable optimization problems, numerous methods (or algorithms) have been reported in the literature, each with some unique advantage over the others. These algorithms are classified into gradient-based and non–gradient-based algorithms. Some popular algorithms are discussed in this chapter.

As an example, a single-variable objective function could be

$$f(x) = 2x^2 - 2x + 8$$

This is an unconstrained optimization problem where x has to be determined, which results in minimization of $f(x)$. If we have to restrict x within $a \leq x \leq b$, where a and b are real numbers, then it becomes a constrained optimization problem. If the function $f(x)$ is either continuously increasing or decreasing between two points a and b, then it is referred to as a *monotonic* function (see Figure 2.1). In a *unimodal* function, the function is monotonic on either side of its minimum point (x^*). The function $f(x) = 2x^2 - 2x + 8$ is plotted in Figure 2.2, in which we observe that $f(x)$ is a unimodal function. Using the property of the unimodal function that it continuously decreases or increases on either side of the minimum point, the single-variable search algorithms can be devised in such a way that they eliminate certain regions of the function where the minimum is not located.

In the next section, a test problem in a solar energy system is defined. Both gradient-based and direct search methods are discussed and tested for this problem. Subsequently, these solution techniques will also be tested on some more standard optimization problems. The performances of these methods are compared toward the end of the chapter. The road map of this chapter is given in Figure 2.3.

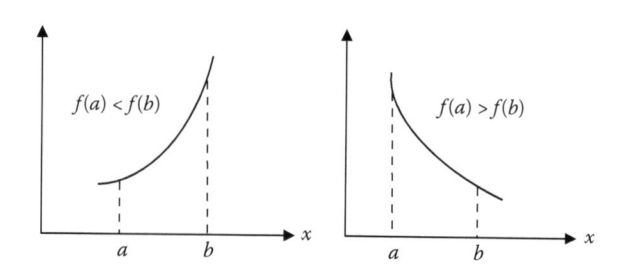

FIGURE 2.1
Monotonic increasing and decreasing functions.

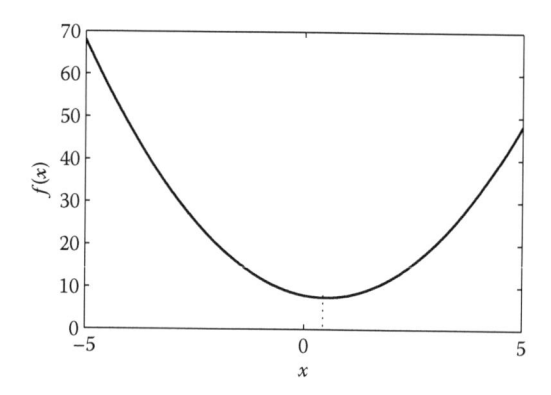

FIGURE 2.2
Unimodal function.

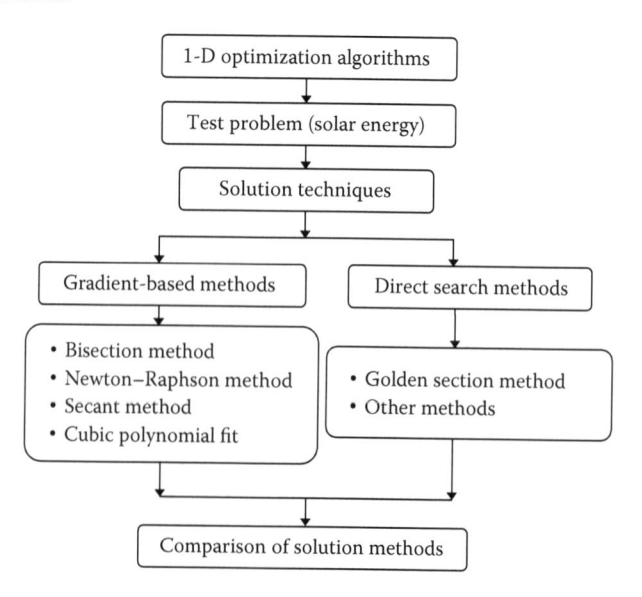

FIGURE 2.3
Road map of Chapter 2.

2.2 Test Problem

Before we discuss the optimization algorithms, let us set a problem on which we will be testing these algorithms. The solar energy problem is defined in Problem 8 of Chapter 1. In this cost minimization problem, the cost is a function of the volume of the storage system and the surface area of the collector. The volume and surface area are functions of the design variable temperature T. Let us rewrite the cost function in terms of T alone as

$$U = \frac{204{,}165.5}{330 - 2T} + \frac{10{,}400}{T - 20} \tag{2.1}$$

The variable T is restricted between 40°C and 90°C. The function U is plotted as a function of T in Figure 2.4. The minimum occurs at $T^* = 55.08$ and the minimum value of the function is $U^* = 1225.166$. Observe from the figure that the cost function is unimodal. A MATLAB® code, *exhaustive.m*, is used to plot the cost function by varying the design variable T from 40 to 90 in steps of 0.01. One may ask why, if this method is able to locate the minimum and is also simple, there is a need to discuss other algorithms. It may be noted that the number of function evaluations by this particular method is $(90 - 40)/0.01 = 5000$. For more complex problems, the time required for the function evaluation is at a premium and it may not be practical to evaluate the function so many times. This necessitates exploring new algorithms that require fewer function evaluations to reach the minimum of any function.

On executing this code, the output obtained is

```
Minimum cost = 1225.17
Occurs at T = 55.08
```

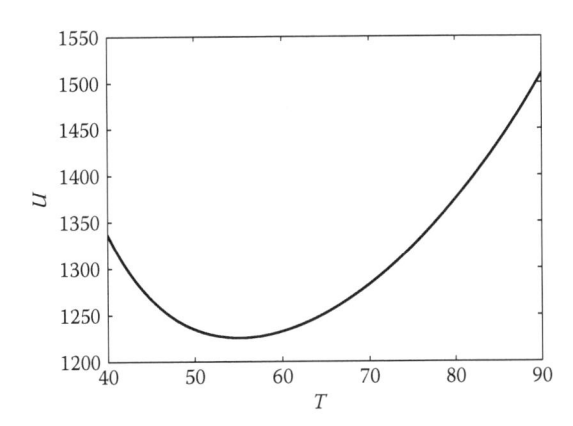

FIGURE 2.4
Cost function for the test problem.

2.3 Solution Techniques

As mentioned previously, the solution techniques for one-dimensional optimization problems can be classified into gradient-based and non–gradient-based algorithms. As the name suggests, gradient-based algorithms require derivative information. These methods find applications to problems in which derivatives can be calculated easily. In the search processes of these algorithms, the derivative of the function is driven to zero. The algorithm is terminated when the derivative of the function is very close to zero and the corresponding x is declared as the point ($x^* = x$) at which minimum of the function occurs. The following gradient-based methods are discussed in this section:

- Bisection method
- Newton–Raphson method
- Secant method
- Cubic polynomial fit

For certain types of optimization problems, the variable x may not be real, but can take only certain discrete values. Recall the pipe size problem discussed in Chapter 1, where pipe size comes in some standard sizes such as 1, 2 inches, and so forth. For such discontinuous functions, gradient information will not be available at all points, and the search algorithm has to proceed using the function evaluations alone to arrive at the minimum of the function. The golden section method is a very effective solution technique for such problems and is discussed later in this section. The golden section method can also be applied to continuous functions. Some other direct search methods such as dichotomous search, the interval halving method, and the Fibonacci method are also briefly discussed.

2.3.1 Bisection Method

In Chapter 1, we discussed that at the maximum or minimum of a function, $f'(x) = 0$. Because in these problems we are considering a unimodal function of minimization type, the condition that the gradient vanishes at the minimum point still holds. The gradient function changes sign near the optimum point. If $f'(x_1)$ and $f'(x_2)$ are the derivatives of a function computed at points x_1 and x_2, then the minimum of the function is located between x_1 and x_2 if

$$f'(x_1)f'(x_2) < 0 \qquad (2.2)$$

Based on this condition, certain regions of the search space can be eliminated. The algorithm is described in Table 2.1.

1-D Optimization Algorithms

TABLE 2.1
Algorithm for the Bisection Method

Step 1: Given a, b, ε, and Δx
Step 2: Compute $\alpha = \dfrac{a+b}{2}$, $f'(a)$ and $f'(\alpha)$
    If $f'(a)f'(\alpha) < 0$
        then $b = \alpha$
        else $a = \alpha$
    If $|a - b| > \varepsilon$
        then goto Step 2
        else goto Step 3
Step 3: Converged. Print $x^* = a$, $f(x^*) = f(a)$

In this algorithm a and b are the bounds of the function, and Δx is used in the central difference formula for computing the derivative and ε is a small number required for terminating the algorithm when $|a - b| < \varepsilon$. See Figure 2.5, which gives physical insight into this method. The algorithm is coded in MATLAB (*bisection.m*). The objective function is coded in MATLAB file (*func.m*). Users can change the function in this file to minimize another objective function that may be of interest to them. In doing so, they also need to give appropriate bounds for the function, given by a and b in the main program (*bisection.m*).

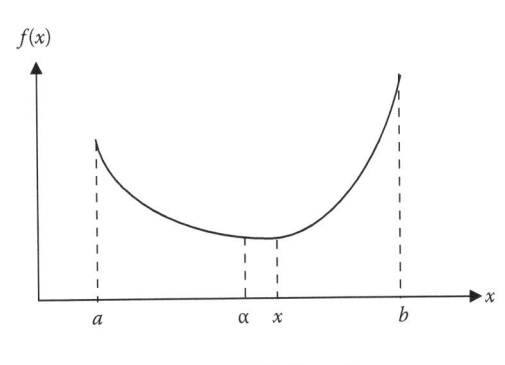

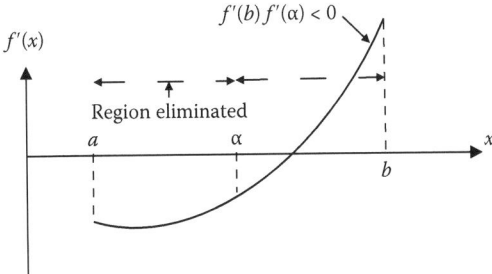

FIGURE 2.5
Bisection method.

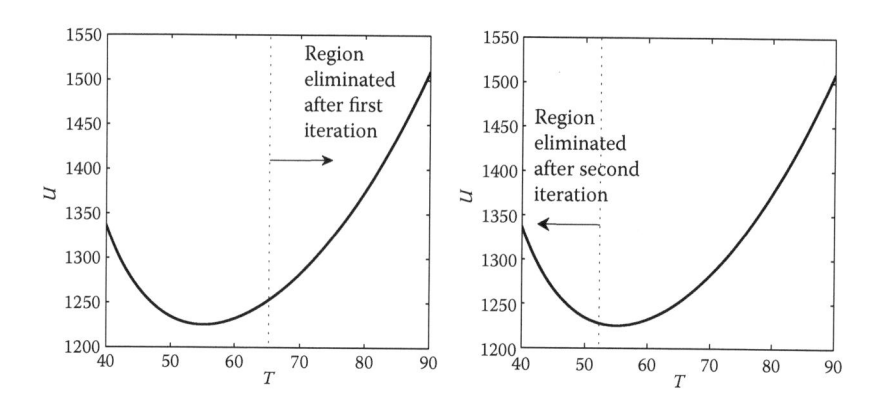

FIGURE 2.6
Region elimination with iterations (bisection method).

On executing the code for the test problem, the output obtained is

```
     a                b
---------------------------
  40.000           90.000
  40.000           65.000
  52.500           65.000
  52.500           58.750
  52.500           55.625
  54.063           55.625
  54.844           55.625
  54.844           55.234
  55.039           55.234
  55.039           55.137
  55.039           55.088
  55.063           55.088
  55.076           55.088
---------------------------
x* = 55.082  Minimum = 1225.166
Number of function calls = 52
```

The minimum obtained from this method matches very closely with the exhaustive search method. But the number of function evaluations in the bisection method is only 52 as compared to 5000 in the exhaustive search method. For this test problem, Figure 2.6 shows the regions that are eliminated in the first two iterations.

2.3.2 Newton–Raphson Method

Isaac Newton evaluated the root of an equation using a sequence of polynomials. The method in the present form was given by Joseph Raphson in 1960, with successive approximation to x given in an iteration form. The

1-D Optimization Algorithms

Newton–Raphson method is a root finding technique in which the root of the equation $f'(x) = 0$ is evaluated. Using the Taylor series, the function $f'(x)$ can be approximated as

$$f'(x_k) + f''(x_k)\Delta x \qquad (2.3)$$

where the gradient is approximated at point x_k. Setting Equation 2.3 to zero, the next approximation point can then be given as

$$x_{k+1} = x_k - \frac{f'(x_k)}{f''(x_k)} \qquad (2.4)$$

Figure 2.7 illustrates the steps of this method. The method shows quadratic convergence. That is, if x^* is the root of the equation, then

$$\frac{\|x_{k+1} - x^*\|}{\|x_k - x^*\|^2} \leq c, \quad c \geq 0 \qquad (2.5)$$

The Newton–Raphson algorithm is described in Table 2.2.

The algorithm is coded in MATLAB (*newtonraphson.m*). On executing the code, the output obtained is

```
    x         f(x)           Deriv.        Second deriv.
-------------------------------------------------------
45.000      1266.690         -9.551            1.449
51.590      1229.340         -2.485            0.800
54.697      1225.214         -0.249            0.650
55.079      1225.166         -0.003            0.636
55.084      1225.166         -0.000            0.635
-------------------------------------------------------
Number of function calls = 25
```

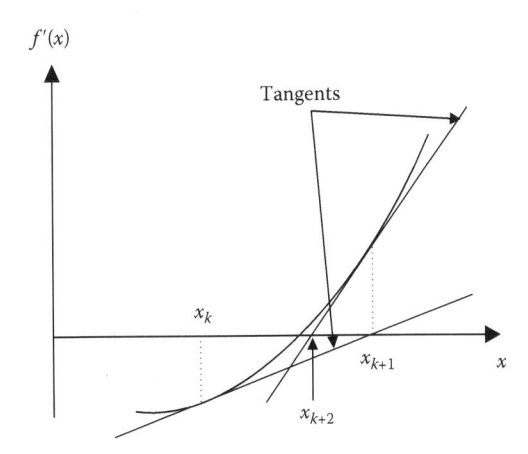

FIGURE 2.7
Newton–Raphson method.

TABLE 2.2

Algorithm for the Newton–Raphson Method

Step 1: Given x and Δx
Step 2: Compute, $f'(x)$ and $f''(x)$
  Store, $xprev = x$
  Update $x = xprev - \dfrac{f'(x)}{f''(x)}$
  If $|x - xprev| > \varepsilon$
    then goto Step 2
    else goto Step 3
Step 3: Converged. Print $x^* = x, f(x^*) = f(x), f'(x^*), f''(x^*)$

The minimum obtained by this method is in agreement with the earlier methods. The number of function evaluations in this method is 25 as compared to those in the bisection method, for which 52 function evaluations were required. The Newton–Raphson method has the following disadvantages:

- The convergence is sensitive to the initial guess. For certain initial guesses, the method can also show divergent trends. For example (Dennis and Schnabel 1983), the solution to the function $\tan^{-1} x$ converges when the initial guess, $|x| < a$, diverges when $|x| > a$ and cycle indefinitely if the initial guess is taken as $|x| = a$, where $a = 1.3917452002707$.
- The convergence slows down when the gradient value is close to zero.
- The second derivative of the function should exist.

2.3.3 Secant Method

In the bisection method, the sign of the derivative was used to locate zero of $f'(x)$. In the secant method, both the magnitude and the sign of the derivative are used to locate the zero of $f'(x)$. The first step in the secant method is the same as in the bisection method, That is, if $f'(x_1)$ and $f'(x_2)$ are the derivatives of a function computed at point x_1 and x_2, then the minimum of the function is located between x_1 and x_2 if

$$f'(x_1)f'(x_2) < 0 \qquad (2.6)$$

Further, it is assumed that $f'(x)$ varies linearly between points x_1 and x_2. A secant line is drawn between the two points x_1 and x_2. The point α where the secant line crosses the x-axis is taken as the improved point in the next iteration (see Figure 2.8).

One of the points, x_1 or x_2, is then eliminated using the aforementioned derivative condition. Thus, either the (x_1, α) or the (α, x_2) region is retained

1-D Optimization Algorithms

$f'(x)$

$f'(b)$

Secant

a

α

x^*

b

x

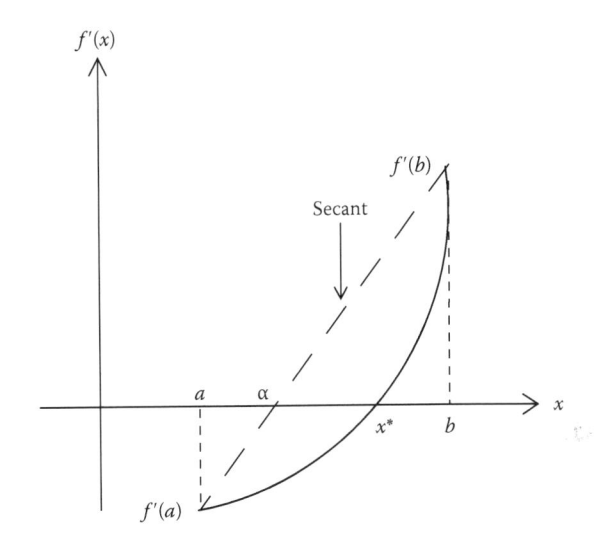

$f'(a)$

FIGURE 2.8
Secant method.

for the next iteration. The iteration continues until $f'(\alpha)$ is close to zero. The algorithm is coded in MATLAB (*secant.m*) and is described in Table 2.3.

On executing the code for the test problem, the output obtained is

Alpha	Deriv.
65.000	5.072
59.832	2.675
57.436	1.402
56.265	0.726
55.680	0.373
55.385	0.190
55.237	0.097
55.161	0.049
55.123	0.025
55.104	0.013
55.094	0.006
55.089	0.003
55.086	0.002

x* = 55.085   Minimum = 1225.166
Number of function calls = 82

The secant method is able to locate the minimum of the function, but with a higher number of function evaluations as compared to other gradient-based methods.

TABLE 2.3
Algorithm for the Secant Method

Step 1: Given a, b, ε, and Δx, flag = 0;

Step 2: Compute $\alpha = \dfrac{a+b}{2}$, $f'(a)$ and $f'(\alpha)$

    If $f'(a)f'(\alpha) < 0$
        then $b = \alpha$
                  set flag = 1(zero is bracketed)
        else $a = \alpha$
    If flag = 1
        then goto Step 3
        else goto Step 2

Step 3: Compute $\alpha = x_2 - \dfrac{f'(x_2)}{\left(f'(x_2)-f'(x_1)\right)/(x_2-x_1)}$

    If $f'(\alpha) > 0$
        then $b = \alpha$
        else $a = \alpha$
    If $|f'(\alpha)| < \varepsilon$
        then goto Step 4
        else goto Step 3

Step 4: Converged. Print $x^* = \alpha$, $f(x^*) = f(\alpha)$

2.3.4 Cubic Polynomial Fit

In this method, the function $f(x)$ to be minimized is approximated by a cubic polynomial $P(x)$ as

$$P(x) = a_0 + a_1 x + a_2 x^2 + a_3 x^3 \tag{2.7}$$

If the function $f(x)$ is evaluated at four different points, then the polynomial coefficients a_0, a_1, a_2, and a_4 can be evaluated by solving four simultaneous linear equations. Alternatively, if the value of the function and its derivatives are available at two points, the polynomial coefficients can still be evaluated. Once a polynomial is approximated for the function, the minimum point can be evaluated using the polynomial coefficients.

The first step in this search method is to bracket the minimum of the function between two points, x_1 and x_2, such that the following conditions hold:

$$f'(x_1)f'(x_2) < 0 \tag{2.8}$$

Using the information of $f(x_1)$, $f'(x_1)$, $f(x_2)$, and $f'(x_2)$, the minimum point of the approximating cubic polynomial can be given as

$$\bar{x} = \begin{cases} x_2 & \text{if } \mu < 0 \\ x_2 - \mu(x_2 - x_1) & \text{if } 0 \le \mu \le 1 \\ x_1 & \text{if } \mu > 0 \end{cases} \tag{2.9}$$

1-D Optimization Algorithms

where

$$\mu = \frac{f'(x_2) + w - z}{f'(x_2) - f'(x_1) + 2w} \tag{2.10}$$

$$z = \frac{3(f(x_1) - f(x_2))}{x_2 - x_1} + f'(x_1) + f'(x_2) \tag{2.11}$$

$$w = \frac{x_2 - x_1}{|x_2 - x_1|} + \sqrt{z^2 - f'(x_1)f'(x_2)} \tag{2.12}$$

The algorithm for this method is coded in MATLAB (*cubic.m*) and is described in Table 2.4.

On executing the code for the test problem, the output obtained is

```
      a                b
------------------------------
   40.000           65.000
   54.109           65.000
   54.109           55.120
------------------------------
x* =  55.084   Minimum = 1225.166
Number of function calls = 28
```

This method is able to capture the minimum point of the function with the number of function evaluations comparable to that in the Newton–Raphson method.

TABLE 2.4

Algorithm for Cubic Polynomial Fit

Step 1: Given x, ε, and Δx
Step 2: Compute $\alpha = \frac{a+b}{2}$, $f'(a)$ and $f'(\alpha)$
  If $f'(a) f'(\alpha) < 0$
    then $b = \alpha$
    else $a = \alpha$
Step 3: Repeat Step 2 until $f'(a) f'(\alpha) < 0$
Step 4: Using $f(a), f'(a), f(b), f'(b)$, compute μ, z, and w
Step 5: Compute $\bar{x}$
  If $|f'(\bar{x})| < \varepsilon$ goto Step 6
  If $f'(a) f'(\bar{x}) < 0$
    then $b = \bar{x}$
    else $a = \bar{x}$
      goto Step 4
Step 6: Converged. Print $x^* = \bar{x}$, $f(x^*) = f(\bar{x})$

2.3.5 Golden Section Method

Two numbers, p and q, are in a golden ratio if

$$\frac{p+q}{p} = \frac{p}{q} = \tau \tag{2.13}$$

Equation 2.13 can be written as

$$1 + \frac{q}{p} = \tau \tag{2.14}$$

or

$$1 + \frac{1}{\tau} = \tau \tag{2.15}$$

On solving the quadratic equation

$$\tau^2 - \tau - 1 = 0 \tag{2.16}$$

we get

$$\tau = \frac{1 + \sqrt{5}}{2} = 1.618033 \tag{2.17}$$

τ is called the golden number, which has a significance in aesthetics (e.g., the Egyptian pyramids).

Gradient information was required in the search methods that were discussed earlier. In the golden section method, the search is refined by eliminating certain regions based on function evaluations alone. No gradient computation is required in the golden section method. This method has two significant advantages over other region elimination techniques:

- Only one new function evaluation is required at each step.
- There is a constant reduction factor at each step.

The algorithm is coded in MATLAB (*golden.m*) and is described in Table 2.5.

1-D Optimization Algorithms

TABLE 2.5
Algorithm for the Golden Section Method

Step 1: Given x, ε, and τ
Step 2: Compute
$$\alpha_1 = a(1 - \tau) + b\tau$$
$$\alpha_2 = a\tau + b(1 - \tau)$$
Step 3: If $f(\alpha_1) > f(\alpha_2)$
  then $a = \alpha_1, \alpha_1 = \alpha_2, \alpha_2 = a\tau + b(1 - \tau)$
  else $a = \alpha_2, \alpha_2 = \alpha_1, \alpha_1 = a(1 - \tau) + b\tau$
Step 4: Repeat Step 3 until $|f(\alpha_1) - f(\alpha_2)| < \varepsilon$
Step 5: Converged. Print $x^* = \alpha_1, f(x^*) = f(\alpha_1)$

On executing the code for the test problem, output obtained is

```
       a                b
-----------------------------
    40.000           90.000
    40.000           70.902
    40.000           59.098
    47.295           59.098
    51.803           59.098
    51.803           56.312
    53.526           56.312
    54.590           56.312
    54.590           55.654
    54.590           55.248
    54.841           55.248
    54.996           55.248
    54.996           55.152
    55.056           55.152
    55.056           55.115
    55.056           55.092
-----------------------------
x* = 55.077  Minimum = 1225.166
Number of function calls = 18
```

2.3.6 Other Methods

In addition to the golden section method, there are other direct search methods that can be used to solve the one-dimensional optimization problems, including

- Dichotomous search
- Interval halving method
- Fibonacci method

In the dichotomous search, a function is evaluated at two points, close to the center of the interval of uncertainty. Let these two points be x_a and x_b given by

$$x_a = \frac{L}{2} - \frac{\delta}{2} \qquad (2.18)$$

$$x_b = \frac{L}{2} + \frac{\delta}{2} \qquad (2.19)$$

where δ is a small number and L is the region of uncertainty. Depending on the computed value of the function at these points, a certain region is eliminated. In Figure 2.9, the region toward the right-hand side of x_b is eliminated. In this method, the region of uncertainty after n function evaluations is given by

$$\frac{L}{2^{n/2}} + \delta\left(1 - \frac{1}{2^{n/2}}\right) \qquad (2.20)$$

In the interval halving method, half of the region of uncertainty is deleted in every iteration. The search space is divided into four equal parts and function evaluation is carried out at x_1, x_2, and x_3. Again, a certain region gets eliminated based on the value of the functions computed at three points. In Figure 2.10, the region toward the right-hand side of x_2 is eliminated. In this method, the region of uncertainty after n function evaluations is given by

$$\left(\frac{1}{2}\right)^{\frac{n-1}{2}} L \qquad (2.21)$$

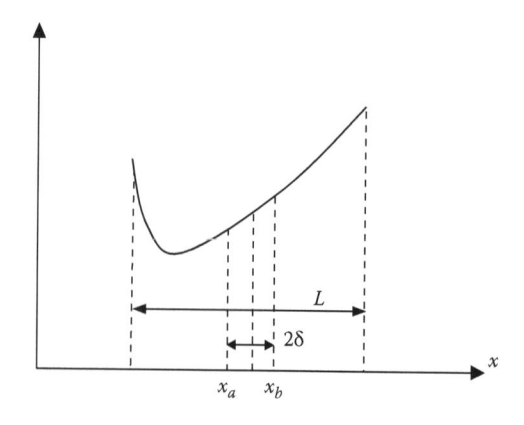

FIGURE 2.9
Dichotomous search.

1-D Optimization Algorithms

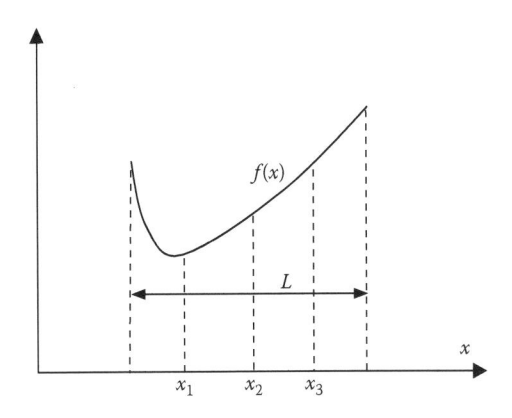

FIGURE 2.10
Interval halving method.

A Fibonacci sequence is given by

$$F_n = F_{n-1} + F_{n-2} \tag{2.22}$$

where

$$F_0 = F_1 = 1 \tag{2.23}$$

In the Fibonacci method, the functions are evaluated at points

$$x_a = a + L^* \tag{2.24}$$

$$x_b = b - L^* \tag{2.25}$$

where $[a, b]$ define the region of uncertainty and L^* is given by

$$L^* = \frac{F_{n-2}}{F_n} L \tag{2.26}$$

In this method n has to be defined before the start of the algorithm.

2.4 Comparison of Solution Methods

Having defined a number of solution methods to find the minimum of a function, it is natural to ask the question of which solution method to use for a given problem. The answer is quite straightforward: no single method can be used for all types of problems. Different methods may have to be tried for different problems.

Let us evaluate the efficiency of each of the methods for the test case problem that we discussed in an earlier section. One way of defining efficiency of an optimization method could be to show how x approaches x^* with increasing iterations. Because the number of function evaluations in each iteration is different for different methods, we can plot $|x - x^*|$ versus number of function evaluations for a meaningful comparison. Figure 2.11 shows this plot for different solution methods for the solar energy test problem. It is observed from this figure that the cubic polynomial fit and Newton–Raphson approach x^* with 25 number of function evaluations. The bisection and secant methods take a much larger number of function evaluations to reach the minimum. The golden section method takes a minimum number of function evaluations.

Let us further evaluate these methods for some well-known test problems (Philips et al. 1976; Reklaitis et al. 1983). Table 2.6 summarizes the number of function evaluations required by each of the methods in reaching the minimum of the function. The golden section, cubic polynomial fit, and Newton–Raphson methods perform well for all the test problems except for the function

$$2(x-3)^2 + e^{0.5x^2} \quad 0 \leq x \leq 100$$

which is highly skewed. The Newton–Raphson method requires a good initial guess for convergence. It takes 275 function evaluations for convergence with an initial guess of $x = 5$. The method takes fewer function evaluations for convergence with $x < 5$. However, the method diverges for $x > 10$. The cubic polynomial fit did not converge for this particular function. The golden

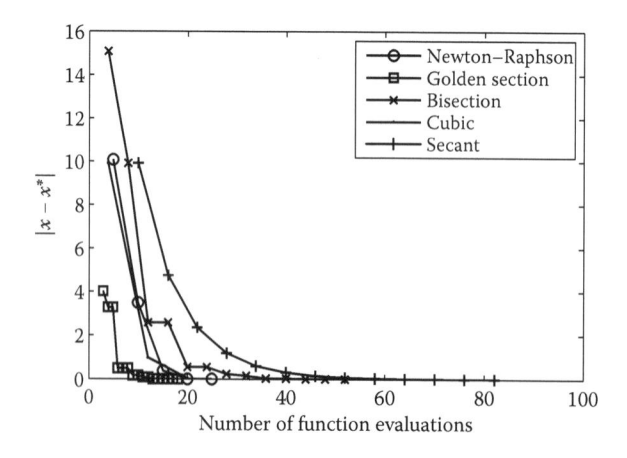

FIGURE 2.11
Comparing different solution methods.

1-D Optimization Algorithms

TABLE 2.6
Comparing Different Solution Techniques for Different Problems

			Number of Function Evaluations				
Minimize	x^*	$f(x^*)$	Golden	Bisection	Cubic	Newton	Secant
$3x^4 + (x-1)^2$ $0 \le x \le 4$	0.451	0.426	16	36	36	35	346
$-4x \sin x$ $0 \le x \le \pi$	2.029	−7.28	14	36	24	20	32
$2(x-3)^2 + e^{0.5x^2}$ $0 \le x \le 100$	1.591	7.516	14	36	–	275	–
$3x^2 + \dfrac{12}{x^3} - 5$ $0.5 \le x \le 2.5$	1.431	5.238	14	32	28	20	604
$2x^2 + \dfrac{16}{x}$ $1 \le x \le 5$	1.587	15.12	12	36	28	25	70

and bisection methods converged for all the test functions. The solution to these problems is obtained by modifying the *func.m* routine and executing the code for the corresponding method.

Chapter Highlights

- The one-dimensional (1-D) optimization problems refer to an objective function that has one variable. 1-D optimization algorithms form the basic building blocks for the multivariable algorithms.
- If a function is either continuously increasing or decreasing between two points, then it is referred as a monotonic function.
- In a unimodal function, the function is monotonic on either side of its minimum point.
- The solution techniques for one-dimensional optimization problems can be classified into gradient-based and non–gradient-based algorithms. Some popular gradient-based algorithms are bisection, cubic polynomial fit, secant, and Newton–Raphson methods. The golden section algorithm does not require derivative information of the function.
- The Newton–Raphson method requires the second derivative of the function, and convergence of this method is strongly dependent on a good initial guess.
- In the bisection method, the sign of the derivative is used to locate the zero of $f'(x)$. In the secant method, both magnitude and sign of the derivative are used to locate the zero of $f'(x)$.

- In the golden section method, the search is refined by eliminating certain regions based on function evaluations only. No gradient computation is required in the golden section method. This method derives its name from the number 1.61803, referred to as the golden number, which has significance in aesthetics.

Formulae Chart

Newton–Raphson method:

$$x_{k+1} = x_k - \frac{f'(x_k)}{f''(x_k)}$$

Secant method:

$$\alpha = x_2 - \frac{f'(x_2)}{(f'(x_2) - f'(x_1))/(x_2 - x_1)}$$

Problems

1. For a lifting body, lift (L) to drag (D) ratio varies with angle of attack (α) as

$$\frac{L}{D} = -0.004\alpha^2 + 0.16\alpha + 0.11$$

   where α lies between 0 and 35 degrees. Find the α at which L/D is maximum. Use different algorithms presented in this chapter to arrive at the optimum.

2. Use golden section, cubic polynomial fit, bisection, and secant methods to minimize the following functions:

   a. $3e^x - x^3 + 5x$ $-3 \leq x \leq 3$

   b. $-x^3 + 4x^2 - 3x + 5$ $-2 \leq x \leq 2$

   c. $e^{x^2} - 2x^3$ $-0.5 \leq x \leq 2$

   d. $2x^2 + \frac{10}{x}$ $0 \leq x \leq 4$

3. Find the maximum value of the function

$$f(x) = \frac{1}{2}\left[1 - \frac{1}{1+x^2} - x\tan^{-1}\left(\frac{1}{2x}\right)\right] \quad 0 \le x \le 3$$

4. Find the maximum value of the function

$$f(x) = 5x^2 - e^x \quad 0 \le x \le 5$$

5. Find the maximum and minimum of the function

$$f(x) = \ln(\cos x^{\cos x} + 1) \quad 0 \le x < \frac{\pi}{2}$$

6. The strength of a beam varies as the product of its breadth and square of its depth. Find the dimension of the strongest beam that can be cut from a circular log of diameter 1 m.

7. A car burns petrol at the rate of $\left(\dfrac{300}{x} + \dfrac{x}{3}\right)$ liters per 100 km where x is the speed in km/h. The cost of petrol is one dollar per liter and the chauffeur is paid $7 per hour. Find the steady speed that will minimize the total cost of the trip of 600 km.

8. A swimmer in the sea is at a distance of 5 km from the closest point C on the shore on a straight line. The house of the swimmer is on the shore at a distance of 7 km from point C. He can swim at a speed of 2 km/h and run at a speed of 6 km/h. At what spot on the shore should he land so that he reaches his house in the shortest possible time?

9. The following data are given for an aircraft that is flying at an altitude of 5 km:

Weight $= W = 700{,}000$ N

Reference area $= S = 140$ m²

Aspect ratio $= AR = 8$

Efficiency factor $= e = 0.82$

Drag coefficient $= C_D = 0.018$

Atmospheric density $= \rho = 0.73612$ kg/m³

The thrust (T) of the aircraft is related to its velocity (v) by the equation

$$T = \frac{1}{2}\rho v^2 S C_D + \frac{2W^2}{\rho v^2 S} \frac{1}{\pi e AR} \quad 100 \le v \le 300 \text{ m/s}$$

Find the velocity of the aircraft at which the thrust requirement is minimum.

10. Plot the function

$$f(x) = x^4 + x^3 - x^2 - 5 \quad -2 \leq x \leq 2$$

and identify the region where the function is concave and convex. Identify the local and global minima for this function.

11. The consumer demand function is given by

$$f(x) = \frac{k}{p_2}x - \frac{p_1}{p_2}x^2$$

where $k = 90$, $p_1 = 10$, and $p_2 = 5$. Maximize the function $f(x)$.

12. Minimize the function

$$f(x) = 2(\sqrt{x} - 3) + \frac{\sqrt{x}}{100} \quad 0.3 \leq x \leq 0.6$$

13. A cone-shaped biscuit cup is to be designed for minimum surface area so that it can hold 130 mL of ice cream. Determine the dimensions of the cone.

14. Microorganisms such as bacteria have an elongated shape (see Figure 2.12). The frictional coefficient τ relates the force on a particle and its velocity when moving through a viscous fluid:

$$\tau = \frac{4\pi\rho a}{\ln\left(\frac{2a}{b}\right) - \frac{1}{2}}$$

where ρ is fluid viscosity (for water this value is 1 (μg/s)/μm). For a short axis of $b = 1$ μm, find the value of a that corresponds to the minimum in the friction coefficient in water (King and Mody 2011).

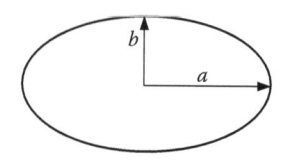

FIGURE 2.12
Elliptical shape of a bacterium.

3
Unconstrained Optimization

3.1 Introduction

The solution techniques for unconstrained optimization problems with multiple variables are dealt in this chapter. In practice, optimization problems are constrained, and unconstrained optimization problems are few. One example of an unconstrained optimization problem is data fitting, where one fits a curve on the measured data. However, the algorithms presented in this chapter can be used to solve constrained optimization problems as well. This is done by suitably modifying the objective function, which includes a penalty term in case constraints are violated.

The solution methods for unconstrained optimization problems can be broadly classified into gradient-based and non–gradient-based search methods. As the name suggests, gradient-based methods require gradient information in determining the search direction. The gradient-based methods discussed in this chapter are steepest descent, Davidon–Fletcher–Powell (DFP), Broyden–Fletcher–Goldfarb–Shanno (BFGS), Newton, and Levenberg–Marquardt methods. The search direction computed by these methods uses the gradient information, Hessian information, or a combination of these two. Some methods also make an approximation of the Hessian matrix. Once the search direction is identified, one needs to evaluate how much to move in that direction so as to minimize the function. This is a one-dimensional problem. We will be using the golden section method, as discussed in Chapter 2, for solving the one-dimensional problem. The non–gradient-based method does not require derivatives or second derivative information in finding the search direction. The search direction is guided by the function evaluations as well as the search directions computed from earlier iterations. Powell's conjugate direction method, a non–gradient-based method, is elaborated in this chapter as it is much superior (shows quadratic convergence) to other non-gradient methods such as simplex and pattern search methods. The simplex method (Nelder–Mead algorithm) is also discussed in Section 3.4.9 on the direct search method. In the last section, Powell's method is used to solve a complicated motion design problem of a robot. The road map of this chapter is shown in Figure 3.1.

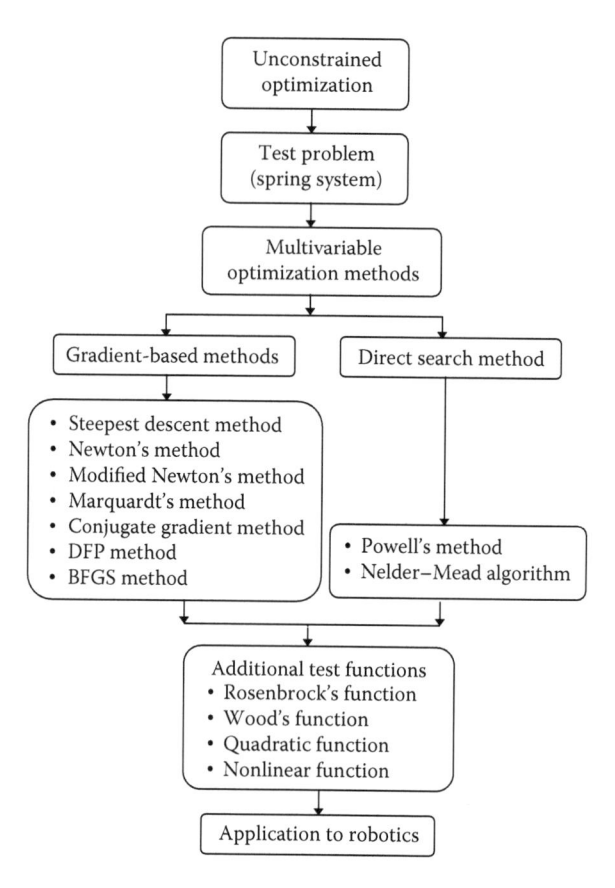

FIGURE 3.1
Road map of Chapter 3.

For a single-variable function, it was discussed earlier that the derivative of the function vanishes at the optimum and the second derivative of the function is greater than zero at the minimum of the function. The same can be extended to a multivariable function. The necessary conditions for x^* to be a minimum are that

$$\nabla f(x^*) = 0 \tag{3.1}$$

and $x^T H x$ is positive definite ($x^T H x > 0$). To ensure this, eigenvalues of H are to be positive. Consider a two-variable function

$$f(x) = x_1^2 + x_2^2 - 2x_1 \tag{3.2}$$

Unconstrained Optimization

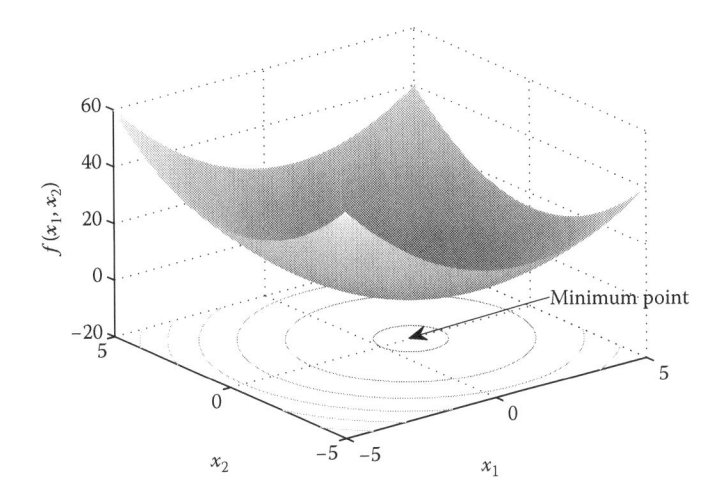

FIGURE 3.2
Surface-contour plot of the function.

The gradient is

$$\nabla f(x) = \begin{bmatrix} 2x_1 - 2 \\ 2x_2 \end{bmatrix} \quad (3.3)$$

Equating the gradient to zero, the optimum is at (1, 0). For this function $x^T Hx > 0$. Hence, the point (1, 0) is the minimum of $f(x)$. The surface-contour plot of this function is shown in Figure 3.2.

For a two-variable function

$$f(x) = x_1^2 - x_2^2 \quad (3.4)$$

the optimum is at (0, 0) from the first-order condition. Checking the second-order condition, we find that $x^T Hx = 0$. Therefore, the point (0, 0) represents saddle point (see Figure 3.3).

3.2 Unidirectional Search

The unidirectional search refers to minimizing the value of a multivariable function along a specified direction. For example, if x_i is the initial starting point of the design variables for minimizing a multivariable function and S_i

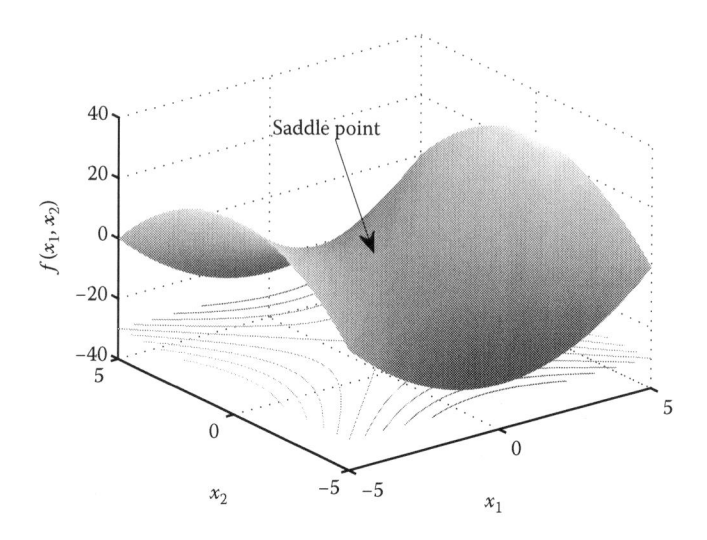

FIGURE 3.3
Surface-contour plot of the function with saddle point.

is the search direction, then we need to determine a scalar quantity α such that the function

$$f(\alpha) = x_i + \alpha S_i \tag{3.5}$$

is minimized. The value of α at which this function reaches a minimum is given by α^*. This is a one-dimensional optimization problem and we can use the golden section technique to minimize this function. The golden section method is modified to handle multivariable functions and the MATLAB® code *golden_funct1.m* is given.

Let us perform a unidirectional search on the Rosenbrock function given by

$$f(x) = 100\left(x_2 - x_1^2\right)^2 + (1 - x_1)^2 \tag{3.6}$$

with different starting values of x and with different search directions. The results are summarized in Table 3.1. It is observed from this table that

TABLE 3.1
Unidirectional Search for a Multivariable Function

x_i	$f(x_i)$	S_i	α^*	$f(\alpha^*)$
(3, 0.5)	7229	(2, 1)	−1.3731	88.45
(3, 0.5)	7229	(2, 3)	−1.1249	1181.7
(1, 1)	0	(2, 2)	0	0
(2, 2)	401	(1, 1)	−1	0

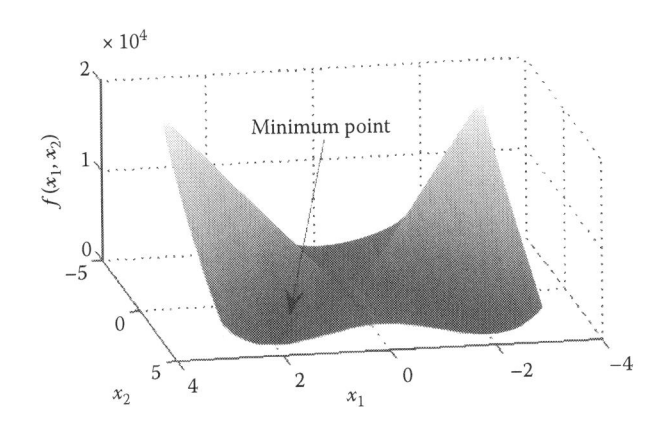

FIGURE 3.4
Rosenbrock function.

performing a linear search in the direction (2, 1) from the starting point (3, 0.5) results in $f(\alpha^*) = 88.45$ as compared to initial function value of 7229. This can be easily shown on the MATLAB command prompt as

```
>> x = [3 0.5];
>> search = [2 1];
>> [alpha1,falpha1] = golden_funct1(x,search)

alpha1 =
  -1.3731

falpha1 =
  88.4501
```

The function has to be appropriately coded in *func_multivar.m*. Note that this function has a minimum at (1, 1) and the minimum value of the function is zero. If we are at minimum point, then any search direction should not improve the function value. It is the reason why search in the direction (2, 2) from the point (1, 1) results in $f(\alpha^*) = 0$ with $\alpha^* = 0$. Similarly, search in the direction (1, 1) from the point (2, 2) results in $f(\alpha^*) = 0$ with $\alpha^* = -1$. This function is plotted in Figure 3.4 and is constructed by executing the MATLAB code (*rosenbrock.m*).

3.3 Test Problem

Let us define a spring system as a test problem on which we will apply multivariable optimization algorithms such as the steepest descent, DFP, BFGS,

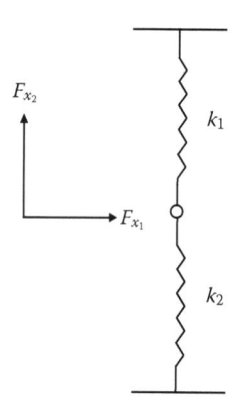

FIGURE 3.5
Spring system.

Newton, and Levenberg–Marquardt methods. Consider two springs of unit length and with stiffness k_1 and k_2, joined at the origin. The other two ends of the springs are fixed on a wall (see Figure 3.5). On applying a force, the spring system will deflect to an equilibrium position, which we are interested in determining. The potential of the spring system is given by

$$U = k_1 \left(\sqrt{x_1^2 + (x_2 + 1)^2} - 1\right)^2 + k_2 \left(\sqrt{x_1^2 + (x_2 - 1)^2} - 1\right)^2 - (F_{x_1} x_1 + F_{x_2} x_2) \quad (3.7)$$

where (F_{x_1}, F_{x_2}) is the force applied at the origin due to which it moves to a position (x_1, x_2). Assuming $k_1 = 100$ N/m, $k_2 = 90$ N/m, and $(F_{x_1}, F_{x_2}) = (20, 40)$, our aim is to evaluate (x_1, x_2) such that U is minimized.

A MATLAB code (*springsystem.m*) is used to find the minimum of the potential function by varying the design variables from –1 to 1 in steps of 0.01. On executing this code, the output obtained is

```
Minimum Potential   = -9.6547
occurs at x1,x2     = 0.5000     0.1200
```

3.4 Solution Techniques

Similar to 1-D optimization algorithms, solution techniques for multivariable, unconstrained optimization problems can be grouped into gradient- and non–gradient-based methods. Gradient-based methods require derivative information of the function in constituting a search. The first and

Unconstrained Optimization

second derivatives can be computed using the central difference formula as given below.

$$\frac{\partial f}{\partial x_i} = \frac{f(x_i + \Delta x_i) - f(x_i - \Delta x_i)}{2\Delta x_i} \tag{3.8}$$

$$\frac{\partial^2 f}{\partial x_i^2} = \frac{f(x_i + \Delta x_i) - 2f(x_i) + f(x_i - \Delta x_i)}{\Delta x_i^2} \tag{3.9}$$

$$\frac{\partial^2 f}{\partial x_i \partial x_j} = \Big[f(x_i + \Delta x_i, x_j + \Delta x_j) - f(x_i + \Delta x_i, x_j - \Delta x_j)$$
$$- f(x_i - \Delta x_i, x_j + \Delta x_j) + f(x_i - \Delta x_i, x_j - \Delta x_j) \Big] / (4\Delta x_i \Delta x_j) \tag{3.10}$$

The computation of first derivative requires two function evaluations with respect to each variable. So for an n variable problem, $2n$ function evaluations are required for computing the gradient vector. The computation of the Hessian matrix requires $O(n^2)$ function evaluations. Note that in the Hessian matrix

$$\frac{\partial^2 f}{\partial x_i \partial x_j} = \frac{\partial^2 f}{\partial x_j \partial x_i} \tag{3.11}$$

Alternatively, one can also compute the derivative of a function using complex variables as

$$f'(x) = \frac{\text{Imaginary}\left[f(x + i\Delta x)/\Delta x \right]}{\Delta x} \tag{3.12}$$

The gradient-based methods such as steepest descent, DFP, BFGS, Newton, and Levenberg–Marquardt methods are discussed next followed by Powell's conjugate direction method, which is a direct search method. The efficiency of solution methods can be gauged by three criteria:

- Number of function evaluations.
- Computational time.
- Rate of convergence. By this we mean how fast the sequence $x_i, x_{i+1}, \ldots$ converges to x^*. The rate of convergence is given by the parameter n in the equation

$$\frac{\|x_{i+1} - x^*\|}{\|x_i - x^*\|^n} \leq c, \quad c \geq 0, \quad n \geq 0 \qquad (3.13)$$

- For $n = 1$ and $0 \leq c \leq 1$ the method is said to have *linear convergence*. For $n = 2$, the method is said to have *quadratic convergence*. When the rate of convergence is higher, the optimization method is better. A method is said to have *superlinear convergence* if

$$\lim_{i \to \infty} \left(\frac{\|x_{i+1} - x^*\|}{\|x_i - x^*\|^n} \right) \leq c, \quad c \geq 0, \quad n \geq 0 \qquad (3.14)$$

3.4.1 Steepest Descent Method

The search direction S_i that reduces the function value is a descent direction. It was discussed earlier that along the gradient direction, there is the maximum change in the function value. Thus, along the negative gradient direction, the function value decreases the most. The negative gradient direction is called the steepest descent direction. That is,

$$S_i = -\nabla f(x_i) \qquad (3.15)$$

In successive iterations, the design variables can be updated using the equation

$$x_{i+1} = x_i - \alpha \nabla f(x_i) \qquad (3.16)$$

where α is a positive scalar parameter that can be determined using the line search algorithm such as the golden section method.

The steepest descent method ensures a reduction in the function value at every iteration. If the starting point is far away from the minimum, the gradient will be higher and the function reduction will be maximized in each iteration. Because the gradient value of the function changes and decreases to a small value near the optimum, the function reduction is uneven and the method becomes sluggish (slow convergence) near the minimum. The method can therefore be utilized as a starter for other gradient-based algorithms. The algorithm for the steepest descent method is described in Table 3.2 and a MATLAB code of its implementation is given in *steep_des.m*.

On executing the code with a starting value of x as (–3, 2), following output is produced for the test problem. After the first iteration, the function value decreases from 1452.2619 to –2.704. Notice from the output that as the gradient value decreases, the reduction in function value at each iteration also

Unconstrained Optimization

TABLE 3.2
Algorithm for the Steepest Descent Method

Step 1: Given x_i (starting value of design variable)
　　　　ε_1 (tolerance of function value from previous iteration)
　　　　ε_2 (tolerance on gradient value)
　　　　Δx (required for gradient computation)
Step 2: Compute $f(x_i)$ and $\nabla f(x_i)$ (function and gradient vector)
　　　　$S_i = -\nabla f(x_i)$　　　　(search direction)
　　　　Minimize $f(x_{i+1})$ and determine α (use golden section method)
　　　　$x_{i+1} = x_i + \alpha S_i$　　　　(update the design vector)
　　　　If $|f(x_{i+1}) - f(x_i)| > \varepsilon_1$ or $\|\nabla f(x_i)\| > \varepsilon_2$
　　　　　　then　　goto Step 2
　　　　　　else　　goto Step 3
Step 3: Converged. Print $x^* = x_{i+1}, f(x^*) = f(x_{i+1})$

decreases. The steepest descent algorithm converges to the minimum of the test problem in 15 iterations. Observe the sluggishness of the algorithm as it approaches the minimum point. The convergence history is shown pictorially in Figure 3.6 along with the function contours of the test problem. The function contours can be plotted using the MATLAB code *contour_testproblem.m*.

```
          Initial function value = 1452.2619
No.        x-vector              f(x)        Deriv.

 1       0.095     0.023        -2.704      1006.074
 2       0.170     0.141        -5.278        37.036
 3       0.318     0.048        -7.369        23.451
 4       0.375     0.138        -8.773        26.656
 5       0.448     0.092        -9.375        14.021
 6       0.470     0.127        -9.583        10.071
 7       0.491     0.114        -9.639         4.403
 8       0.497     0.123        -9.652         2.509
 9       0.501     0.120        -9.655         1.050
10       0.503     0.122        -9.656         0.554
11       0.504     0.122        -9.656         0.236
12       0.504     0.122        -9.656         0.125
13       0.504     0.122        -9.656         0.047
14       0.504     0.122        -9.656         0.027
15       0.504     0.122        -9.656         0.016
```

3.4.2 Newton's Method

The search direction in this method is based on the first and second derivative information and is given by

$$S_i = -[H]^{-1}\nabla f(x_i) \tag{3.17}$$

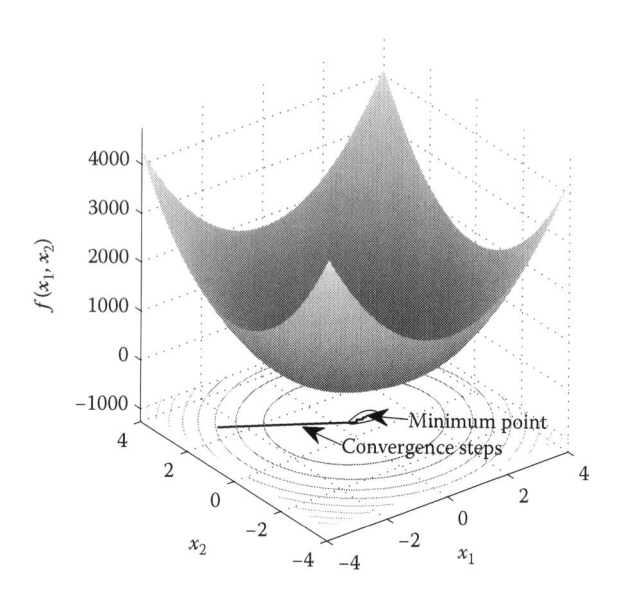

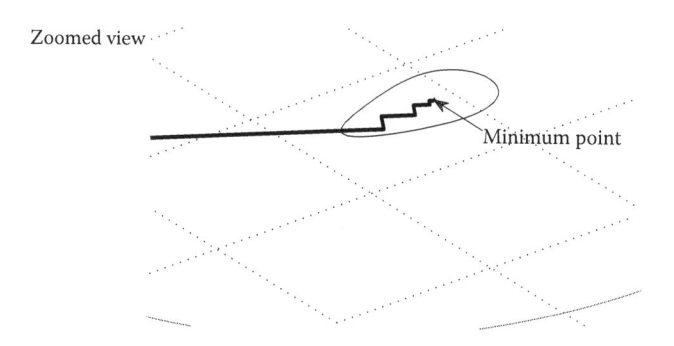

FIGURE 3.6
Function contours of the test problem and convergence history.

where $[H]$ is the Hessian matrix. If this matrix is positive definite, then S_i will be a descent direction. The same can be assumed true near the vicinity of the optimum point. However, if the initial starting point is far away from the optimum, the search direction may not always be descent. Often a restart is required with a different starting point to avoid this difficulty. Though the Newton's method is known for converging in a single iteration for a quadratic function, seldom do we find functions in practical problems that are quadratic. However, Newton's method is often used as a hybrid method in conjunction with other methods.

The algorithm for the Newton's method is described in Table 3.3 and a MATLAB code of its implementation is given in *newton.m*. A MATLAB code that computes Hessian matrix is given in *hessian.m*.

TABLE 3.3
Algorithm for Newton's Method

Step 1: Given x_i (starting value of design variable)
        ε_1 (tolerance of function value from previous iteration)
        ε_2 (tolerance on gradient value)
        Δx (required for gradient computation)
Step 2: Compute $f(x_i)$, $\nabla f(x_i)$, and $[H]$ (function, gradient, and Hessian)
        $S_i = -[H]^{-1}\nabla f(x_i)$ (search direction)
        $x_{i+1} = x_i + S_i$ (update the design vector)
        If $|f(x_{i+1}) - f(x_i)| > \varepsilon_1$ or $\|\nabla f(x_i)\| > \varepsilon_2$
            then   goto Step 2
            else   goto Step 3
Step 3: Converged. Print $x^* = x_{i+1}$, $f(x^*) = f(x_{i+1})$

On executing the code with a starting value of x as (–3, 2), the following output is displayed in the command window for the test problem. Note that in some iteration, the search direction is not a descent as the function value increases instead of monotonically decreasing. The method, however, converges to the minimum point.

Initial function value = 1452.2619

No.	x-vector		f(x)	Deriv.
1	-0.754	0.524	44.244	1006.074
2	-0.362	-0.010	8.398	116.281
3	0.094	0.125	-3.920	50.597
4	11.775	0.324	22007.14	21.420
5	1.042	0.093	14.533	4076.916
6	0.640	0.142	-8.479	102.745
7	0.524	0.122	-9.635	18.199
8	0.505	0.122	-9.656	2.213
9	0.504	0.122	-9.656	0.059
10	0.504	0.122	-9.656	0.000

Let us restart the method with x as (1, 1). The output is given below. If the starting value is closer to the minimum, the function value reduces monotonically in all the iterations and eventually converges to the minimum.

Initial function value = 92.7864

No.	x-vector		f(x)	Deriv.
1	0.818	0.041	-1.428	202.492
2	0.569	0.138	-9.386	56.085
3	0.510	0.122	-9.655	8.516
4	0.504	0.122	-9.656	0.602
5	0.504	0.122	-9.656	0.004

TABLE 3.4

Algorithm for Modified Newton's Method

Step 1: Given x_i (starting value of design variable)
        ε_1 (tolerance of function value from previous iteration)
        ε_2 (tolerance on gradient value)
        Δx (required for gradient computation)
Step 2: Compute $f(x_i)$, $\nabla f(x_i)$, and $[H]$ (function, gradient, and Hessian)
        $S_i = -[H]^{-1} \nabla f(x_i)$ (search direction)
        Minimize $f(x_{i+1})$ and determine α (use golden section method)
        $x_{i+1} = x_i + \alpha S_i$ (update the design vector)
        If $|f(x_{i+1}) - f(x_i)| > \varepsilon_1$ or $\|\nabla f(x_i)\| > \varepsilon_2$
           then   goto Step 2
           else   goto Step 3
Step 3: Converged. Print $x^* = x_{i+1}$, $f(x^*) = f(x_{i+1})$

3.4.3 Modified Newton's Method

The method is similar to Newton's method with a modification that a unidirectional search is performed in the search direction S_i of the Newton method. The algorithm for the modified Newton method is described in Table 3.4 and a MATLAB code of its implementation is given in *modified_newton.m*.

On executing the code with a starting value of x as (−3, 2), the following output is displayed in the command window for the test problem. For the same starting point, the modified Newton's method converges to the minimum point in just six iterations as compared to Newton's method, which converges in ten iterations.

```
     Initial function value = 1452.2619
No.      x-vector              f(x)           Deriv.

1     0.006      0.025       -1.010         1006.074
2     0.498      0.026       -8.227           36.392
3     0.496      0.121       -9.653           29.839
4     0.504      0.122       -9.656            0.873
5     0.504      0.122       -9.656            0.018
6     0.504      0.122       -9.656            0.003
```

3.4.4 Levenberg–Marquardt Method

The advantage of the steepest descent method is that it reaches closer to the minimum of the function in a few iterations even when the starting guess is far away from the optimum. However, the method shows sluggishness near the optimum point. On the contrary, Newton's method shows a faster convergence if the starting guess is close to the minimum point. Newton's method may not converge if the starting point is far away from the optimum point.

Unconstrained Optimization

The Levenberg–Marquardt method is a kind of hybrid method that combines the strength of both the steepest descent and Newton's methods. The search direction in this method is given by

$$S_i = -[H + \lambda I]^{-1}\nabla f(x_i) \tag{3.18}$$

where I is an identity matrix and λ is a scalar that is set to a high value at the start of the algorithm. The value of λ is altered during every iteration depending on whether the function value is decreasing or not. If the function value decreases in the iteration, λ it decreases by a factor (less weightage on steepest descent direction). On the other hand, if the function value increases in the iteration, λ it increases by a factor (more weightage on steepest descent direction). The algorithm for the Levenberg–Marquardt method is described in Table 3.5 and a MATLAB code of its implementation is given in *levenbergmarquardt.m*.

On executing the code with a starting value of x as $(-3, 2)$, following output is displayed at the command window for the test problem.

```
         Initial function value = 1452.2619
No.        x-vector              f(x)          Deriv.

1       -2.384    1.604         815.738        1006.074
2       -1.680    1.139         325.925         733.709
3       -1.104    0.705         102.059         429.113
4       -0.740    0.327          28.673         201.554
5       -0.444    0.133           8.324          86.884
6       -0.164    0.105           1.186          34.005
7        0.546    0.091          -9.390          20.542
8        0.508    0.122          -9.655          11.361
9        0.505    0.122          -9.656           0.409
10       0.504    0.122          -9.656           0.016
```

TABLE 3.5

Algorithm for the Levenberg–Marquardt Method

Step 1: Given x_i (starting value of design variable)
        ε_1 (tolerance of function value from previous iteration)
        ε_2 (tolerance on gradient value)
        Δx (required for gradient computation)
Step 2: Compute $f(x_i)$, $\nabla f(x_i)$, and $[H]$ (function, gradient, and Hessian)
        $S_i = -[H + \lambda I]^{-1}\nabla f(x_i)$ (search direction)
        $x_{i+1} = x_i + S_i$ (update the design vector)
        If $f(x_{i+1}) < f(x_i)$
            then   change the value of λ as $\lambda/2$
            else   change the value of λ as 2λ
        If $|f(x_{i+1}) - f(x_i)| > \varepsilon_1$ or $\|\nabla f(x_i)\| > \varepsilon_2$
            then   goto Step 2
            else   goto Step 3
Step 3: Converged. Print $x^* = x_{i+1}$, $f(x^*) = f(x_{i+1})$

3.4.5 Fletcher–Reeves Conjugate Gradient Method

The Levenberg–Marquardt method uses the strengths of both steepest descent and Newton's method for accelerating the convergence to reach the minimum of a function. The method is a second-order method, as it requires computation of the Hessian matrix. On the other hand, the conjugate gradient method is a first-order method, but shows the property of quadratic convergence and thus has a significant advantage over the second-order methods. Two directions, S_1 and S_2, are said to be conjugate if

$$S_1^T H S_2 = 0 \qquad (3.19)$$

where H is a symmetric matrix. For example, orthogonal directions are conjugate directions. In Figure 3.7, starting from point x_{1a}, the search direction S_1 results in the minimum point x_a^*. Similarly, starting from point x_{1b}, the search direction S_1 results in the minimum point x_b^*. The line joining x_a^* and x_b^* is the search direction S_2. Then, S_1 and S_2 are conjugate directions.

The steepest descent method was modified by Fletcher and Reeves in the conjugate gradient method. Starting with the search direction

$$S_1 = -\nabla f(x_1) \qquad (3.20)$$

the subsequent search direction is taken as a linear combination of S_1 and $-\nabla f(x_2)$. That is,

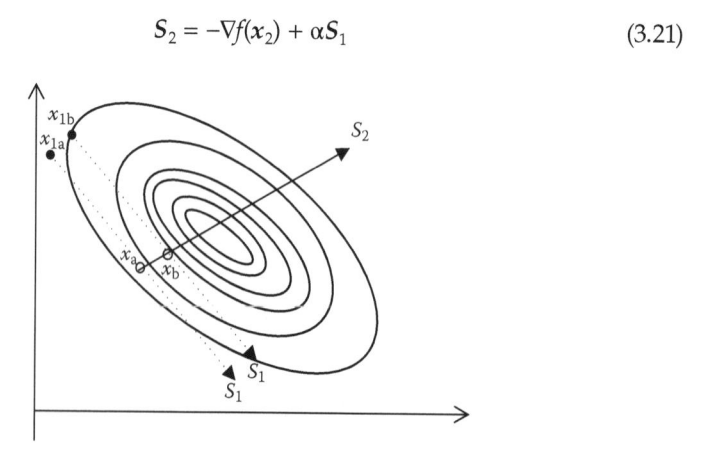

$$S_2 = -\nabla f(x_2) + \alpha S_1 \qquad (3.21)$$

FIGURE 3.7
Conjugate directions.

Unconstrained Optimization

Using the property $S_1^T HS_2 = 0$ of conjugate directions, α can be evaluated as

$$\alpha = \frac{\|\nabla f(x_{i+1})\|^2}{\|\nabla f(x_i)\|^2} \tag{3.22}$$

Starting with $S_1 = -\nabla f(x_1)$, the search direction in every iteration is calculated using the equation

$$S_{i+1} = -\nabla f(x_i) + \frac{\|\nabla f(x_{i+1})\|^2}{\|\nabla f(x_i)\|^2} S_i \tag{3.23}$$

The algorithm for the conjugate gradient method is described in Table 3.6 and a MATLAB code of its implementation is given in *conjugate.m*.

On executing the code with a starting value of x as $(-3, 2)$, the following output is displayed at the command window in the test problem. The efficiency of conjugate gradient method can be seen from Figure 3.8, where it

TABLE 3.6

Algorithm for Fletcher–Reeves's Conjugate Gradient Method

Step 1: Given x_i (starting value of design variable)
ε_1 (tolerance of function value from previous iteration)
ε_2 (tolerance on gradient value)
Δx (required for gradient computation)

Step 2: Compute $f(x_i)$ $\nabla f(x_i)$ (function and gradient)
$S_i = -\nabla f(x_i)$ (search direction)
Minimize $f(x_{i+1})$ and determine α (use the golden section method)
$x_{i+1} = x_i + \alpha S_i$ (update the design vector)

Step 3: $S_{i+1} = -\nabla f(x_{i+1}) + \dfrac{\|\nabla f(x_{i+1})\|^2}{\|\nabla f(x_i)\|^2} S_i$

Minimize $f(x_{i+2})$ and determine α (use the golden section method)
$x_{i+2} = x_{i+1} + \alpha S_{i+1}$
Minimize $f(x_{i+2})$ and determine α (use the golden section method)
If $|f(x_{i+2}) - f(x_{i+1})| > \varepsilon_1$ or $\|\nabla f(x_{i+1})\| > \varepsilon_2$
then goto Step 3
else goto Step 4

Step 4: Converged. Print $x^* = x_{i+2}$, $f(x^*) = f(x_{i+2})$

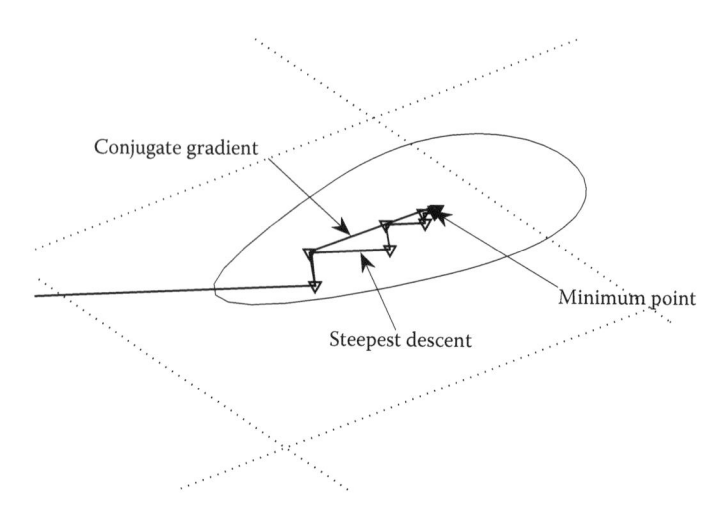

FIGURE 3.8
Convergence plot of conjugate gradient/steepest descent method.

is compared with the first-order, steepest descent method. The conjugate method does not show sluggishness in reaching the minimum point.

```
          Initial function value = 1452.2619
No.        x-vector            f(x)          Deriv.

1        0.095    0.023        -2.704        1006.074
2        0.178    0.145        -5.404          37.036
3        0.507    0.136        -9.627          23.958
4        0.510    0.123        -9.655           4.239
5        0.505    0.121        -9.656           0.605
6        0.504    0.122        -9.656           0.340
7        0.504    0.122        -9.656           0.023
```

3.4.6 DFP Method

In the DFP method, the inverse of the Hessian is approximated by a matrix $[A]$ and the search direction is given by

$$S_i = -[A]\nabla f(x_i) \qquad (3.24)$$

The information stored in the matrix $[A]$ is called as the metric and because it changes with every iteration, the DFP method is known as the variable metric method. Because this method uses first-order derivatives and has the

Unconstrained Optimization

property of quadratic convergence, it is referred to as a *quasi-Newton* method. The inverse of the Hessian matrix can be approximated as

$$[A]_{i+1} = [A]_i + \frac{\Delta x \Delta x^T}{\Delta x^T \nabla g} - \frac{[A]_i \nabla g \nabla g^T [A]_i}{\nabla g^T [A]_i \nabla g} \tag{3.25}$$

where

$$\Delta x = \Delta x_i - \Delta x_{i-1} \tag{3.26}$$

$$\nabla g = \nabla g_i - \nabla g_{i-1} \tag{3.27}$$

The matrix $[A]$ is initialized to the identity matrix. The algorithm for the DFP method is described in Table 3.7 and a MATLAB code of its implementation is given in *dfp.m*.

On executing the code with a starting value of x as $(-3, 2)$ the following output is displayed in the command window for the test problem. Observe that in the second and the third iterations, search points are similar in this method and the conjugate gradient method, indicating that search directions were similar. In further iterations, however, the search direction is different. Further, on typing *inv(A)* in the MATLAB command prompt and then

TABLE 3.7

Algorithm for the DFP Method

Step 1: Given x_i (starting value of design variable)
    ε_1 (tolerance of function value from previous iteration)
    ε_2 (tolerance on gradient value)
    Δx (required for gradient computation)
    $[A]$ (initialize to identity matrix)
Step 2: Compute $f(x_i)$ and $\nabla f(x_i)$ (function and gradient vector)
    $S_i = -\nabla f(x_i)$ (search direction)
    $x_{i+1} = x_i + \alpha S_i$ (update the design vector)
    Minimize $f(x_{i+1})$ and determine α (use the golden section method)
Step 3: Compute Δx and ∇g

$$[A]_{i+1} = [A]_i + \frac{\Delta x \Delta x^T}{\Delta x^T \nabla g} - \frac{[A]_i \nabla g \nabla g^T [A]_i}{\nabla g^T [A]_i \nabla g}$$

    $S_{i+1} = -[A]_{i+1} \nabla f(x_{i+1})$
    $x_{i+2} = x_{i+1} + \alpha S_{i+1}$
    Minimize $f(x_{i+2})$ and determine α (use the golden section method)
    If $|f(x_{i+2}) - f(x_{i+1})| > \varepsilon_1$ or $\|\nabla f(x_{i+1})\| > \varepsilon_2$
        then   goto Step 3
        else   goto Step 4
Step 4: Converged. Print $x^* = x_{i+2}$, $f(x^*) = f(x_{i+2})$

printing the Hessian matrix at the converged value of x, it is observed that [A] approaches $[H]^{-1}$.

```
         Initial function value = 1452.2619
No.         x-vector              f(x)           Deriv.

 1        0.095      0.023       -2.704        1006.074
 2        0.179      0.145       -5.418          37.036
 3        0.508      0.145       -9.576          23.983
 4        0.501      0.122       -9.656           7.004
 5        0.504      0.122       -9.656           0.396
 6        0.504      0.122       -9.656           0.053
 7        0.504      0.122       -9.656           0.038
 8        0.504      0.122       -9.656           0.028
 9        0.504      0.122       -9.656           0.005

>> A
A =
    0.0091    0.0005
    0.0005    0.0033
>> inv(hessian(x,delx,n_of_var))
ans =
    0.0091    0.0005
    0.0005    0.0033
```

3.4.7 BFGS Method

In the BFGS method, the Hessian is approximated using the variable metric matrix [A] given by the equation

$$[A]_{i+1} = [A]_i + \frac{g \nabla g^T}{\nabla g^T \Delta x} + \frac{\nabla f(x_i) \nabla f(x_i)^T}{\nabla f(x_i)^T S_i} \tag{3.28}$$

It is important to note that whereas the matrix [A] converges to the inverse of the Hessian in the DFP method, the matrix [A] converges to the Hessian itself in the BFGS method. As the BFGS method needs fewer restarts as compared to the DFP method, it is more popular than the DFP method. The algorithm for the BFGS method is described in Table 3.8 and a MATLAB code of its implementation is given in BFGS.m.

On executing the code with a starting value of x as (−3, 2) the following output is displayed in the command window for the test problem. Again, it is observed that in the second and third iterations, search points are similar to this method as compared to DFP and the conjugate gradient methods, indicating that search directions were similar. Further, on typing A in the

TABLE 3.8
Algorithm for the BFGS Method

Step 1: Given x_i (starting value of design variable)
  ε_1 (tolerance of function value from previous iteration)
  ε_2 (tolerance on gradient value)
  Δx (required for gradient computation)
  $[A]$ (initialize to identity matrix)
Step 2: Compute $f(x_i)$ and $\nabla f(x_i)$ (function and gradient vector)
  $S_i = -\nabla f(x_i)$ (search direction)
  $x_{i+1} = x_i + \alpha S_i$ (update the design vector)
  Minimize $f(x_{i+1})$ and determine α (use golden section method)
Step 3: Compute Δx and ∇g

$$[A]_{i+1} = [A]_i + \frac{g\nabla g^T}{\nabla g^T \Delta x} + \frac{\nabla f(x_i)\nabla f(x_i)^T}{\nabla f(x_i)^T S_i}$$

  $S_{i+1} = -[[A]_{i+1}]^{-1}\nabla f(x_{i+1})$
  $x_{i+2} = x_{i+1} + \alpha S_{i+1}$
  Minimize $f(x_{i+2})$ and determine α (use the golden section method)
  If $|f(x_{i+2}) - f(x_{i+1})| > \varepsilon_1$ or $\|\nabla f(x_{i+1})\| > \varepsilon_2$
    then goto Step 3
    else goto Step 4
Step 4: Converged. Print $x^* = x_{i+2}, f(x^*) = f(x_{i+2})$

MATLAB command prompt and then printing the Hessian matrix at the converged value of x, it is observed that $[A]$ approaches $[H]$.

```
      Initial function value = 1452.2619
No.       x-vector           f(x)          Deriv.

1      0.095     0.023      -2.704        1006.074
2      0.179     0.145      -5.418          37.036
3      0.508     0.145      -9.578          24.017
4      0.501     0.122      -9.655           6.900
5      0.504     0.122      -9.656           0.471
6      0.504     0.122      -9.656           0.077
7      0.504     0.122      -9.656           0.056
8      0.504     0.122      -9.656           0.040
9      0.504     0.122      -9.656           0.007

>> A
A =
    110.5001   -16.9997
    -16.9997   306.7238
>> hessian(x,delx,n_of_var)
ans =
    111.0981   -15.9640
    -15.9640   308.5603
```

3.4.8 Powell Method

The Powell method is a direct search method (no gradient computation is required) with the property of quadratic convergence. Previous search directions are stored in this method and they form a basis for the new search direction. The method makes a series of unidirectional searches along these search directions. The last search direction replaces the first one in the new iteration and the process is continued until the function value shows no improvement. A MATLAB code (*powell.m*) is written in which this method is implemented and the algorithm is described in Table 3.9.

On executing the code with a starting value of *x* as (–3, 2), following output is displayed at the command window for the test problem.

```
Initial function value = 1452.2619
No.         x-vector           f(x)

1         0.504     0.122     -9.656
2         0.505     0.122     -9.656
3         0.504     0.122     -9.656
4         0.504     0.122     -9.656
5         0.505     0.122     -9.656
```

TABLE 3.9

Algorithm for the Powell Method

Step 1: Given x_i (starting value of design variable)
  ε (tolerance of function value from previous iteration)
  S_i (linearly independent vectors)
  $f(X_{prev}) = f(x_i)$
Step 2: $X = x_i + \alpha S_i$
  Minimize $f(X)$ and determine α (use the golden section method)
Step 3: Set $Y = X$, $i = 1$
  do
  Minimize $f(X)$ and determine α (use the golden section method)
  $X = X + \alpha S_i$
  $i = i + 1$
  while $i <$ (number of variable) + 1
  If $|f(X) - f(X_{prev})| < \varepsilon$
    then   goto Step 4
    else   continue
  $S_i = X - Y$
  $X = X + \alpha S_i$
  $f(X_{prev}) = f(X)$
  goto Step 3
Step 4: Converged. Print $x^* = X$, $f(x^*) = f(X)$

3.4.9 Nelder–Mead Algorithm

Simplex refers to a geometric figure formed by $n + 1$ points in an n dimension space. For example, in a two-dimensional space, the figure formed is a triangle. The Nelder–Mead algorithm is a direct search method and uses function information alone (no gradient computation is required) to move from one iteration to another. The objective function is computed at each vertex of the simplex. Using this information, the simplex is moved in the search space. Again, the objective function is computed at each vertex of the simplex. The process of moving the simplex is continued until the optimum value of the function is reached. Three basic operations are required to move the simplex in the search space: reflection, contraction, and expansion.

In an optimization problem with two dimensions, the simplex will be a triangle, whose vertices are given by (say) x_1, x_2, and x_3. Of these, let the worst value of the objective function be at $x_3 = x_{worst}$. If the point x_{worst} is reflected on the opposite face of the triangle, the objective function value is expected to decrease. Let the new reflected point be designated as x_r. The new simplex (see Figure 3.9) is given by the vertices x_1, x_2, and x_r. The centroid point x_c is computed using all the points but with the exclusion of x_{worst}. That is,

$$x_c = \frac{1}{n} \sum_{\substack{i=1 \\ i \neq worst}}^{n+1} x_i \tag{3.29}$$

The reflected point is computed as

$$x_r = x_c + \alpha(x_c - x_{worst}) \tag{3.30}$$

where α is a predefined constant. Typically, $\alpha = 1$ is taken in the simulations. If the reflected value does not show improvement, the second worst value is taken and the process as discussed earlier is repeated. Sometimes reflection

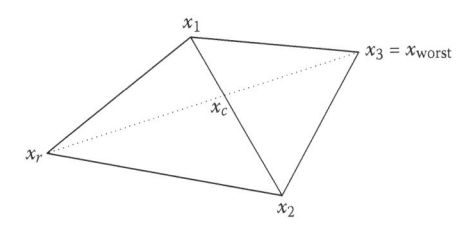

FIGURE 3.9
Reflection operation.

can lead to cycling with no improvement in the objective function value. Under such conditions, a contraction operation is performed.

If x_r results in a new minimum point, then it is possible to further expand the new simplex (see Figure 3.10) in the hope of further reducing the objective function value. The expanded point is computed as

$$x_e = x_c + \gamma(x_c - x_{worst}) \tag{3.31}$$

where γ is a predefined constant. Typically, $\gamma = 2$ is taken in the simulations. If x_e results in the new minimum point, it replaces the x_{worst} point. Else, x_r replaces the x_{worst} point.

The contraction operation is used when it is certain that the reflected point is better than the second worst point ($x_{second\ worst}$). The contracted point is computed as

$$x_{contr} = x_c + \rho(x_c - x_{worst}) \tag{3.32}$$

where ρ is a predefined constant. Typically, $\rho = -0.5$ is taken in the simulations.

The preceding operations are continued until the standard deviation of the functions computed at the vertices of the simplex becomes less than ε. That is,

$$\sum_{i=1}^{n+1} \frac{\left[f(x_i) - f(x_c)\right]^2}{n+1} \leq \varepsilon \tag{3.33}$$

The Nelder–Mead algorithm is described in Table 3.10 and a MATLAB code (*neldermead.m*) is written in which this method is implemented.

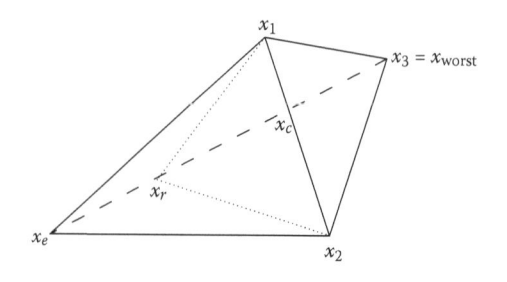

FIGURE 3.10
Expansion operation.

TABLE 3.10

Nelder–Mead Algorithm

Step 1: Given x_i (randomly select starting value of design variables)
  $\alpha, \gamma, \rho, \sigma, \varepsilon$ (value of constants)
  Compute $f(x_i)$, $f(x_{best}) \leq \ldots \leq f(x_{second\ worst}) \leq f(x_{worst})$
Step 2: Compute the centroid as
$$x_c = \frac{1}{n} \sum_{\substack{i=1 \\ i \neq worst}}^{n+1} x_i$$
Step 3: Reflection
  $x_r = x_c + \alpha(x_c - x_{worst})$
  If $f(x_{best}) \leq \ldots \leq f(x_r) \leq f(x_{second\ worst})$ then replace x_{worst} with x_r and goto Step 1
Step 3: Expansion
  If $f(x_r) \leq f(x_{best})$ then
  $x_e = x_c + \gamma(x_c - x_{worst})$
  If $f(x_e) \leq f(x_r)$ then replace x_{worst} with x_e and goto Step 1
  Else
  replace x_{worst} with x_r
  Else
  goto Step 5
Step 4: Contraction
  $x_{contr} = x_c + \rho(x_c - x_{worst})$
  If $f(x_{contr}) \leq f(x_{worst})$ then replace x_{worst} with x_c and goto Step 1
Step 5: If
$$\sum_{i=1}^{n+1} \frac{\left[f(x_i) - f(x_c)\right]^2}{n+1} \leq \varepsilon$$
  then converged,
  else
  goto Step 1

On executing the code with a random value of x, the following output is displayed at the command window for the test problem.

Iteration	Deviation	f(x)
1	72.2666	-0.733
2	36.7907	-0.733
3	6.8845	-0.733
4	9.7186	-8.203
5	5.0965	-8.203
6	3.8714	-8.426
7	1.3655	-8.426
8	0.7944	-9.351
9	0.6497	-9.509
10	0.2242	-9.509
11	0.1083	-9.509
12	0.1068	-9.641
13	0.0794	-9.641
14	0.0299	-9.641

15	0.0173	-9.641
16	0.0126	-9.653
17	0.0079	-9.653
18	0.0034	-9.654
19	0.0025	-9.654
20	0.0021	-9.656
21	0.0011	-9.656
22	0.0003	-9.656
23	0.0004	-9.656
24	0.0003	-9.656

```
xc =
    0.5028    0.1219
```

3.5 Additional Test Functions

Different solution techniques were applied to the test problem on the spring system in the previous section. In this section, some additional test problems such as Rosenbrock's function, Wood's function, quadratic function, and so forth are taken, on which different solution methods will be tested. The performance of each method is compared in terms of the computational time. The MATLAB functions *tic* and *toc* can be used to estimate the computational time.

3.5.1 Rosenbrock Function

The two-variable function is given by

$$f(x) = 100\left(x_2 - x_1^2\right)^2 + (1 - x_1)^2 \tag{3.34}$$

The minimum of this "banana valley" function is zero (see Figure 3.11 where the minimum is marked with *) and occurs at (1, 1). Different solution methods are applied from the same starting point (–1.5, 1.5) and their performances are summarized in Table 3.11. All methods are able to track the minimum of the function. The steepest descent method takes a maximum computational time as compared to all other methods. The computational time required by other methods is comparable. The convergence history of the steepest descent method is plotted in Figure 3.12 and marked with °. Because of the particular nature of the problem, the method dwells in the region with a low gradient value. The Nelder–Mead method is not compared here as it uses more than one starting point.

Unconstrained Optimization

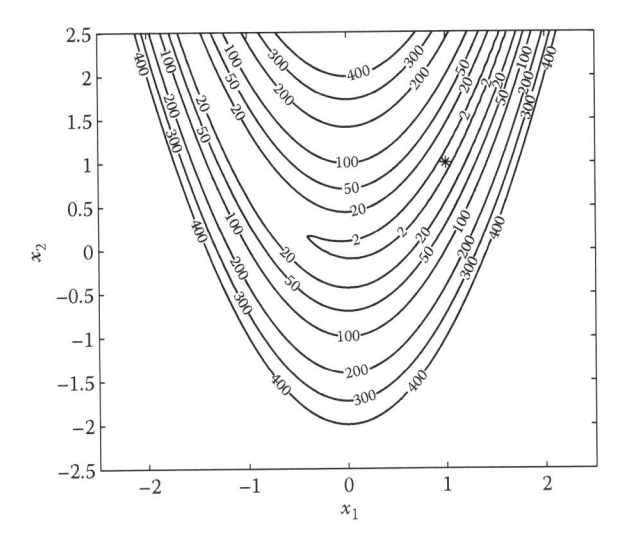

FIGURE 3.11
Contours of Rosenbrock function.

TABLE 3.11

Performance Comparison of Different Solution Methods for Rosenbrock's Function

Method	Computational Time (ms)
Steepest descent	49.7
Newton	8.04
Modified Newton	11.9
Marquardt	9.4
Conjugate gradient	18.8
DFP	11.23
BFGS	10.34
Powell	10.52

3.5.2 Quadratic Function

The two-variable function is given by

$$f(x) = (1 - x_1)^2 + (2 - x_2)^2 \tag{3.35}$$

The minimum of this function is zero (see Figure 3.13, where the minimum is marked with *) and occurs at (1, 2). Different solution methods are applied from a starting point (2, −3) and their performances are summarized in Table 3.12. All methods are able to track the minimum of the function. The

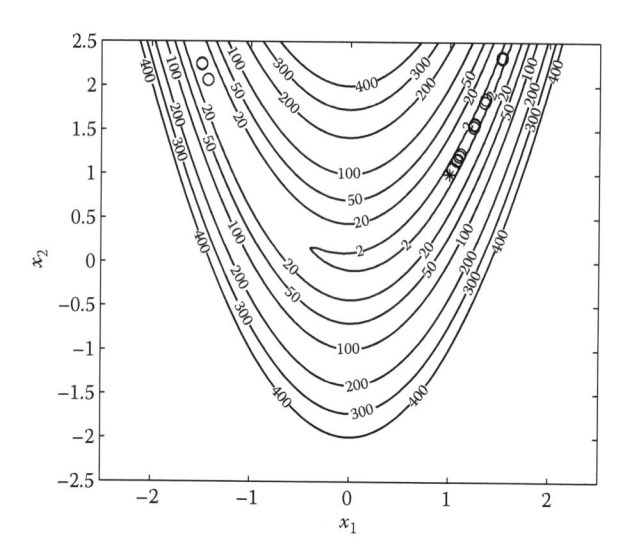

FIGURE 3.12
Behavior of steepest descent method on Rosenbrock function.

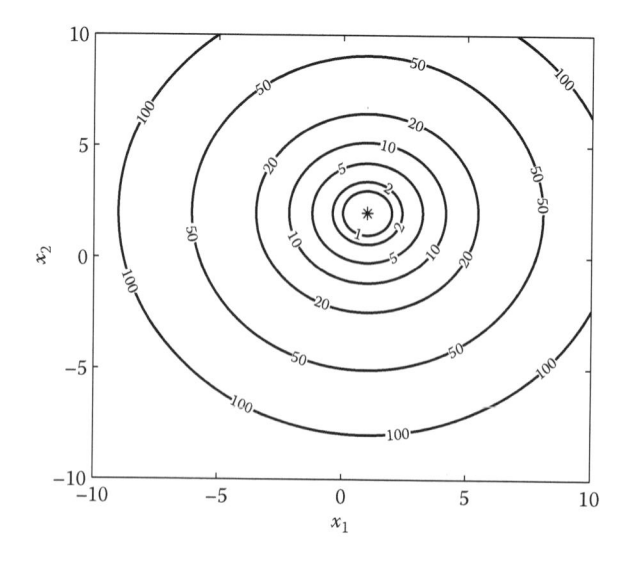

FIGURE 3.13
Contours of a quadratic function.

Unconstrained Optimization

TABLE 3.12

Performance Comparison of Different Solution Methods for a Quadratic Function

Method	Computational Time (ms)
Steepest descent	6.06
Newton	7.5
Modified Newton	10.38
Marquardt	9.72
Conjugate gradient	5.79
DFP	7.69
BFGS	7.45
Powell	7.26

conjugate gradient method takes minimum computational time compared to other solution methods.

3.5.3 Nonlinear Function

The two-variable function is given by

$$f(x) = 4x_1^2 - 4x_1x_2 + 3x_2^2 + x_1 \tag{3.36}$$

The minimum of this function is −0.09375 (see Figure 3.14, where the minimum is marked with *) and occurs at (−3/16, −1/8). Different solution methods are applied from a starting point (4, 3) and their performances are

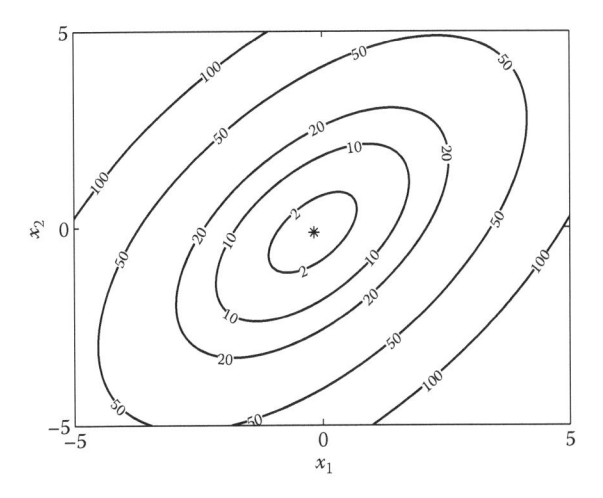

FIGURE 3.14
Contours of a nonlinear function.

TABLE 3.13
Performance Comparison of Different Solution Methods for a Nonlinear Function

Method	Computational Time (ms)
Steepest descent	11.19
Newton	7.67
Modified Newton	10.51
Marquardt	10.0
Conjugate gradient	6.27
DFP	7.85
BFGS	7.70
Powell	8.32

summarized in Table 3.13. All methods are able to track the minimum of the function. The conjugate gradient method takes minimum computational time compared to other solution methods.

3.5.4 Wood's Function

The two-variable function is given by

$$f(x) = \frac{1}{10}\left(12 + x_1^2 + \frac{1+x_1^2}{x_1^2} + \frac{100 + x_1^2 x_2^2}{(x_1 x_2)^4}\right) \tag{3.37}$$

The minimum of this function is 1.744 (see Figure 3.15, where the minimum is marked with *) and occurs at (1.743, 2.03). Different solution methods are

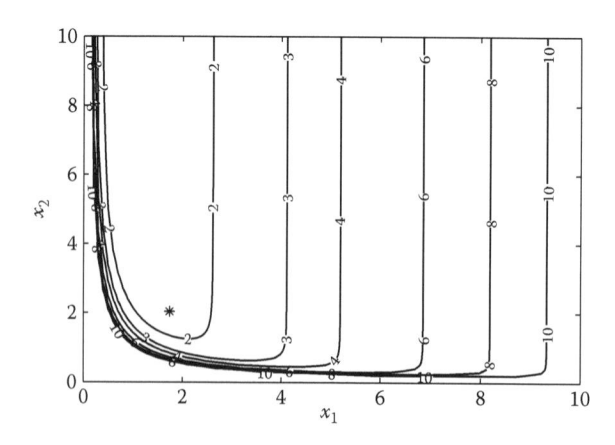

FIGURE 3.15
Contours of Wood's function.

Unconstrained Optimization

TABLE 3.14

Performance Comparison of Different Solution Methods for Wood's Function

Method	Computational Time (ms)
Steepest descent	7.16
Newton	9.46
Modified Newton	12.1
Marquardt	10.6
Conjugate gradient	6.22
DFP	9.54
BFGS	8.33
Powell	12.75

applied from a starting point (0.5, 0.5) and their performances are summarized in Table 3.14. All methods are able to track the minimum of the function. The conjugate gradient method takes minimum computational time compared to other solution methods.

3.6 Application to Robotics

An industrial robot typically comprises a number of mechanical links with one end fixed and the other *end-effector* free to move. If the joint angles (θ_1, θ_2, and θ_3) are known, then the trajectory of the end-effector can be calculated using kinematic relationships. Often a predefined motion of the end-effector is given for which we have to evaluate the joint angles. This can be stated as an unconstrained optimization problem (Andreas 2007).

The design variables for the optimization problem are

$$x = \begin{bmatrix} \theta_1 \\ \theta_2 \\ \theta_3 \end{bmatrix} \tag{3.38}$$

The kinematic equations are

$$f_1(x) = c_1(a_2 c_2 + a_3 c_{23} - d_4 s_{23}) - d_3 s_1 - p_x \tag{3.39}$$

$$f_2(x) = s_1(a_2c_2 + a_3c_{23} - d_4s_{23}) + d_3s_1 - p_y \qquad (3.40)$$

$$f_3(x) = d_1 - a_2s_2 - a_3s_{23} - d_4c_{23} - p_z \qquad (3.41)$$

where

$$c_1 = \cos(\theta_1)$$

$$c_2 = \cos(\theta_2)$$

$$c_{23} = \cos(\theta_2 + \theta_3)$$

$$s_1 = \sin(\theta_1)$$

$$s_2 = \sin(\theta_2)$$

$$s_{23} = \sin(\theta_2 + \theta_3)$$

$d_1 = 66.04$, $d_3 = 14.91$, $d_4 = 43.31$, $a_2 = 43.18$, $a_3 = 2.03$

The desired trajectory equation is given by

$$\begin{bmatrix} p_x \\ p_y \\ p_z \end{bmatrix} = \begin{bmatrix} 30\cos t \\ 100\sin t \\ 10t + 66.04 \end{bmatrix} \qquad (3.42)$$

The unconstrained optimization problem is

$$\text{Minimize } \sum_{i=1}^{3} f_i^2(x) \qquad (3.43)$$

Here $-\pi \leq t \leq \pi$. t is divided into 100 parts. It means there are 100 variables for θ_1, 100 variables for θ_2, and 100 variables for θ_3. The unconstrained problem thus has 300 variables that need to be determined. The optimization problem is solved using the Powell method.

Go to the *Robotics* directory in Chapter 3 and type *powell* in the command prompt. Then generate the optimized trajectory by executing the MATLAB code *generate_optimized_traj(x)*. Give a *hold on* command and then execute

Unconstrained Optimization

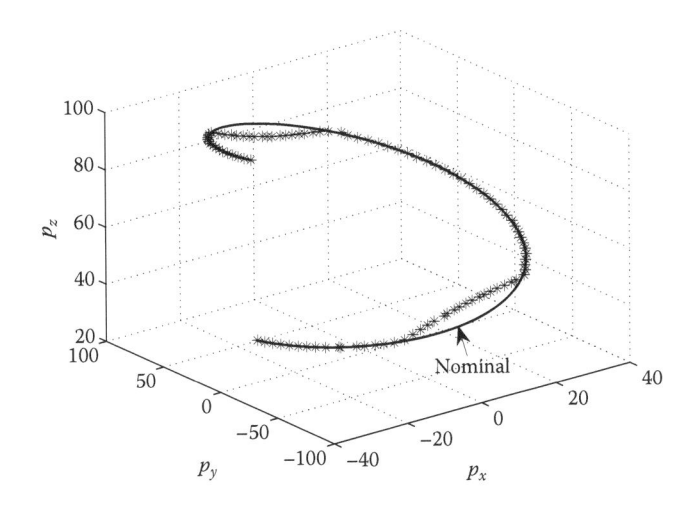

FIGURE 3.16
Comparison of manipulator's trajectories (optimized with nominal).

robotics_nominal_traj.m. The desired (nominal shown by solid line) and optimized (shown by *) trajectories are compared in Figure 3.16. It is observed that in some regions, the motion of the end-effector is not exactly matched with the desired profile. Similar results are also seen in Andreas (2007), where the reason for the difference is attributed to "beyond manipulators reach."

Chapter Highlights

- The unidirectional search refers to minimizing the value of a multi-variable function along a specified direction.
- Solution techniques for multivariable, unconstrained optimization problems can be grouped into gradient- and non–gradient-based methods.
- The negative gradient direction is addressed as the steepest descent direction.
- The steepest descent method ensures a reduction in the function value at every iteration. If the starting point is far away from the minimum, the gradient will be higher and function reduction will be maximum in each iteration. Because the gradient value of the function decreases near the optimum, the method becomes sluggish (slow convergence) near the minimum.

- Newton's method requires computation of the Hessian matrix, which is computationally expensive. Newton's method is known for converging in one iteration for a quadratic function. The method requires a restart if the starting point is far away from optimum.
- In the modified Newton method, a line search is performed in the search direction computed by the Newton method.
- The Levenberg–Marquardt method is a sort of hybrid method that combines the strength of both the steepest descent and Newton methods.
- The conjugate gradient method is a first-order method, but shows the property of quadratic convergence and thus has a significant advantage over the second-order methods.
- DFP and BFGS methods are called the variable metric methods.
- It is important to note that whereas the matrix [A] converges to the inverse of the Hessian in the DFP method, it converges to the Hessian itself in the BFGS method.
- The Powell method is a direct search method (no gradient computation is required) with the property of quadratic convergence.
- In the Nelder–Mead algorithm, the simplex is moved using reflection, expansion, and contraction.

Formulae Chart

Necessary conditions for minimum of a function:

$$\nabla f(x^*) = 0$$

$$\nabla^2 f(x^*) \geq 0$$

Unidirectional search:

$$f(\alpha) = x_i + \alpha S_i$$

Search direction in steepest descent method:

$$S_i = -\nabla f(x_i)$$

Search direction in the Newton method:

$$S_i = -[H]^{-1} \nabla f(x_i)$$

Unconstrained Optimization

Search direction in the Levenberg–Marquardt method:

$$S_i = -[H + \lambda I]^{-1}\nabla f(x_i)$$

Search direction in the conjugate gradient method:

$$S_{i+1} = -\nabla f(x_i) + \frac{\|\nabla f(x_{i+1})\|^2}{\|\nabla f(x_i)\|^2} S_i$$

Search direction in the DFP method:

$$S_i = -[A]\nabla f(x_i)$$

$$[A]_{i+1} = [A]_i + \frac{\Delta x \Delta x^T}{\Delta x^T \nabla g} - \frac{[A]_i \nabla g \nabla g^T [A]_i}{\nabla g^T [A]_i \nabla g}$$

Search direction in the BFGS method:

$$S_i = -[A]^{-1}\nabla f(x_i)$$

$$[A]_{i+1} = [A]_i + \frac{g \nabla g^T}{\nabla g^T \Delta x} + \frac{\nabla f(x_i) \nabla f(x_i)^T}{\nabla f(x_i)^T S_i}$$

Problems

1. Find the steepest descent direction for the function

$$f(x) = x_1^2 + 3x_1 x_2 + 2x_2^2$$

at point (1, 2).

2. Minimize the function

$$f(x) = 10{,}000 x_1 x_2 + e^{-x_1} + e^{-x_2} - 2.0001$$

from a starting value of (2, 2) using the BFGS, DFP, and steepest descent methods.

3. Minimize the function

$$f(x) = \left(x_1^2 + x_2 - 11\right)^2 + \left(x_2^2 + x_1 - 7\right)^2$$

from a starting value of (2, 3) using the following methods:
   i. Steepest descent
   ii. Newton
   iii. Modified Newton
   iv. Levenberg–Marquardt
   v. DFP
   vi. BFGS
   vii. Powell
   viii. Nelder–Mead
4. Show that in the DFP method, the variable metric [A] approaches the inverse of the Hessian matrix for the following function which needs to be minimized.

$$f(x) = x_1^2 + 3x_1x_2 + 5x_2^2$$

Take starting value as (1, 1).

5. Show that in the BFGS method, the variable metric [A] approaches the Hessian matrix for the following function which needs to be minimized.

$$f(x) = x_1^2 + 3x_1x_2 + 5x_2^2$$

Take the starting value as (1, 1).

6. Minimize the function using the DFP method with a starting value of (1, 1)

$$f(x) = e^{x_1^2 + x_2^2} + x_1 + x_2 - 3 - \sin(3(x_1 + x_2))$$

7. Minimize the function

$$f(x) = 100(x_3 - 10\theta)^2 + 100\left(\sqrt{x_1^2 + x_2^2} - 1\right)^2 + x_3^2$$

where

$$2\pi\theta = \tan^{-1}\left(\frac{x_2}{x_1}\right) \quad x_1 > 0$$

$$2\pi\theta = \pi + \tan^{-1}\left(\frac{x_2}{x_1}\right) \quad x_1 < 0$$

Take the starting value as (–1, 0, 0).

8. Instead of using the central difference formula for computing the derivative of a function, use the complex variable formula

$$f'(x) = \frac{\text{Imaginary}\left[f(x+i\Delta x)/\Delta x\right]}{\Delta x}$$

The MATLAB code *grad_vec.m* can be modified as

```
%%%%%%%%%%%%%%%%%%%%%%%%%%%%%%%%%%%%%%%%%
% MATLAB code grad_vec.m
%%%%%%%%%%%%%%%%%%%%%%%%%%%%%%%%%%%%%%%%%
%
function deriv = grad_vec_complex(x,delx,n_of_var)
xvec = x;
h = 1e-14;
for j = 1:length(x)
xvec = x;
c = complex(xvec(j),h);
xvec(j) = c;
deriv(j) = imag(func_multivar(xvec)/h);
end
%
%%%%%%%%%%%%%%%%%%%%%%%%%%%%%%%%%%%%%%%%%
```

Now use the steepest descent method to optimize the test function given in the main text.

9. Compare the accuracy of the derivative computation using the central difference formula and the complex variable formula against the analytical value of the derivative of the test function

$$f(x) = \sin x + \ln x$$

at = 0.1.

10. Use the line search algorithm to minimize the function

$$f(x) = \left(x_1^2 + x_2 - 11\right)^2 + \left(x_2^2 + x_1 - 7\right)^2$$

starting from different initial points and different search directions:
  i. Starting point (1, 1) and search direction (2, 4)
  ii. Starting point (0, 0) and search direction (1, 2)
  iii. Starting point (3, 2) and search direction (1, 1)

11. Minimize the function

$$f(x) = \frac{1}{2}\left(x_1^2 + 8x_2^2\right)$$

from the starting point (1, 2) using the steepest descent method. Observe the sluggishness of this method. Again, solve the function by the conjugate gradient method and compare the performance with the steepest descent method.

12. A manufacturing firm wants to divide its resources suitably between capital (x_1) and labor (x_2) so as to maximize the profit function given by

$$f(x) = p\{\ln(1 + x_1) + \ln(1 + x_2)\} - wx_2 - vx_1$$

where p is the unit price of the product, w is the wage rate of labor, and v is the unit cost of capital.

   i. By computing the gradient vector of the above function with respect to x_1 and x_2 and then equating it to zero, compute the design variables x_1 and x_2 as a function of p, v, and w.

   ii. Using the second-order condition, check whether the solution corresponds to a maximum of the function.

   iii. Compute numerical values of x_1 and x_2 by assuming suitable values of p, v, and w ($p > w, v$).

   iv. Starting with an initial guess of (0, 0) and using the values of p, v, and w as assumed in (iii), find the maximum of the function using the steepest descent method. Compare the values of x_1 and x_2 with those obtained from (iii).

13. The stable equilibrium configuration (Haftka and Gurdal 1992) of a two-bar unsymmetrical shallow truss (Figure 3.17) can be obtained

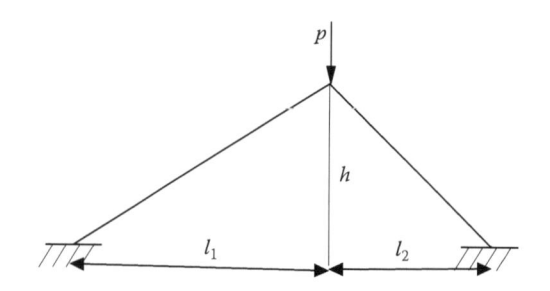

FIGURE 3.17
Two-bar truss.

by minimizing the potential energy function of the nondimensional displace variables x_1 and x_2 as

$$f(x) = \frac{1}{2}m\gamma\left(-\alpha_1 x_1 + \frac{1}{2}x_1^2 + x_2\right)^2 + \frac{1}{2}\left(-\alpha_1 x_1 + \frac{1}{2}x_1^2 - \frac{x_2}{\gamma}\right)\gamma^4 - \bar{p}\gamma x_1$$

where m, γ, α, and $\bar{p}$ are the nondimensional quantities defined as

$$m = \frac{A_1}{A_2} \quad \gamma = \frac{l_1}{l_2} \quad \alpha_1 = \frac{h}{l_1} \quad \bar{p} = \frac{p}{EA_2}$$

where E is the elastic modulus and A_1 and A_2 are the cross-sectional areas of the bars. Take $m = 5$, $\gamma = 4$, $\alpha = 0.02$, and $\bar{p} = 0.00002$. Staring with an initial guess of (0, 0), minimize the function using the DFP and BFGS methods.

4

Linear Programming

4.1 Introduction

Linear programming refers to an optimization problem that has the objective and the constraints as a linear function of the design variables. The constraints could be of an equality or inequality type or both. Mathematically, a linear function satisfies the following properties:

$$f(x + y) = f(x) + f(y) \tag{4.1}$$

$$f(kx) = kf(x) \tag{4.2}$$

where x and y are the variables and k is a scalar. A practical linear programming problem (LPP) might contain hundreds of design variables and constraints and thus require special solution techniques that are different from the methods that were described in the previous chapters. A number of applications of LPP can be found in the literature, some of which include

- An airline company would like to assign crews to different flights in an optimal way so that total cost is minimized while covering its entire network.
- In a portfolio optimization problem, an investor would like to know the investment allocation to different assets that would maximize the return.
- An oil company blends different qualities of oil to produce different grades of gasoline, which need to be shipped to users who are located in different places. The quantity of gasoline that can be produced is fixed at a certain maximum and so is the input oil quantity. The company would like to maximize its profit.
- A company produces a number of products and this requires a number of processes on different machines. The profit from each product is known and the maximum time available for each machine

is fixed. The company would like to determine the manufacturing policy that would maximize its profit.

- A government-run bus company has to cover different places in a metro city. As a government company, it has an obligation to cover all parts of the city, irrespective of whether a particular route is profitable or not. The company would like to find the number of routes and allocate a number of buses for each of these routes in such a way that it can maximize its profit.

The next section discusses the solution to LPP using the graphical method and its limitations. The need to convert an LPP into the standard form along with procedural details is discussed next. Basic definitions of linear programming such as feasible solutions, basic solutions, basic feasible solutions, and optimal solution are further introduced. The simplex method is discussed in detail for solving LPPs. The degeneracy problem in the simplex method and how it can be overcome is also discussed. The importance of converting a primal problem into a dual problem is explained followed by the dual-simplex method to solve such problems. In the simplex method, the algorithm moves from one feasible point to another feasible point. For a large LPP, this can be time consuming. As an alternate, interior point methods move inside the feasible region to reach the optimum. The road map of this chapter is given in Figure 4.1.

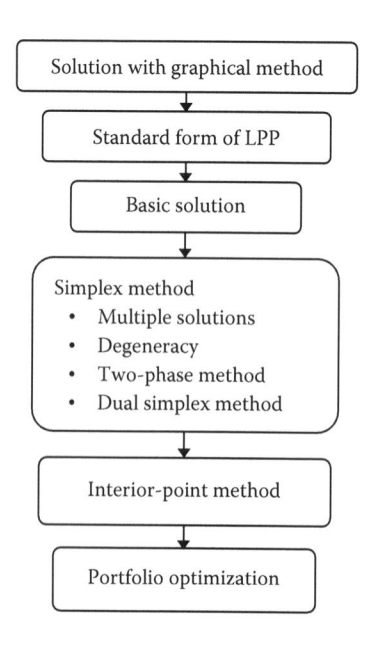

FIGURE 4.1
Road map of Chapter 4.

4.2 Solution with the Graphical Method

The graphical method is a simple technique for locating the optimal solution of problems with up to two to three design variables only. Beyond three variables and with many constraints, the representation of the optimization problem through graphs becomes complex. Consider the LPP

Maximize

$$z = x + 2y \tag{4.3}$$

subject to

$$2x + y \geq 4 \tag{4.4}$$

$$-2x + 4y \geq -2 \tag{4.5}$$

$$-2x + y \geq -8 \tag{4.6}$$

$$-2x + y \leq -2 \tag{4.7}$$

$$y \leq 6 \tag{4.8}$$

The intersection of five constraints leads to a feasible region $ABCDE$ as shown in Figure 4.2. To make this plot, first type MuPad in the MATLAB® command prompt. Open a new window in MuPad and then type the following commands:

```
k := [{2*x + y >= 4, -2*x + 4*y >= -2, -2*x + y >= -8,
-2*x + y <= -2, y <= 6}, x + 2*y]:
g := linopt::plot_data(k, [x, y]):
plot(g, Color = RGB::Grey)
```

The coordinate value of the vertex is given in the brackets. The values of the objective function at points A, B, C, D, and E are given as 9, 13/5, 7/2, 16 and 19 respectively. In an LPP, the optimal value of the objective function occurs at the edge of the convex polyhedron. Thus, the maximum value of the objective function is 19 and the values of the variables x and y are 7 and 6 respectively at the optimal point. Note that the objective function $z = x + 2y$, also referred to as the cost equation represents, a family of parallel lines (shown by the dashed line in Figure 4.2) called equicost lines. The value of the objective function is constant along this line.

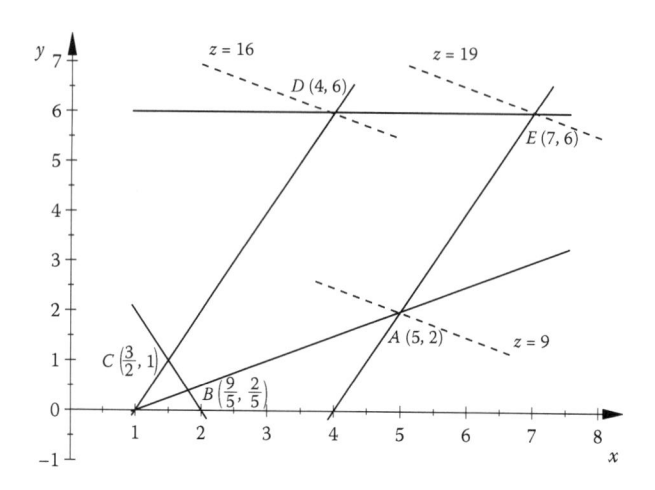

FIGURE 4.2
Feasible region (*ABCDE*) for the LPP.

An LPP need not have a unique solution. For example, if we change the previous LPP to

Minimize

$$z = 2x + y \tag{4.9}$$

subject to

$$2x + y \geq 4 \tag{4.10}$$

$$-2x + 4y \geq -2 \tag{4.11}$$

$$-2x + y \geq -8 \tag{4.12}$$

$$-2x + y \leq -2 \tag{4.13}$$

$$y \leq 6 \tag{4.14}$$

Open a new window in MuPad and then type the following command and observe the plot in Figure 4.3.

```
k := [{2*x + y >= 4, -2*x + 4*y >= -2, -2*x + y >= -8,
-2*x + y <= -2, y <= 6}, -2*x - y]:
g := linopt::plot_data(k, [x, y]):
plot(g, Color = RGB::Grey)
```

Linear Programming

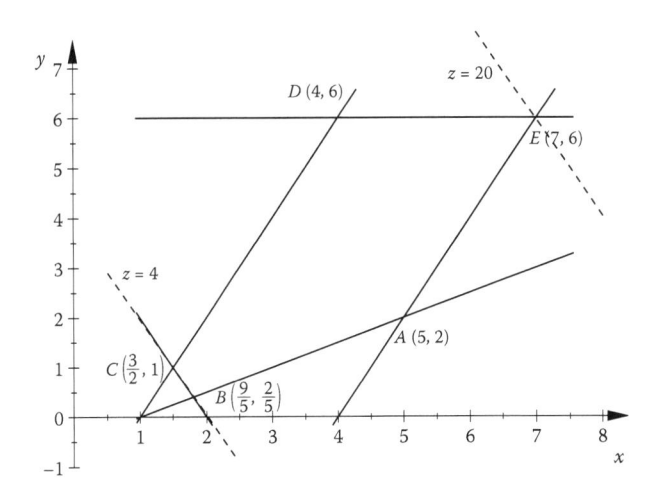

FIGURE 4.3
Infinite solutions for the LPP.

The equicost line $z = 4$ passes through points B and C. The minimum value of the objective function is 4 and occurs at $(3/2, 1)$ and $(9/5, 2/5)$. In fact, for infinite number of points in the line joining points B and C, the objective function value is 4. That is, in the given LPP, the solution is not unique.

Now consider the LPP in which one of the constraints is removed. The LPP is given by

Maximize

$$z = x + 2y \qquad (4.15)$$

subject to

$$2x + y \geq 4 \qquad (4.16)$$

$$-2x + 4y \geq -2 \qquad (4.17)$$

$$-2x + y \geq -8 \qquad (4.18)$$

$$-2x + y \leq -2 \qquad (4.19)$$

The constraints are plotted in Figure 4.4. Observe that the value of the objective function can be increased to an infinitely large value, without leaving the feasible region. The solution of the LPP, in this case, is said to be *unbounded*.

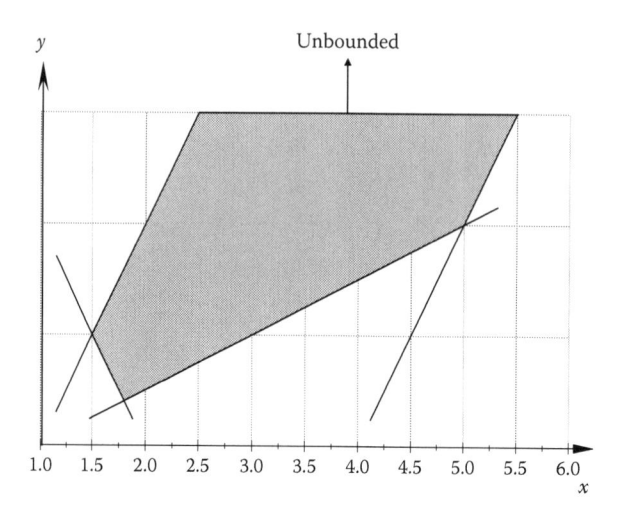

FIGURE 4.4
Unbounded solution for the LPP.

In addition, there can be inconsistent constraints in a LPP or the constraints may be such that no feasible solution exists for the problem. The solution of the LPP, in this case, is said to be *infeasible*. From the discussion so far, we can say that an LPP can have

- A unique solution
- Infinite solutions
- An unbounded solution
- An infeasible solution

4.3 Standard Form of an LPP

In the previous section, the graphical method was used to find the optimal solution of a two-variable LPP. In practice, LPP would contain several variables and constraints. Thus, there is a need to put LPP in a standard form. For an n variable LPP, the scalar form is given as

Minimize

$$z = c_1 x_1 + c_2 x_2 + \cdots + c_n x_n \tag{4.20}$$

Linear Programming

subject to

$$a_{11}x_1 + a_{12}x_2 + \cdots + a_{1n}x_n = b_1 \quad (4.21)$$

$$a_{21}x_1 + a_{22}x_2 + \cdots + a_{2n}x_n = b_2 \quad (4.22)$$

$$\vdots$$

$$a_{m1}x_1 + a_{m2}x_2 + \cdots + a_{mn}x_n = b_m \quad (4.23)$$

$$x_j, b_m \geq 0 \quad (4.24)$$

where $a_{ij}(i = 1, 2, \cdots, m; j = 1, 2, \cdots, n)$, b_j, c_j are constants and x_j are the design variables. LPP can also be put in matrix form as

Minimize

$$z = c^T x \quad (4.25)$$

subject to

$$Ax = b \quad (4.26)$$

$$x, b \geq 0 \quad (4.27)$$

where A is an $m \times n$ constraint matrix given by

$$A = \begin{bmatrix} a_{11} & a_{12} & \cdots & a_{1n} \\ a_{21} & a_{22} & \cdots & a_{2n} \\ \vdots & & & \\ a_{m1} & a_{m2} & \cdots & a_{mn} \end{bmatrix} \quad (4.28)$$

and b, c, and x are column vectors given by

$$b = \begin{bmatrix} b_1 \\ b_2 \\ \vdots \\ b_m \end{bmatrix}, \quad c = \begin{bmatrix} c_1 \\ c_2 \\ \vdots \\ c_n \end{bmatrix}, \quad x = \begin{bmatrix} x_1 \\ x_2 \\ \vdots \\ x_n \end{bmatrix}$$

The following important points are to be noted when an LPP is written in standard form.

- The objective function needs to be in the minimization type.
- All of the design variables should be nonnegative.
- All of the components of the vector b are to be nonnegative.
- All of the constraints are of the equality type.

If the objective function is the maximization type, it can be converted to the minimization type by multiplying the cost coefficients by –1. For example, if the objective function is

$$\text{Maximize} \quad z = x_1 + 2x_2$$

Then it can be converted to the minimization type as

$$\text{Minimize} \quad -z = -x_1 - 2x_2$$

If a ≤ type constraint is present, then it can be converted into an equality constraint by adding a *slack variable*. For example, the inequality constraint

$$4x_1 - 5x_2 + 6x_3 + 9x_4 \leq 20$$

can be converted to an equality constraint by the addition of the slack variable s_1

$$4x_1 - 5x_2 + 6x_3 + 9x_4 + s_1 = 20$$

where $s_1 \geq 0$. If a ≥ type constraint is present, then it can be converted into an equality constraint by subtracting it with a *surplus variable*. For example, the inequality constraint

$$2x_1 + 4x_2 - 6x_3 + 7x_4 \geq 8$$

can be converted to an equality constraint by subtracting it with a surplus variable e_1

$$2x_1 + 4x_2 - 6x_3 + 7x_4 - e_1 = 8$$

An *unrestricted* or *free* variable (without any specified bounds) can be replaced by a pair of nonnegative variables. If x_1 is an unrestricted variable, then it can be replaced by

$$x_1 = x_1' - x_1''$$

with $x_1' \geq 0$ and $x_1'' \geq 0$.

Example 4.1

Transform the following LPP into the standard form.

Maximize

$$z = -4x_1 - 2x_2 + x_3 - 3x_4$$

subject to

$$2x_1 + 3x_2 - x_3 - 3x_4 = 5$$

$$-5x_1 - 2x_2 + 4x_3 - 7x_4 \leq 8$$

$$4x_1 - x_2 - 2x_3 + 5x_4 \leq -6$$

$$x_1 \geq -1,\ 0 \leq x_2 \leq 3,\ x_3 \geq 0,\ x_4 \text{ free}$$

Since the objective function is of the maximization type, it needs to be converted into the minimization type. This can be done by multiplying the objective function by -1, that is,

Minimize

$$-z = 4x_1 + 2x_2 - x_3 + 3x_4$$

The right-hand side of the third constraint is negative (-6). In standard form, this has to be positive. Hence, the third constraint has to be multiplied by -1 throughout. Third constraint thus becomes

$$-4x_1 + x_2 + 2x_3 - 5x_4 \geq 6$$

Note that inequality type also changes during this operation. Now transforming the variables

$$x_1' = x_1 + 1$$

$$x_4 = x_4' - x_4''$$

and substituting these variables in the LPP, we get

Minimize

$$z' = 4x_1' + 2x_2 - x_3 + 3x_4' - 3x_4'' - 4$$

subject to

$$2x_1' + 3x_2 - x_3 - 3x_4' + 3x_4'' = 7$$

$$-5x_1' - 2x_2 + 4x_3 - 7x_4' + 7x_4'' \le 3$$

$$-4x_1' + x_2 + 2x_3 - 5x_4' + 5x_4'' \ge 2$$

$$x_2 \le 3$$

$$x_1', x_2, x_3, x_4', x_4'' \ge 0$$

Using the slack and surplus variables, inequality constraints can be converted into equality constraints. Thus, the LPP problem converted into standard form is

Minimize

$$z'' = z' + 4 = 4x_1' + 2x_2 - x_3 + 3x_4' - 3x_4''$$

subject to

$$2x_1' + 3x_2 - x_3 - 3x_4' + 3x_4'' = 7$$

$$-5x_1' - 2x_2 + 4x_3 - 7x_4' + 7x_4'' + s_2 = 3$$

$$-4x_1' + x_2 + 2x_3 - 5x_4' + 5x_4'' - e_3 = 2$$

$$x_2 + s_4 = 3$$

$$x_1', x_2, x_3, x_4', x_4'', s_2, e_3, s_4 \ge 0$$

In matrix form, the LPP in standard form can be written as

Minimize

$$z'' = c^T x$$

subject to

$$Ax = b$$

$$x \ge 0$$

where

$$A = \begin{bmatrix} 2 & 3 & -1 & -3 & 3 & 0 & 0 & 0 \\ -5 & -2 & 4 & -7 & 7 & 1 & 0 & 0 \\ -4 & 1 & 2 & -5 & 5 & 0 & -1 & 0 \\ 0 & 1 & 0 & 0 & 0 & 0 & 0 & 1 \end{bmatrix} ; c = \begin{bmatrix} 4 \\ 2 \\ -1 \\ 3 \\ -3 \\ 0 \\ 0 \\ 0 \end{bmatrix} ; b = \begin{bmatrix} 7 \\ 3 \\ 2 \\ 2 \end{bmatrix} ; x = \begin{bmatrix} x_1' \\ x_2 \\ x_3 \\ x_4' \\ x_4'' \\ s_2 \\ e_3 \\ s_4 \end{bmatrix}$$

4.4 Basic Solution

Consider an LPP in the standard form

Minimize

$$z = c^T x \qquad (4.29)$$

subject to

$$Ax = b \qquad (4.30)$$

$$x, b \geq 0 \qquad (4.31)$$

with n variables and m constraints. If $m = n$, then the solution is given by satisfying the constraint equations $Ax = b$ and there is no need for optimization. For $m > n$, there will be $m - n$ redundant equations. The case $m < n$ will correspond to an underdetermined system of linear equations that will have infinite solutions. The solution technique of LPP is to determine the optimal solution among many solutions.

A solution that satisfies the constraints is called the *feasible solution*. If we set $n - m$ variables to zero and solve the constraint equations $Ax = b$, we get the *basic solution*. The corresponding variable x obtained from the basic solution is termed the *basis*. A basic solution that also satisfies $x \geq 0$ is called the *basic feasible solution*. It may be noted that every basic feasible solution is an extreme point of the convex set of feasible solutions. If the basic feasible solution is optimal then it is said to be the *optimal basic solution*.

Example 4.2

Find all the basic solutions for the system of equations:

$$3x_1 - 4x_2 + 2x_3 + x_4 = 0$$

$$x_1 + 3x_2 + 2x_3 + x_4 = 500$$

$$7x_1 + x_2 + x_3 - x_4 = 700$$

Writing the above equations in matrix form

$$Ax = b$$

where

$$A = \begin{bmatrix} 3 & -4 & 2 & 1 \\ 1 & 3 & 2 & 1 \\ 7 & 1 & 1 & -1 \end{bmatrix}; \quad x = \begin{bmatrix} x_1 \\ x_2 \\ x_3 \\ x_4 \end{bmatrix}; \quad b = \begin{bmatrix} 0 \\ 500 \\ 700 \end{bmatrix}$$

Let x_1, x_2, and x_3 be the basic variables and x_4 be the nonbasic variable. Since the nonbasic variable(s) take the value zero in the basic solution, we can rewrite the matrix equation as

$$Bx = b$$

where

$$B = \begin{bmatrix} 3 & -4 & 2 \\ 1 & 3 & 2 \\ 7 & 1 & 1 \end{bmatrix}; \quad x = \begin{bmatrix} x_1 \\ x_2 \\ x_3 \end{bmatrix}; \quad b = \begin{bmatrix} 0 \\ 500 \\ 700 \end{bmatrix}$$

The matrix B corresponds to the basic variable columns of A. If B is invertible, then we can evaluate x as

$$x = B^{-1}b$$

$$x = \begin{bmatrix} x_1 \\ x_2 \\ x_3 \end{bmatrix} = \begin{bmatrix} 3 & -4 & 2 \\ 1 & 3 & 2 \\ 7 & 1 & 1 \end{bmatrix}^{-1} \begin{bmatrix} 0 \\ 500 \\ 700 \end{bmatrix} = \begin{bmatrix} 6800/89 \\ 8300/89 \\ 6400/89 \end{bmatrix}$$

Since x_1, x_2, and x_3 are all greater than zero, the solution obtained is a basic feasible solution.

Similarly, we can take x_1, x_2, and x_4 as the basic variables and x_3 as the nonbasic variable. Then,

$$x = \begin{bmatrix} x_1 \\ x_2 \\ x_4 \end{bmatrix} = \begin{bmatrix} 3 & -4 & 1 \\ 1 & 3 & 1 \\ 7 & 1 & -1 \end{bmatrix}^{-1} \begin{bmatrix} 0 \\ 500 \\ 700 \end{bmatrix} = \begin{bmatrix} 100 \\ 100 \\ 100 \end{bmatrix}$$

Again, the basic variables obtained have a value greater than zero, corresponding to a basic feasible solution.

If we take x_1, x_3, and x_4 as the basic variables and x_2 as the nonbasic variable, then

$$x = \begin{bmatrix} x_1 \\ x_3 \\ x_4 \end{bmatrix} = \begin{bmatrix} 3 & 2 & 1 \\ 1 & 2 & 1 \\ 7 & 1 & -1 \end{bmatrix}^{-1} \begin{bmatrix} 0 \\ 500 \\ 700 \end{bmatrix} = \begin{bmatrix} -250 \\ 3200/3 \\ -4150/3 \end{bmatrix}$$

Since some of the basic variables are negative, the basic solution is not feasible.

Now take x_2, x_3, and x_4 as the basic variables and x_1 as the nonbasic variable. Then,

$$x = \begin{bmatrix} x_2 \\ x_3 \\ x_4 \end{bmatrix} = \begin{bmatrix} -4 & 2 & 1 \\ 3 & 2 & 1 \\ 1 & 1 & -1 \end{bmatrix}^{-1} \begin{bmatrix} 0 \\ 500 \\ 700 \end{bmatrix} = \begin{bmatrix} 500/7 \\ 6400/21 \\ -6800/21 \end{bmatrix}$$

Since some of the basic variables are negative, the basic solution is not feasible.

4.5 Simplex Method

In the previous example, we examined four basic solutions for a system of equations with four variables and three constraints. The number of basic solutions that need to be inspected for an n variable problem with m constraints is given by

$$\frac{n!}{(n-m)!m!}$$

For a large LPP, the number of basic solutions could be very high. For example, for a 15-variable problem with 10 constraints, number of basic solutions is 3003.

In the simplex method, all the basic solutions are not evaluated. Rather, this is an iterative method that moves from one basic feasible solution to another until the basis becomes optimal. To begin with, the simplex method requires an initial basic feasible solution for the problem. This can be accomplished by the introduction of artificial variables in the problem. The coefficient matrix associated with the artificial variables will be an identity matrix. The artificial variables can provide initial bases since the columns of an identity matrix are linearly independent.

Consider an LPP

Maximize

$$z = 6x_1 + 7x_2$$

subject to

$$3x_1 + x_2 \leq 10$$

$$x_1 + 2x_2 \leq 8$$

$$x_1 \leq 3$$

$$x_1, x_2 \geq 0$$

Writing the problem in standard form

Minimize

$$z = -6x_1 - 7x_2$$

subject to

$$3x_1 + x_2 + x_3 = 10$$

$$x_1 + 2x_2 + x_4 = 8$$

$$x_1 + x_5 = 3$$

$$x_1, x_2, x_3, x_4, x_5 \geq 0$$

Linear Programming

Writing the constraints in matrix form

$$Ax = b$$

where

$$A = \begin{bmatrix} 3 & 1 & 1 & 0 & 0 \\ 1 & 2 & 0 & 1 & 0 \\ 1 & 0 & 0 & 0 & 1 \end{bmatrix}; \quad x = \begin{bmatrix} x_1 \\ x_2 \\ x_3 \\ x_4 \\ x_5 \end{bmatrix}; \quad b = \begin{bmatrix} 10 \\ 8 \\ 3 \end{bmatrix}$$

Taking x_3, x_4, and x_5 as basic variables, we can evaluate them as

$$x_B = \begin{bmatrix} x_3 \\ x_4 \\ x_5 \end{bmatrix} = B^{-1}b = \begin{bmatrix} 1 & 0 & 0 \\ 0 & 1 & 0 \\ 0 & 0 & 1 \end{bmatrix}^{-1} \begin{bmatrix} 10 \\ 8 \\ 3 \end{bmatrix} = \begin{bmatrix} 10 \\ 8 \\ 3 \end{bmatrix}$$

The solution obtained is a basic feasible solution since all the elements of x are positive. This is not surprising since all the elements of the vector b are positive (problem already written in the standard form) and B is an identity matrix. In this way, a basic feasible solution is ensured at the start of the simplex method. Now we need to check whether the basic feasible solution is optimal.

Since nonbasic variables (x_1 and x_2) have zero values, the objective function

$$z = -6x_1 - 7x_2$$

takes the value zero. That is,

$$z = 0$$

The value of z will decrease if x_1 or x_2 is increased from zero. Thus, the current basis is not optimal. In the simplex method, we can add or remove only one variable from the basis. So we can bring either x_1 or x_2 into the basis. Since the coefficient of x_2 is most negative ($-7 < -6$), we bring x_2 into the basis. The idea is that z decreases more rapidly when x_2 is brought into the basis.

Keeping the other nonbasic variable x_1 equal to zero, let us write the basic variable equations in terms of x_2 as

$$x_3 = 10 - x_2$$

$$x_4 = 8 - 2x_2$$

$$x_5 = 3$$

To maintain nonnegativity of the basic variables x_3, x_4, and x_5, the variable x_2 can take a maximum value of 10 in the equation

$$x_3 = 10 - x_2$$

and x_2 can take a maximum value of 4 in the equation

$$x_4 = 8 - 2x_2$$

Though we would like to take x_2 as large as possible to minimize the objective function, it can take a maximum value of 4 without making x_4 negative. Thus, x_4 is the leaving basic variable. Note that if the coefficients of x_2 on the left-hand side of the constraint equations were negative, then x_2 can be increased to any larger value without violating the nonnegativity constraint of the basic variable. The preceding discussion can be put in a *ratio test* for determining the leaving basic variable. In this test we compute the minimum of the ratios

$$\frac{b_i}{a_{ij}}; \quad a_{ij} > 0$$

In the present example, since x_2 is the new basic variable, j becomes 2. So the ratios are

$$\frac{10}{1} = 10; \quad \frac{8}{2} = 4$$

Since the second row has the minimum value (4 < 10), the second basic variable (x_4) leaves the basis. The new basic feasible solution is

$$x_B = \begin{bmatrix} x_2 \\ x_3 \\ x_5 \end{bmatrix} = \begin{bmatrix} 4 \\ 6 \\ 3 \end{bmatrix}$$

Rewriting the objective function in terms of the new nonbasic variables x_1 and x_4 by using the second constraint equation, we get

$$z = -28 - \frac{5}{2}x_1 + \frac{7}{2}x_4$$

In the first iteration of the simplex method, the objective function is minimized to -28 and since the coefficient of the nonbasic variable x_1 is negative, the basic feasible solution obtained is not optimal. The steps described

Linear Programming

earlier are to be repeated until the cost coefficients associated with nonbasic variables in the objective function are all nonnegative. Based on the discussion, let us write the algorithm for the simplex method (see Table 4.1) and corresponding MATLAB code (*simplex.m*).

The initial simplex tableau is written as follows:

- A vector of basic variables
- A vector of nonbasic variables
- The matrix $[B \quad N \quad \hat{b}]$, and
- The cost coefficients $[c_B^T \quad c_N^T]$

The simplex tableau at the end of each iteration is written as follows:

- A vector of basic variables
- A vector of nonbasic variables
- The matrix $[I \quad B^{-1}N \quad B^{-1}b]$, and
- The cost coefficients $[c_B^T \quad \hat{c}_N^T \quad -z]$

TABLE 4.1

Algorithm for Simplex Method

Step 1: Write the LPP in the canonical form

$$\text{Minimize} \quad z = c^T x$$
$$\text{subject to} \quad Ax = b$$
$$x \geq 0$$

Compute matrices B and N from matrix A, corresponding to basic and nonbasic variable sets. c_N and c_B are the cost coefficients. Print the initial simplex tableau.

Step 2: Compute the minimum(c_N), which gives *i*th entering basic variable

Compute the entering column as $\hat{A}_i = B^{-1} A_i$

Step 3: For all components of $\hat{A}_i$ that are greater than zero, compute the ratios $\dfrac{b_i}{\hat{A}_i}$. From the minimum of these ratios decide the leaving basic variable.

Step 4: Using the updated basic and nonbasic variable sets, update the B and N matrix along with c_N and c_B.

Step 5: Compute

$$x_B = \hat{b} = B^{-1}b$$
$$y^T = c_B^T B^{-1}$$
$$\hat{c}_N^T = c_N^T - y^T N$$
$$z = -c_B^T x_B$$

If $\hat{c}_N^T \geq 0$
  then   goto Step 6
  else   print simplex tableau and goto Step 2

Step 6: Print the optimal basis, value of basic variables, and the objective function value.

The components of the vector y are called *simplex multipliers*.

Let us execute the code for the LPP

Minimize

$$z = -6x_1 - 7x_2$$

subject to

$$3x_1 + x_2 + x_3 = 10$$

$$x_1 + 2x_2 + x_4 = 8$$

$$x_1 + x_5 = 3$$

$$x_1, x_2, x_3, x_4, x_5 \geq 0$$

The output obtained is

```
basic_set =      3    4    5
nonbasic_set =   1    2
Initial_Table =
    1    0    0    3    1    10
    0    1    0    1    2     8
    0    0    1    1    0     3
Cost =
    0    0    0   -6   -7     0

basic_set =      3    2    5        (x3, x2, x5)
nonbasic_set =   1    4             (x1, x4)
Table =
    1    0    0    5/2  -1/2    6
    0    1    0    1/2   1/2    4
    0    0    1    1     0      3
Cost =
    0    0    0   -5/2   7/2   28

basic_set =      1    2    5
nonbasic_set =   3    4
Table =
    1    0    0    2/5  -1/5   12/5
    0    1    0   -1/5   3/5   14/5
    0    0    1   -2/5   1/5    3/5
Cost =
    0    0    0    1     3     34
---  SOLUTION  ---
basic_set = 1    2    5             (x1, x2, x5)
xb =
    12/5
    14/5
    3/5
zz =
    -34
```

Linear Programming

Since the reduced cost coefficients (1 and 3) of the nonbasic variables are nonnegative, the basis is optimal. The basis for the optimal solution is x_1, x_2, and x_5 and their values are

$$x_1 = \frac{12}{5}$$

$$x_2 = \frac{14}{5}$$

$$x_5 = \frac{3}{5}$$

and the objective function is minimized to

$$z = -34$$

Let us graph the constraints (see Figure 4.5) for this problem. The initial basic feasible solution in the simplex method corresponds to point A (0, 0). In the first iteration, the method moved to point B (0, 4) as the next basic feasible solution where objective function value was reduced to −28. The basis here is x_3, x_2, and x_5 and their values are

$$x_2 = 4$$

$$x_3 = 6$$

$$x_5 = 3$$

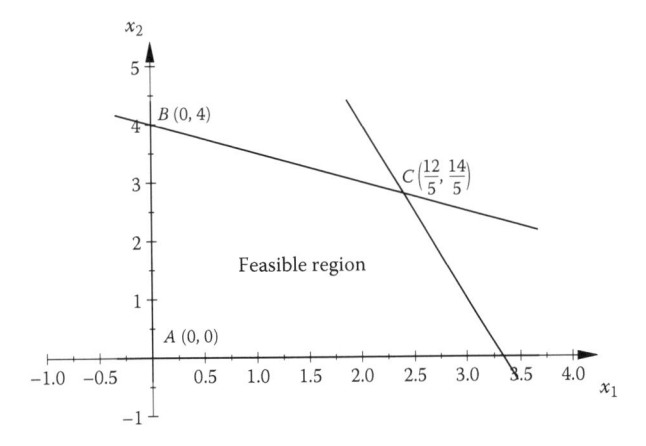

FIGURE 4.5
Feasible region for the problem solved by the simplex method.

In the second iteration, the simplex method moved to point $C\left(\dfrac{12}{5}, \dfrac{14}{5}\right)$ as the next basic feasible solution where the objective function value was reduced to −34. Since all the cost coefficients corresponding to the nonbasic variable were nonnegative, the basic feasible solution was optimal and further iterations were terminated.

4.5.1 Multiple Solutions

Let us modify the objective function of the previous problem and rewrite LPP as

Minimize

$$z = -6x_1$$

subject to

$$3x_1 + x_2 + x_3 = 10$$

$$x_1 + 2x_2 + x_4 = 8$$

$$x_1 + x_5 = 3$$

$$x_1, x_2, x_3, x_4, x_5 \geq 0$$

The MATLAB code (*simplex.m*) is executed again with the following modification:

```
c = [-6;0;0;0;0];
```

The output obtained is

```
basic_set =
        3    4    5
nonbasic_set =
        1    2
Initial_Table =
1    0    0    3    1    10
0    1    0    1    2    8
0    0    1    1    0    3
Cost =
0    0    0    -6    0    0
```

Linear Programming

```
basic_set =
     3      4      1
nonbasic_set =
     2      5
Table =
  1    0    0    1   -3    1
  0    1    0    2   -1    5
  0    0    1    0    1    3
Cost =
  0    0    0    0    6   18
------SOLUTION------
basic_set =
     3      4      1
xb =
     1
     5
     3
zz =
   -18
```

The simplex method converges to the optimal solution in one iteration and the minimum value of the objective function is –18. Observe from the output that at the end of the first iteration the cost coefficient corresponding to the nonbasic variable x_2 is zero as compared to another nonbasic variable x_5 that has a value of 6. Allow the MATLAB code (*simplex.m*) to be executed for one more iteration by commenting the terminating criterion as follows:

```
%   if cn_cap >=0
%       break;
%   end
```

The output obtained in the second iteration is

```
basic_set =              2      4      1
nonbasic_set =           3      5
Table =
  1    0    0    1   -3    1
  0    1    0   -2    5    3
  0    0    1    0    1    3
Cost =
  0    0    0    0    6   18
------SOLUTION------
basic_set =
     2      4      1
xb =
     1
     3
     3
zz =
   -18
```

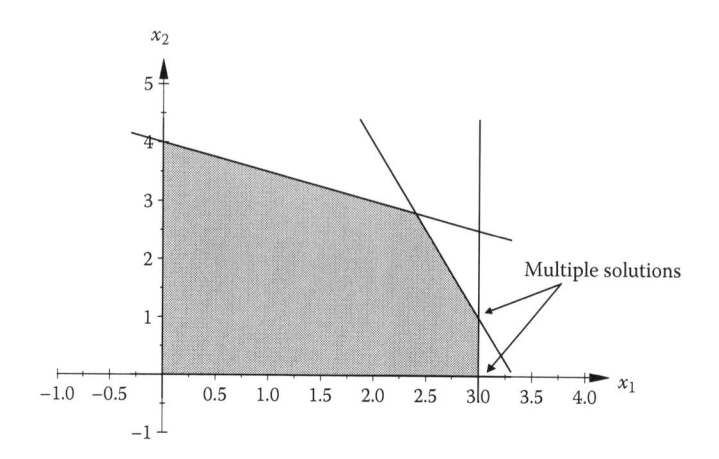

FIGURE 4.6
Concept of multiple solutions.

Observe that this basis is also optimal. An LPP is said to have *multiple solutions* when the cost coefficient of a nonbasic variable is zero in the optimal basis. This is also shown in Figure 4.6, where these points correspond to an edge of the convex polyhedron.

4.5.2 Degeneracy

Sometimes, during the course of the simplex procedure, the method can become cyclic with no further improvement in the objective function. This occurs when the entering basic variable becomes zero in a basis. That is, component of vector b_i becomes zero during the iteration. Let us show this with an example for the following LPP:

Minimize

$$z = -3x_1 - 3x_2$$

subject to

$$x_1 \leq 4$$

$$x_1 + 2x_2 \leq 4$$

$$x_1, x_2 \geq 0$$

Linear Programming

Modify the *simplex.m* code with the following changes:

```
A = [1  0  1  0;
     1  2  0  1 ];
b = [4;4];
c = [-3;-3;0;0];
basic_set = [3 4];
nonbasic_set = [1 2];
```

On executing the code, the following output is displayed on the command window.

```
basic_set =           3    4
nonbasic_set =        1    2
Initial_Table =
        1       0       1       0       4
        0       1       1       2       4
Cost =
        0       0      -3      -3       0

basic_set =           1    4
nonbasic_set =        2    3
Table =
        1       0       0       1       4
        0       1       2      -1       0
Cost =
        0       0      -3       3      12

basic_set =           1    2
nonbasic_set =        3    4
Table =
        1       0       1       0       4
        0       1    -1/2     1/2       0
Cost =
        0       0     3/2     3/2      12
```

Note that in the first iteration, the basic variable x_4 becomes zero. The value of the objective function does not improve during the second iteration. The problem can be avoided by adding a small perturbation on the b vector and the same can be implemented in the *simplex.m* code as

```
b = [4;4];
pertb = [1e-2;1e-3];
b = b+pertb;
```

On executing the modified code, the following output is displayed on the command window.

```
basic_set =            3    4
nonbasic_set =         1    2
Initial_Table =
        1      0      1      0      401/100
        0      1      1      2      4001/1000
Cost =
        0      0     -3     -3      0

basic_set =            3    1
nonbasic_set =         2    4
Table =
        1      0     -2     -1      9/1000
        0      1      2      1      4001/1000
Cost =
        0      0      3      3      3997/333

basic_set =            3    2
nonbasic_set =         1    4
Table =
        1      0      1      0      401/100
        0      1     1/2    1/2     4001/2000
Cost =
        0      0    -3/2    3/2     4003/667
------SOLUTION------
basic_set =            3    2
xb =
        401/100
        4001/2000
zz =
        -4003/667
```

Note that by making a small perturbation on b, we are able to achieve the minimum value of the objective function as −6.0015 at $(x_1, x_2) = (0, 2.0005)$. The exact minimum value of the objective function is −6 and occurs at $(x_1, x_2) = (0, 2)$.

4.5.3 Two-Phase Method

As discussed earlier, to start a simplex method, a basic feasible solution is required. A basic feasible solution may not be readily available for an LPP. For example, the addition of a negative slack variable in a ≥ type constraint will not lead to the canonical form of equations. By addition of artificial variables, this problem can be overcome. The original LPP gets modified as a result of the introduction of the artificial variables. In phase I of the simplex method, we solve the modified LPP to get a basic feasible solution. Once a basic feasible solution is available from phase I, phase II involves solving the original LPP. Let us explain the two-phase simplex method with an example.

Consider a LPP

Minimize

$$z = 3x_1 + 4x_2$$

subject to

$$3x_1 + 2x_2 = 10$$
$$2x_1 - 4x_2 \geq 3$$
$$3x_1 + 4x_2 \leq 16$$
$$x_1, x_2 \geq 0$$

Writing the LPP with slack (x_3, x_4) and artificial variables (y_1, y_2) as

Minimize

$$z = 3x_1 + 4x_2$$

subject to

$$3x_1 + 2x_2 + y_1 = 10$$
$$2x_1 - 4x_2 - x_3 + y_2 = 3$$
$$3x_1 + 4x_2 + x_4 = 16$$
$$x_1, x_2, x_3, x_4, y_1, y_2 \geq 0$$

The objective function in the phase I problem is

Minimize

$$z' = \sum y_i = y_1 + y_2 \qquad (4.32)$$

The constraints of the phase I problem remain same as in the original LPP. The variables y_1, y_2, and x_4 can be taken as the basic variables. The objective function in the phase I problem is not a function of the nonbasic variables.

Writing the modified objective function in terms of the nonbasic variables using the first and second constraint equations:

$$y_1 + y_2 = 13 - 5x_1 + 2x_2 + x_3$$

This can also be done using the formula

$$\hat{c}_N^T = c_N^T - y^T N \qquad (4.33)$$

where

$$y^T = c_B^T B^{-1} \qquad (4.34)$$

By executing the MATLAB code (*initial_cost.m*), the cost coefficients for the nonbasic variables can be obtained as

(-5, 2, 1)

With a minor modification of the MATLAB code (*simplex.m*), phase I code is written in *phase1.m*. On executing the code, the following output is displayed on the command window.

```
basic_set =           5      6     4
nonbasic_set =        1      2     3
Initial_Table =
1      0      0      3      2      0      10
0      1      0      2     -4     -1       3
0      0      1      3      4      0      16
Cost =
0      0      0     -5      2      1     -13

basic_set =           5      1     4
nonbasic_set =        2      3     6
Table =
1      0      0      8     3/2   -3/2    11/2
0      1      0     -2    -1/2    1/2     3/2
0      0      1     10     3/2   -3/2    23/2
Cost =
0      0      0     -8    -3/2    5/2   -11/2
```

The variable number 6 (y_2) has left the basis and so can be removed from the basis. A MATLAB code (*remove_variable.m*) removes the user-specified column from the nonbasic set. Note that this variable corresponds to the third column in the nonbasic set. On executing the code, the following output is displayed on the command window.

Linear Programming

```
----Table after removing artificial variable------
basic_set =         5      1      4
nonbasic_set =      2      3
Initial_Table =
1     0     0       8      3/2    11/2
0     1     0      -2     -1/2    3/2
0     0     1      10      3/2    23/2
Cost =
0     0     0      -8     -3/2   -11/2
```

Now rerun the phase I code without initializing the A and b matrix. This can be done by modifying the code *phase1.m* to *phase1_without_initialization.m*. On executing the code *phase1_without_initialization.m*, the following output is displayed on the command window.

```
basic_set =         2      1      4
nonbasic_set =      3      5
Table =
1     0     0      3/16    1/8    11/16
0     1     0     -1/8     1/4    23/8
0     0     1     -3/8    -5/4    37/8
Cost =
0     0     0       0       1       0
```

Again, the variable number 5 (y_1) has left the basis and so can be removed from the basis. This corresponds to the second column in the nonbasic set. Make the following modification in the code *remove_variable.m* and then rerun this code.

```
remove_column = 2;

----Table after removing artificial variable------
basic_set =         2      1      4
nonbasic_set =      3
Initial_Table =
    1     0     0       0    3/16    11/16
    0     1     0       0   -1/8     23/8
    0     0     1       1   -3/8     37/8
Cost =
    0     0     0       0     0       0
```

This basis does not involve any artificial variable and the value of the objective function is zero. So this is the feasible solution for the original problem. In case the objective function value was greater than zero, the solution would be infeasible. This is the end of phase I. The objective function of the original problem is

$$z = 3x_1 + 4x_2$$

The objective function has to be written in terms of the nonbasic variable. Again using the code *initial_cost.m*, with the following modifications, the cost coefficients for the nonbasic variable can be computed.

```
cb = [3 4 0];
cn = [0];
N = [3/16;
     -1/8;
     -3/8];
B = [0 1 0;1 0 0;0 0 1];
```

In phase II of the simplex method, execute the *phase1.m* code with following modifications.

```
A = [0 1 3/16 0;
     1 0 -1/8 0;
     0 0 -3/8 1];
b = [11/16;23/8;37/8];
c = [0;0; -3/8; 0];
basic_set = [2 1 4];
nonbasic_set = [3];
zz1 = 91/8;
```

On executing the *phase1.m* code the following output is printed on the command window.

basic_set =	2	1	4		
nonbasic_set =	3				
Initial_Table =					
1	0	0	3/16	11/16
0	1	0	-1/8	23/8
0	0	1	-3/8	37/8
Cost =				
0	0	0	-3/8	-91/8

basic_set =	3	1	4		
nonbasic_set =	2				
Table =					
1	0	0	16/3	11/3
0	1	0	2/3	10/3
0	0	1	2	6
Cost =				
0	0	0	2	-10

Since all the cost coefficients are nonnegative, the basis is optimal. The minimum value of the objective function is 10 and occurs at $(x_1, x_2) = \left(\dfrac{10}{3}, 0\right)$.

4.5.4 Dual Simplex Method

Every LPP, called the *primal*, is associated with another LPP, called its *dual*. The optimal solution of the primal problem and its dual remain the same. In the dual problem, the components of the b vector (right-hand side of the constraint equation in the primal problem) become the cost coefficients in the objective function and vice versa. If there are n variables and m constraints in the primal problem, then there will be m variables and n constraints in the dual problem. If the objective function in the primal problem is of the minimization type, then it becomes a maximization type in the dual problem. All constraints are to be written as $\leq$ in the dual problem. An equality constraint $x = b$ can be converted into two $\leq$ constraints by writing it as $x \leq b$ and $-x \leq -b$. For a primal LPP

Maximize
$$z = c^T x$$

subject to
$$Ax \geq b$$
$$x \geq 0$$

Its corresponding dual is

Minimize
$$w = b^T y$$

subject to
$$A^T y \leq c$$
$$y \geq 0$$

The transformation rules from primal to dual problems are given in Table 4.2. In the primal LPP, the simplex method moves from one feasible solution to another. The dual simplex method moves from one primal infeasible solution to another with reduced infeasibility. On reaching the primal feasibility conditions, the method stops as the solution obtained is the optimal one. One may argue the need for a dual problem and its solution. It is observed that some of the LPPs show degeneracy when used with the primal problems. The corresponding dual problems are much easier to solve in

TABLE 4.2

Transformation Rules from Primal to Dual Conversion

Primal	Dual
$a_i x \geq b_i$	$y_i \geq 0$
$a_i x \leq b_i$	$y_i \leq 0$
$a_i x = b_i$	y_i free
$x_i \geq 0$	$a_j^T y \leq c_j$
$x_i \leq 0$	$a_j^T y \geq c_j$
x_i free	$a_j^T y = c_j$

such cases. Further, dual methods are more suited for mixed-integer type problems.

Let us write the algorithm for the dual-simplex method (Table 4.3) and the corresponding MATLAB code is written in the file *dual.m*.

Consider the primal LPP

Maximize

$$z = 3y_1 + 4y_2 + 25y_3 + 26y_4$$

subject to

$$y_1 + 2y_3 + y_4 \leq 9$$

$$y_2 + y_3 + 3y_4 \leq 8$$

$$y_1, y_2, y_3, y_4 \geq 0$$

Its dual is

Minimize

$$z = 9x_1 + 8x_2$$

subject to

$$x_1 \geq 3$$

$$x_2 \geq 4$$

$$2x_1 + x_2 \geq 25$$

$$x_1 + 3x_2 \geq 26$$

$$x_1, x_2 \geq 0$$

TABLE 4.3

Algorithm for the Dual-Simplex Method

Step 1: Write the dual LPP in canonical form

$$\text{Minimize} \quad z = c^T x$$

$$\text{subject to} \quad Ax = b$$

$$x \geq 0$$

Compute matrices B and N from matrix A, corresponding to basic and nonbasic variable sets. c_N and c_B are the cost coefficients of basic and nonbasic variables. Print the initial tableau.
Step 2: Compute minimum(b_i), which gives the ith leaving basic variable.
Compute the pivot row as $\hat{A}_i = B^{-1} A_i$
Step 3: For all components of $\hat{A}_i$ which are less than zero, compute the ratios $-\dfrac{c_j}{\hat{A}_i}$.
Minimum of these ratios decide the entering basic variable.
Step 4: Using the updated basic and nonbasic variable sets, update B and N matrix along with c_N and c_B.
Step 5: Compute

$$x_B = \hat{b} = B^{-1} b$$

$$y^T = c_B^T B^{-1}$$

$$\hat{c}_N^T = c_N^T - y^T N$$

$$z = -c_B^T x_B$$

If $\hat{b} \geq 0$
    then   goto Step 6
    else   print the simplex tableau and goto Step 2
Step 6: Print the optimal basis, value of basic variables and the objective function value.

Since the constraints are of $\geq$ type, the dual problem is not in standard form. Writing the constraints in the canonical form,

$$Ax = b$$

where

$$A = \begin{bmatrix} -1 & 0 & 1 & 0 & 0 & 0 \\ 0 & -1 & 0 & 1 & 0 & 0 \\ -2 & -1 & 0 & 0 & 1 & 0 \\ -1 & -3 & 0 & 0 & 0 & 1 \end{bmatrix}; \quad x = \begin{bmatrix} x_1 \\ x_2 \\ x_3 \\ x_4 \\ x_5 \\ x_6 \end{bmatrix}; \quad b = \begin{bmatrix} -3 \\ -4 \\ -25 \\ -26 \end{bmatrix}$$

On executing the code *dual.m*, the following output is printed on the MATLAB command window.

```
basic_set =       3    4     5     6
nonbasic_set =    1    2
Initial_Table =
1    0    0    0    -1    0     -3
0    1    0    0     0   -1     -4
0    0    1    0    -2   -1    -25
0    0    0    1    -1   -3    -26
Cost =
0    0    0    0     9    8      0

basic_set =       3    4     5     2
nonbasic_set =    1    6
Table =
1    0    0    0    -1     0      -3
0    1    0    0    1/3   -1/3   14/3
0    0    1    0   -5/3   -1/3  -49/3
0    0    0    1    1/3   -1/3   26/3
Cost =
0    0    0    0   19/3   8/3  -208/3

basic_set =       3    4     1     2
nonbasic_set =    5    6
Table =
1    0    0    0   -3/5   1/5    34/5
0    1    0    0    1/5  -2/5     7/5
0    0    1    0   -3/5   1/5    49/5
0    0    0    1    1/5  -2/5    27/5
Cost =
0    0    0    0   19/5   7/5  -657/5
------FINAL SOLUTION------
basic_set =   3    4    1    2
xb =
       34/5
        7/5
       49/5
       27/5
zz =
       657/5
```

Since all the b_i are nonnegative, the basis is optimal. The minimum value of the objective function is $\dfrac{657}{5}$ and occurs at $(x_1, x_2) = \left(\dfrac{49}{5}, \dfrac{27}{5}\right)$.

Linear Programming

4.6 Interior-Point Method

In the simplex method, one moves along the boundary of the feasible region to arrive at the optimum. For an LPP with a large number of constraints, this may be time consuming if the initial guess is far from the optimal. On the other hand, interior-point methods move inside the feasible region to reach the optimal solution. Narenndra Karmarkar proposed a new polynomial-time algorithm (Karmarkar 1984) that claimed to be up to 50 times faster as compared to the simplex method for large LPP. His algorithm did create further interest in such methods. Interior point methods can be classified into

- Barrier function methods
- Potential-reduction methods
- Affine scaling methods

The affine scaling method is very simple to implement and has been successful in solving large LPP. The method is due to Barnes and Vanderbei (Barnes 1986; Vanderbei et al. 1986). In this method, we start with a point inside the feasible region (see Figure 4.7) and then use the projected steepest-descent direction to get the next improved point. Note that if the point (x_c) is close to the central position, a considerable improvement in the objective function can be made. On the other hand, if the point (x_a) is away from the central position, the improvement in the objective function would be less. The affine scaling method transforms LPP to another equivalent problem so that the point is closer to the central position.

Let us write an algorithm for the affine scaling method (Table 4.4) and the corresponding MATLAB code is written in the file *interior.m*.

Let us take the same problem that was solved by the simplex method. The LPP

Maximize
$$z = 6x_1 + 7x_2$$

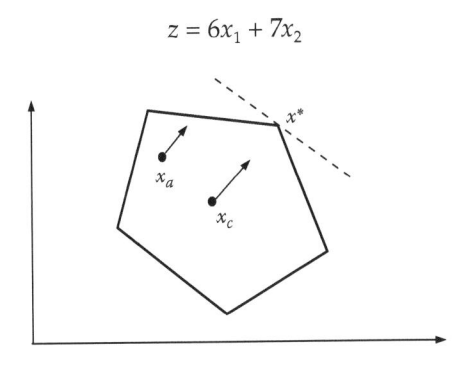

FIGURE 4.7
Interior-point method.

TABLE 4.4

Algorithm for the Interior-Point Method

Step 1: Write the LPP in the form

$$\text{Maximize} \quad z = c^T x$$

$$\text{subject to} \quad Ax = b$$

$$x \geq 0$$

Give inputs A, b, c, x_0 (initial feasible point), γ (accelerating parameter), and ε (tolerance parameter).

Step 2: Compute $v_i = b - Ax_i$

$$D = \text{diag}(v_i)$$

$$h_x = (A^T D^{-2} A)^{-1} c$$

$$h_v = -A h_x$$

$$\alpha = \gamma \cdot \min\left\{ \left. -v_i \middle/ h_v \right| h_v < 0 \right\}$$

$$x_{i+1} = x_i + \alpha h_x$$

Step 3: If $|z(x_{i+1}) - z(x_i)| > \varepsilon$
 then goto Step 2
 else goto Step 4

Step 4: Print x and z.

---

subject to

$$3x_1 + x_2 \leq 10$$

$$x_1 + 2x_2 \leq 8$$

$$x_1 \leq 3$$

$$x_1, x_2 \geq 0$$

On executing the code, the maximum value of the objective function obtained is 34, which occurs at $(x_1, x_2) = (2.4, 2.8)$. This matches with the result that was obtained by the simplex method. The convergence history of the affine scaling method is shown in Figure 4.8.

Linear Programming

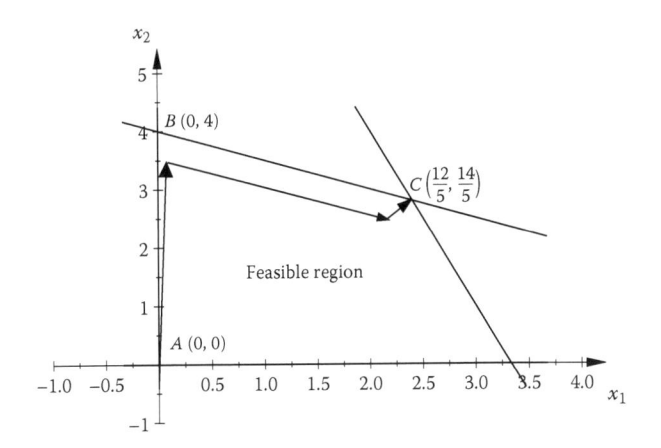

FIGURE 4.8
Convergence history for the affine scaling method.

4.7 Portfolio Optimization

Let us solve a 10-variable portfolio optimization problem using some of the techniques described earlier in this chapter. A company has to invest $600,000 in different financial products such that its earnings are maximized. The expected return on investment for different financial products is given in Table 4.5.

TABLE 4.5
Portfolio Optimization Problem Description

Financial Product	Market	Return (in %)
x_1	Trucks—Germany	9.5
x_2	Cars—Japan	11.2
x_3	Laptops—USA	10.5
x_4	Computers—USA	11.9
x_5	Appliances—Australia	11.7
x_6	Appliances—Europe	13.2
x_7	Insurance—Germany	10.5
x_8	Insurance—USA	10.9
x_9	Currency carry trade	5.5
x_{10}	Others	5.1

The following constraints are specified on the investment.

- No more than $140,000 in the transport segment
- No more than $160,000 in the computer segment
- No more than $120,000 in the appliances segment
- No more than $230,000 in the German segment
- No more than $220,000 in the USA segment

The LPP can be written mathematically as

Maximize

$$z = 0.095x_1 + 0.112x_2 + 0.105x_3 + 0.119x_4 + 0.117x_5 + 0.132x_6 + 0.105x_7 + 0.109x_8 + 0.055x_9 + 0.051x_{10}$$

subject to

$$x_1 + x_2 \leq 140$$

$$x_3 + x_4 \leq 160$$

$$x_5 + x_6 \leq 120$$

$$x_1 + x_7 \leq 230$$

$$x_3 + x_4 + x_8 \leq 220$$

$$x_1 + x_2 + x_3 + x_4 + x_5 + x_6 + x_7 + x_8 + x_9 + x_{10} = 600$$

$$x_1, x_2, x_3, x_4, x_5, x_6, x_7, x_8, x_{91}, x_{10} \geq 0$$

Writing the problem in standard form

Minimize

$$z = -0.095x_1 - 0.112x_2 - 0.105x_3 - 0.119x_4 - 0.117x_5 - 0.132x_6 - 0.105x_7 - 0.109x_8 - 0.055x_9 - 0.051x_{10}$$

subject to

$$x_1 + x_2 + x_{11} = 140$$

Linear Programming

$$x_3 + x_4 + x_{12} = 160$$

$$x_5 + x_6 + x_{13} = 120$$

$$x_1 + x_7 + x_{14} = 230$$

$$x_3 + x_4 + x_8 + x_{15} = 220$$

$$x_1 + x_2 + x_3 + x_4 + x_5 + x_6 + x_7 + x_8 + x_9 + x_{10} + x_{16} = 600$$

$$x_1, x_2, x_3, x_4, x_5, x_6, x_7, x_8, x_9, x_{10}, x_{11}, x_{12}, x_{13}, x_{14}, x_{15}, x_{16} \geq 0$$

The variables x_1, x_2, and so on are in thousands of dollars. Update the following input data in the code *simplex.m* and then execute the code.

```
A =[1 1 0 0 0 0 0 0 0 0 1 0 0 0 0 0;
    0 0 1 1 0 0 0 0 0 0 0 1 0 0 0 0;
    0 0 0 0 1 1 0 0 0 0 0 0 1 0 0 0;
    1 0 0 0 0 0 1 0 0 0 0 0 0 1 0 0;
    0 0 1 1 0 0 0 1 0 0 0 0 0 0 1 0;
    1 1 1 1 1 1 1 1 1 1 0 0 0 0 0 1];
b = [140;160;120;230;220;600];
c = [-0.095;-0.112;-0.105;-0.119;-0.117;-0.132;-0.105;-0.109;-0.055;
    -0.051;0;0;0;0;0;0];
basic_set = [11 12 13 14 15 16];
nonbasic_set = [1 2 3 4 5 6 7 8 9 10]
```

The following output is displayed on the command window.

```
basic_set =            11      12      13      14      15      16
nonbasic_set =   1   2   3   4   5   6   7   8   9   10
Initial_Table =
1 0 0 0 0 0 1 1 0 0 0 0 0 0 0 140
0 1 0 0 0 0 0 0 1 1 0 0 0 0 0 160
0 0 1 0 0 0 0 0 0 0 1 1 0 0 0 120
0 0 0 1 0 0 1 0 0 0 0 0 1 0 0 230
0 0 0 0 1 0 0 0 1 1 0 0 0 1 0 0 220
0 0 0 0 0 1 1 1 1 1 1 1 1 1 1 600
Cost =
0 0 0 0 0 0 -19/200 -14/125 -21/200 -119/1000 -117/1000 -33/250
       -21/200     -109/1000              -11/200      -51/1000     0

basic_set = 11      12      6       14              15      16
nonbasic_set = 1   2   3   4    5 7 8 9   10    13
Table =
1 0 0 0 0 0 1 1 0 0 0 0 0 0 0 140
0 1 0 0 0 0 0 0 1 1 0 0 0 0 0 160
```

```
0 0 1 0 0 0 0 1 1 0 0 0 0 0 1 120
0 0 0 1 0 0 1 0 0 0 0 1 0 0 0 0 230
0 0 0 0 1 0 0 0 1 1 0 0 1 0 0 0 220
0 0 0 0 0 1 1 1 1 1 0 1 1 1 1 -1 480
Cost =
0 0 0 0 0 0  -19/200  -14/125  -21/200  -119/1000  3/200   -21/200
 -109/1000        -11/200        -51/1000        33/250    396/25
```

basic_set =	11	4	6	14	15	16
nonbasic_set =	1	2	3	5	7	8 9 10 12 13

```
Table =
1 0 0 0 0 0 1 1 0 0 0 0 0 0 0 140
0 1 0 0 0 0 0 0 1 0 0 0 0 0 1 0 160
0 0 1 0 0 0 0 0 0 1 0 0 0 0 0 1 120
0 0 0 1 0 0 1 0 0 0 1 0 0 0 0 0 230
0 0 0 0 1 0 0 0 0 0 0 1 0 0 -1 0 60
0 0 0 0 0 1 1 1 0 0 1 1 1 1 -1 -1 320
Cost =
0 0 0 0 0 0  -19/200   -14/125  7/500   3/200   -21/200  -109/1000
   -11/200        -51/1000      119/1000       33/250      872/25
```

basic_set =	2	4	6	14	15	16
nonbasic_set =	1	3	5	7	8 9 10 11 12	13

```
Table =
1 0 0 0 0 0 1 0 0 0 0 0 1 0 0 140
0 1 0 0 0 0 0 1 0 0 0 0 0 1 0 160
0 0 1 0 0 0 0 0 1 0 0 0 0 0 1 120
0 0 0 1 0 0 1 0 0 1 0 0 0 0 0 230
0 0 0 0 1 0 0 0 0 0 1 0 0 0 -1 0 60
0 0 0 0 0 1 0 0 0 1 1 1 1 -1 -1 -1 180
Cost =
0 0 0 0 0 0  17/1000  7/500  3/200  -21/200  -109/1000 -11/200
     -51/1000    14/125        119/1000       33/250     1264/25
```

basic_set =	2	4	6	14	8	16
nonbasic_set =	1	3	5	7	9 10 11 12	13	15

```
Table =
1 0 0 0 0 0 1 0 0 0 0 1 0 0 0 140
0 1 0 0 0 0 0 1 0 0 0 0 1 0 0 160
0 0 1 0 0 0 0 0 1 0 0 0 0 1 0 120
0 0 0 1 0 0 1 0 0 1 0 0 0 0 0 230
0 0 0 0 1 0 0 0 0 0 0 0 -1 0 1 60
0 0 0 0 0 1 0 0 0 1 1 1 -1 0 -1 -1 120
Cost =
0 0 0 0 0 0  17/1000  7/500  3/200  -21/200  -11/200  -51/1000
     14/125         1/100         33/250      109/1000    571/10
```

basic_set =	2	4	6	14	8	7
nonbasic_set =	1	3	5	9	10 11 12 13 15	16

```
Table =
1 0 0 0 0 0 1 0 0 0 0 1 0 0 0 0 140
0 1 0 0 0 0 0 1 0 0 0 0 1 0 0 0 160
0 0 1 0 0 0 0 0 1 0 0 0 0 1 0 0 120
0 0 0 1 0 0 1 0 0 -1 -1 1 0 1 1 -1 110
0 0 0 0 1 0 0 0 0 0 0 0 -1 0 1 0  60
0 0 0 0 0 1 0 0 0 1 1 -1 0 -1 -1 1 120
Cost =
0 0 0 0 0 0    17/1000  7/500   3/200   1/20      27/500    7/1000
        1/100              27/1000   1/250         21/200    697/10
------SOLUTION------
basic_set =    2       4       6       14      8       7
xb =
          140
          160
          120
          110
           60
          120
zz =
         -697/10
```

The optimal solution for the LPP is

$x_1, x_2, x_3, x_4, x_5, x_6, x_7, x_8, x_9, x_{10}$ = (0, 140, 0, 160, 0, 120, 120, 60, 0, 0)

Again notice that the variables x_1, x_2, and so on are in thousands of dollars. Since the maximization problem was converted into the minimization problem, the optimal solution has to be multiplied by –1. Thus the maximum earnings are $69,700.

Chapter Highlights

- An optimization problem that has the objective and the constraints as a linear function of the design variables is a linear programming problem.
- The graphical method is a simple technique for locating the optimal solution for problems with up to two or three design variables only.
- Inequalities can be plotted in MATLAB in the *MuPad* command window.
- In an LPP, the optimal value of the objective function occurs at the edge of the convex polyhedron.

- When the objective function can be increased to an infinitely large value, without leaving the feasible region, the solution of the LPP is said to be unbounded.
- In the standard form of an LPP, the objective function is of minimization type, all the design variables should be nonnegative, all the constraints should be of the equality type, and all the numbers on the right-hand side of the constraint equation should be nonnegative.
- A ≤ type constraint can be converted into an equality constraint by adding a slack variable.
- A ≥ type constraint can be converted into an equality constraint by subtracting it with a surplus variable.
- An unrestricted or free variable (without any specified bounds) can be replaced by a pair of nonnegative variables.
- A solution that satisfies the constraints is called a feasible solution.
- The variables x obtained from the basic solution are termed the basis. A basic solution that also satisfies $x \geq 0$ is called the basic feasible solution. It may be noted that every basic feasible solution is an extreme point of the convex set of feasible solutions. If the basic feasible solution is optimal then it is called the optimal basic solution.
- The simplex method is an iterative method that moves from one basic feasible solution to another until the basis becomes optimal. The method requires an initial basic feasible solution for the problem. This can be achieved by the introduction of artificial variables in the problem. The coefficient matrix associated with the artificial variables will be an identity matrix. The artificial variables can provide initial bases because the columns of an identity matrix are linearly independent.
- An LPP is said to have multiple solutions when the cost coefficient of a nonbasic variable is zero in the optimal basis.
- The simplex method can become cyclic with no improvement in the objective function during iterations. This occurs when the entering basic variable becomes zero in a basis.
- The degeneracy problem in the simplex method can be avoided by adding a small perturbation on the b vector.
- In phase I of the simplex method, we solve the modified LPP to get a basic feasible solution. Once a basic feasible solution is available from phase I, phase II involves solving the original LPP.
- Every LPP, called the primal, is associated with another LPP, called its dual. The optimal solution of the primal problem and its dual remain the same.

Linear Programming

- It is observed that some of the LPPs show degeneracy when used with the primal problems. The corresponding dual problems are much easier to solve in such cases.
- Interior-point methods move inside the feasible region to reach the optimal solution.

Formulae Chart

LPP in the standard matrix form:

Minimize

$$z = c^T x$$

subject to

$$Ax = b$$

$$x \geq 0$$

Simplex tableau:

$$x_B = \hat{b} = B^{-1} b$$

$$y^T = c_B^T B^{-1}$$

$$\hat{c}_N^T = c_N^T - y^T N$$

$$z = -c_B^T x_B$$

Problems

1. A manufacturer produces two components, X and Y. Component X requires 2 hours of machining and 3 hours of polishing. Component Y requires 3 hours of machining and 4 hours of polishing. Every week, 42 hours of machining and 48 hours of polishing can be done. The company makes a profit of $5 on X and $7 on Y. Assume that whatever is produced gets sold in the market. Formulate the LPP and solve it using the graphical method.

2. Solve the following LPP using the graphical method.
   i. Minimize
   $$z = 3x_1 - 2x_2$$
   subject to
   $$x_1 + 2x_2 \le 10$$
   $$2x_1 - x_2 \le 5$$
   $$-4x_1 + 3x_2 \ge 5$$
   $$x_1, x_2 \ge 0$$

   ii. Maximize
   $$z = 2x_1 + 5x_2$$
   subject to
   $$3x_1 + x_2 \le 11$$
   $$x_1 + x_2 \ge 6$$
   $$2x_1 + x_2 \le 10$$
   $$x_1, x_2 \ge 0$$

   iii. Maximize
   $$z = 4x_1 + 5x_2$$
   subject to
   $$2x_1 + x_2 \le 20$$
   $$-3x_1 + 2x_2 \le 25$$
   $$-x_1 + x_2 \le 30$$
   $$x_1, x_2 \ge 0$$

iv. Maximize
$$z = -x_1 + 2x_2$$
subject to
$$2x_1 + x_2 \geq 5$$
$$4x_1 + x_2 \geq 10$$
$$2x_1 + 3x_2 \leq 8$$
$$x_1, x_2 \geq 0$$

3. Determine all the basic solutions (feasible and infeasible) for the following system of linear equations.
$$x_1 - 2x_2 - x_3 + 4x_4 = 3$$
$$x_1 + 2x_3 + 2x_4 = 4$$
$$2x_1 - x_2 + x_3 + x_4 = 5$$

4. Find the value of k so that the following LPP has an optimal solution at $\left(\dfrac{-44}{7}, \dfrac{48}{7}\right)$.

Minimize
$$z = -3x_1 + 2x_2$$
subject to
$$-x_1 + 2x_2 \geq 10k$$
$$2x_1 + x_2 \leq 5k$$
$$2x_1 + 3x_2 \leq 4k$$
$$x_1, x_2 \geq 0$$

5. Convert the following LPP into standard form with matrix notations:
   i. Minimize
$$z = 2x_1 + 3x_2 - x_3$$

subject to

$$-x_1 + 2x_2 - 3x_3 \leq 5$$

$$2x_1 - x_2 + 4x_3 \leq -5$$

$$3x_1 - 2x_2 - 5x_3 \geq -7$$

$$x_1, x_2, x_3 \geq 0$$

ii. Maximize

$$z = 2x_1 - 3x_2 + 4x_3$$

subject to

$$3x_1 - 2x_2 - 3x_3 \geq 11$$

$$-4x_1 - 3x_2 + x_3 \geq -6$$

$$x_1 + 2x_2 + x_3 \leq 10$$

$$x_1 \geq 2, x_2 \leq 5, x_3 \text{ free}$$

6. Consider the system of equations $Ax = b, x \geq 0$ where

$$A = \begin{bmatrix} 2 & 3 & 1 & 0 & 0 \\ 2 & 1 & 0 & 1 & 0 \\ 4 & 2 & 0 & 0 & 1 \end{bmatrix}; \quad x = \begin{bmatrix} x_1 \\ x_2 \\ x_3 \\ x_4 \\ x_5 \end{bmatrix}; \quad b = \begin{bmatrix} 7 \\ 8 \\ 5 \end{bmatrix}$$

Find the initial basic solution.

7. Solve the following LPP using the simplex method.

Minimize

$$z = 3x_1 - 2x_2$$

subject to

$$x_1 + 2x_2 \leq 10$$

$$2x_1 - x_2 \leq 5$$

$$-4x_1 + 3x_2 \geq 5$$

$$x_1, x_2 \geq 0$$

8. Using the simplex method check whether the following LPP has multiple solutions

   Minimize

   $$z = x_1 - 2x_2$$

   subject to

   $$2x_1 - 4x_2 \leq 2$$

   $$-x_1 + x_2 \leq 3$$

   $$x_1 \leq 4$$

   $$x_1, x_2 \geq 0$$

9. Use Phase I of the simplex method to find a basic feasible solution for the system of equations

   $$2x_1 - 4x_2 + x_3 \geq 2$$

   $$-3x_1 + 2x_2 + 2x_3 \geq 4$$

   $$x_1, x_2, x_3 \geq 0$$

10. Write the dual of the following LPP:

    Maximize

    $$z = 4y_1 + 5y_2 + 23y_3 + 24y_4$$

    subject to

    $$y_1 + 2y_3 + y_4 \leq 7$$

    $$y_2 + y_3 + 3y_4 \leq 6$$

    $$y_1, y_2, y_3, y_4 \geq 0$$

    Solve the dual problem. Show that the optimal solution is same for the primal and the dual problem.

11. Use the affine scaling method to solve the following LPP.
    Maximize

$$z = x_1 + x_2$$

$$Ax \leq b$$

where

$$A = \begin{bmatrix} 0.1 & 1 \\ 0.2 & 1 \\ 0.4 & 1 \\ 0.6 & 1 \\ 0.8 & 1 \\ 1.0 & 1 \\ 1.2 & 1 \\ 1.4 & 1 \\ 1.6 & 1 \\ 1.8 & 1 \\ 2.0 & 1 \end{bmatrix}; \quad x = \begin{bmatrix} x_1 \\ x_2 \end{bmatrix}; \quad b = \begin{bmatrix} 1.00 \\ 1.01 \\ 1.04 \\ 1.09 \\ 1.16 \\ 1.25 \\ 1.36 \\ 1.49 \\ 1.64 \\ 1.81 \\ 2.00 \end{bmatrix}$$

Take initial x as $(0, 0)$ and $\gamma = 0.9$.

5
Guided Random Search Methods

5.1 Introduction

The solution techniques for unconstrained optimization problems that have been described in earlier chapters invariably use the gradient information to locate the optimum. Such methods, as we have seen, require the objective function to be continuous and differentiable, and the optimal solution depends on the chosen initial conditions. These methods are not efficient in handling discrete variables and are more likely to stay at a local optimum for a multimodal objective function. Gradient-based methods often have to be restarted to ensure that the local optimum reached is indeed the global one.

In this chapter we explore five different types of solution techniques that do not require the objective function to be either continuous or differentiable. The solution techniques are

- Genetic algorithms (GAs)
- Particle swarm optimization (PSO)
- Simulated annealing (SA)
- Ant colony optimization (ACO)
- Tabu search

All these methods are based on random searches in locating the optima. However, these methods are different from pure "random walk" methods in the sense that they use information from the previous iteration in locating the next best point(s). These methods are hence classified under guided random search methods. The guided random search techniques can be subclassified into evolutionary methods. GA and PSO methods fall under the heading of evolutionary methods. Instead of using a single point in the search space, both GA and PSO techniques use population of points in the search space and hence have a better chance of locating the global optima. The GA technique mimics the biological process (genetics) whereas the PSO technique is based on the idea of natural phenomena such as birds flocking together or school of fishes moving together. The SA method is based on

```
┌─────────────────────────────────┐
│  Guided random search methods   │
└─────────────────────────────────┘
                │
                ▼
┌─────────────────────────────────┐
│  Genetic algorithm              │
│    • Initialize population      │
│    • Fitness evaluation         │
│    • Reproduction               │
│    • Crossover                  │
│    • Mutation                   │
│    • Multimodal test function   │
└─────────────────────────────────┘
                │
                ▼
┌─────────────────────────────────┐
│  Simulated annealing            │
└─────────────────────────────────┘
                │
                ▼
┌─────────────────────────────────┐
│  Particle swarm optimization    │
└─────────────────────────────────┘
                │
                ▼
┌─────────────────────────────────┐
│  Other methods                  │
│    • Ant colony optimization    │
│    • Tabu search                │
└─────────────────────────────────┘
```

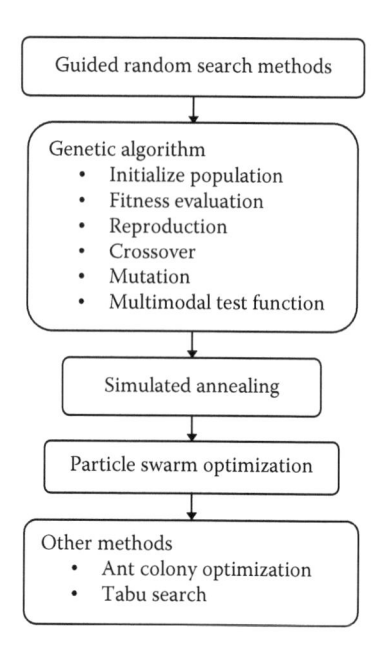

FIGURE 5.1
Road map of Chapter 5.

the physical analogy of the *annealing* process of a material that is heated to a high temperature and then slowly cooled in a controlled manner. The properties of the material get improved through this process. In a similar way, using the SA technique, the transformation is made from the nonoptimal solution for an optimized solution. Some other popular methods such as ant colony optimization and tabu search, which are used for solving combinatorial problems, are briefly discussed in the last section. The road map of this chapter is shown in Figure 5.1.

5.2 Genetic Algorithms

Genetic algorithms (GAs) are search algorithms based on the mechanism of natural selection. They rely on one of the most important principles of Darwin: survival of the fittest. Globally the population is submitted to many transformations. After some generations, when the population is enduring no more, the best individual in the population is assumed to represent the optimal solution. GA mimics the genetic process in which hereditary characteristics are transmitted from a parent to an offspring. The basic unit of inheritance is a gene. Several such genes, encoding specific characteristics (eye color, height, etc.) are present on a chromosome. For example, humans

Guided Random Search Methods

have 23 pairs of chromosomes. One chromosome in each pair is derived from the maternal and and one from the paternal parent. As a result of the crossover operation, some characteristics of each parent can be seen in the offspring. In the natural hereditary process, some genes also randomly mutate. For instance, if the gene corresponding to eye color mutates, the offspring can have blue eyes even if both of the parents' eyes are brown. The mutation in a sense brings variety into the offspring and improves his survivability in a changing environment.

In gradient-based methods, the solution moves from one point to another using the gradient and the Hessian information. In GA, one works with a population of points rather than a single point. The fitness (value of the objective function) of each individual in the population (corresponding to a point in the search space) is then computed. Individuals who have high fitness value undergo crossover and mutation with the hope that they produce better offspring. By better offspring, we mean that they have higher fitness value as compared to their parents. To facilitate the easy working of the genetic operators on the design variables x, these are often coded into binary strings. Once these variables have undergone genetic operations, the new values of the variables can be computed by decoding the binary strings. Using the decoded value of the variables, the fitness of the each individual in the new population is computed. This completes one generation (iteration) of the GA. The working principle of a GA is depicted through a flow chart (Figure 5.2).

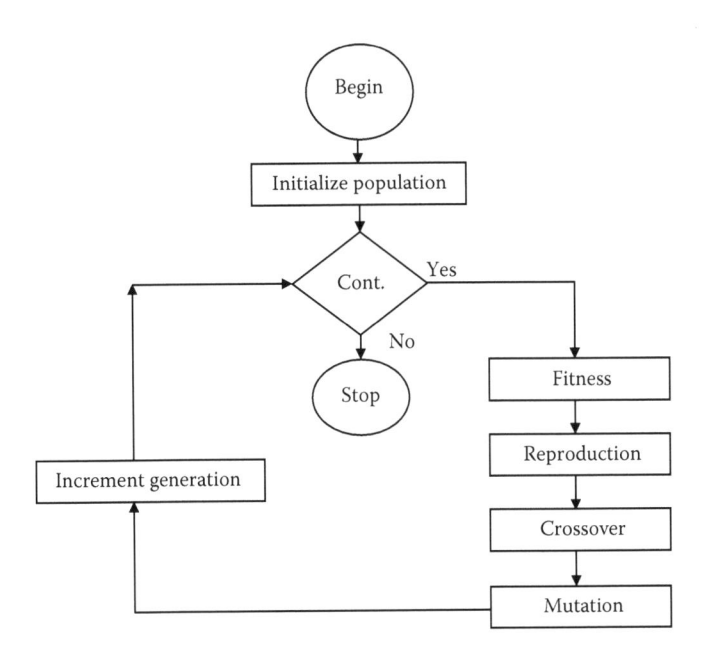

FIGURE 5.2
Working principle of a genetic algorithm.

Let us take the solar energy test problem (see Problem 8 in Chapter 1) in which the following cost function (U) is to be minimized and the variable to be evaluated is temperature T, which is restricted between 40°C and 90°C.

$$U = \frac{204{,}165.5}{330 - 2T} + \frac{10{,}400}{T - 20}$$

Each step of the GA will be explained for the solar energy test problem.

5.2.1 Initialize Population

The variable T has to be restricted within [40, 90]. Since the variable T has to be coded into a binary string, we have to first decide on the number of bits in the string (also called as the string length). Because each bit can take a value of 0 or 1, for a 5-bit string, the minimum value will be 00000 and maximum value will be 11111. This corresponds to a decimal value of 0 and 32 (2^5). If this is linearly mapped into the search space, the variable T will have an accuracy of

$$\frac{90 - 40}{2^5} = 1.5625°C$$

Because we require a finer value of the variable T as 0.001 degrees, the required string length will be 15. The initial population of variables (in binary form) will be generated randomly. A uniform random number generator can be used that generates a random number between 0 and 1. If the random number values are less than 0.5 we take the bit value as 0; else it is taken as 1. To generate a string length of 15, we have to generate the same quantity of random numbers. The following random numbers are generated using the rand command in MATLAB®. The corresponding bit string is mentioned in the second row.

0.81	0.90	0.13	0.91	0.63	0.09	0.28	0.54	0.96	0.15	0.35	0.47	0.74	0.19	0.8
1	1	0	1	1	0	0	1	1	0	0	0	1	0	1

Therefore, the first individual in the population will be 110110011000101. Repeat the step for the number of individuals in the population. For a population size of 10, the following strings are randomly generated:

```
110110011000101
100001010111010
000110101110101
100000110011101
000011100100111
100100101011000
```

Guided Random Search Methods

```
010110100110001
100110101110011
111100100011010
001111100111001
```

In the next step, we decode these strings and compute their fitness.

5.2.2 Fitness Evaluation

The binary string (genotype) has to be decoded to its real value (phenotype) using the equation

$$T_i = T_{i\min} + \frac{(T_{i\max} - T_{i\min})DV(s_i)}{(2^{l_i} - 1)} \qquad (5.1)$$

where $T_{i\min}$ and $T_{i\max}$ are the lower and upper bounds of the variable T_i, $DV(s_i)$ is the decoded value of the string s_i, and l_i is the string length used to code the ith parameter. The binary string 110110011000101 can be decoded as

1	1	0	1	1	0	0	1	1	0	0	0	1	0	1
2^{14}	2^{13}		2^{11}	2^{10}			2^7	2^6				2^2		2^0

Assuming the leftmost bit as the most significant bit, the real value of the string is

$$2^{14} + 2^{13} + 2^{11} + 2^{10} + 2^7 + 2^6 + 2^2 + 2^0 = 27{,}845$$

The value of the variable for this string will be

$$T_i = T_{i\min} + \frac{(T_{i\max} - T_{i\min})DV(s_i)}{(2^{l_i} - 1)} = 40 + \frac{(90 - 40)27{,}845}{(2^{15} - 1)} = 82.4894$$

To get fitness value of this string, simply compute the objective function value corresponding to $T_i = 82.4894$. That is,

$$f_i = \frac{204{,}165.5}{330 - 2T_i} + \frac{10{,}400}{T_i - 20} = 1403.6$$

Table 5.1 summarizes the decoded and fitness value of all 10 strings.

5.2.3 Reproduction

In reproduction, good and bad chromosomes (strings) are identified based on their fitness value. More copies of good chromosomes are made and bad

TABLE 5.1

Fitness Evaluation for Different Strings

Name	String	Decoded Value	Fitness f_i
S1	110110011000101	82.4894	1403.6
S2	100001010111010	66.0659	1257.6
S3	000110101110101	45.2568	1264.3
S4	100000110011101	65.6310	1255.2
S5	000011100100111	42.7940	1291.6
S6	100100101011000	68.6508	1273.3
S7	010110100110001	57.6534	1227.2
S8	100110101110011	70.2545	1284.4
S9	111100100011010	87.3067	1468.4
S10	001111100111001	52.1967	1228.0

ones are eliminated. This can be achieved using *Roulette wheel* or *tournament* selection. In the first approach, roulette wheel slots are sized in proportion to the fitness value of each string. The wheel is spun and the string to which it is pointed is picked up. This is repeated until all the population is filled up. The roulette wheel selection procedure is suited for objective functions of the maximizing type. Because in the test problem the objective function is to be minimized, we have to suitably convert the fitness values so that roulette wheel selection procedure can be used. This is a two-step procedure.

In the first step, identify whether there are any negative values in the fitness value. If the answer is yes, identify the minimum value and scale up the remaining fitness values by that number. For example, if the fitness values are −5, −1, 2, and 7, then the fitness values after scaling will be 0, 4, 7, and 12.

In the second step, convert the fitness values f_i into F_i using the equation

$$F_i = \frac{1}{1+f_i} \tag{5.2}$$

The fitness values 0, 4, 7, and 12 now become 1, 0.2, 0.1429, 0.0833.

The fitness values for the test problem do not have any negative values. So, we can ignore the first step and compute the fitness F_i and some other terms as given in Table 5.2.

Let us make a pie chart with the data corresponding to last column of Table 5.2. The probability of picking strings 7 and 10 (denoted by S-7 and S-10) for the next generation is highest (11%). The next step in the selection process is to make slots (see Figures 5.3 and 5.4) of the roulette wheel using the cumulative values of the data corresponding to last column of the table. Generate 10 random numbers (corresponding to the population size) between

TABLE 5.2
Modified Fitness Evaluation for Different Strings

String	Fitness f_i	$F_i = \dfrac{1}{1+f_i}$	$\dfrac{F_i}{\sum F_i}$
110110011000101	1403.6	0.00071195	0.0924
100001010111010	1257.6	0.00079453	0.1031
000110101110101	1264.3	0.00079033	0.1026
100000110011101	1255.2	0.00079605	0.1033
000011100100111	1291.6	0.00077363	0.1004
100100101011000	1273.3	0.00078474	0.1019
010110100110001	1227.2	0.00081420	0.1057
100110101110011	1284.4	0.00077797	0.1010
111100100011010	1468.4	0.00068055	0.0883
001111100111001	1228.0	0.00081367	0.1056
		$\sum F_i = 0.0077$	

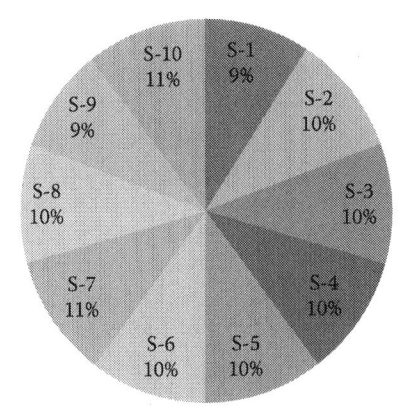

FIGURE 5.3
Pie chart showing probability of a string to be picked up during reproduction.

0 and 1, and select corresponding strings where these random numbers lie in the slots. Thus, two copies each of strings S-2, S-7, and S-8, and one copy each of strings S-4, S-5, S-6, and S-10 are made in the reproduction process. These strings will participate in the crossover and mutation operations.

The convergence rate of GA is determined largely by the *selection pressure* (degree to which better individuals are favored), with larger selection pressure resulting in better convergence. However, if the selection pressure is too high, there are increased chances of GA prematurely converging to a suboptimal solution. Roulette wheel selection methodology is known for providing high selection pressure and this often results in premature convergence. An

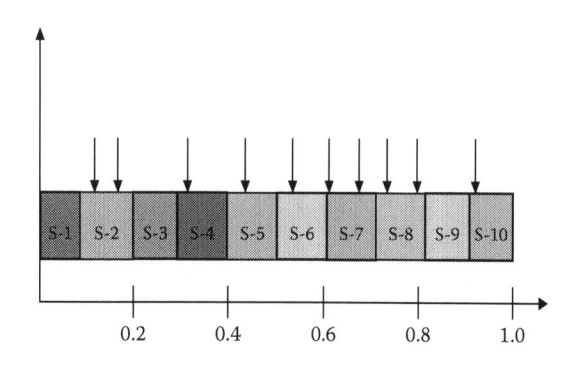

FIGURE 5.4
Slots of roulette wheel.

alternative method is the *tournament selection*, in which a number of individuals (say, 2) are chosen randomly from the population and the best individual (in terms of fitness value) from the group is selected as a parent. The process is repeated once for every individual in the new population. This methodology ensures that the best string is always retained and the worst string always gets eliminated from the selection process. It is important to note that whereas roulette wheel selection is used for maximization type objective functions, tournament selection is used for minimization type objective functions. Because the test problem's objective function is of the minimization type, we had to modify the function suitably for the roulette wheel selection methodology. The modification of the function is not required in the tournament selection methodology because the objective function is already of minimization type.

In the tournament selection methodology, we begin with the first individual of the population. Then any other individual from the population is selected randomly. The fitness values of the two individuals are then compared. The individual with lower fitness value is declared the "winner." The

TABLE 5.3

Tournament Selection

String	Competitor	Fitness Comparison	Winner
110110011000101 (S-1)	S-8	1403.6 < 1284.4 (No)	S-8
100001010111010 (S-2)	S-4	1257.6 < 1255.2 (No)	S-4
000110101110101 (S-3)	S-2	1264.3 < 1257.6 (No)	S-2
100000110011101 (S-4)	S-9	1255.2 < 1468.4 (Yes)	S-4
000011100100111 (S-5)	S-10	1291.6 < 1228.0 (No)	S-10
100100101011000 (S-6)	S-7	1273.3 < 1227.2 (No)	S-7
010110100110001 (S-7)	S-8	1227.2 < 1284.4 (Yes)	S-7
100110101110011 (S-8)	S-1	1284.4 < 1403.6 (Yes)	S-8
111100100011010 (S-9)	S-4	1468.4 < 1255.2 (No)	S-4
001111100111001 (S-10)	S-2	1228.0 < 1257.6 (Yes)	S-10

5.2.4 Crossover and Mutation

In the reproduction step of the GA, we have merely copied the strings that will participate in the crossover and mutation operation. The strings were not altered in the reproduction step. In the crossover operation, two parents are taken randomly from the mating pool (previous step of GA using roulette wheel or tournament selection) and bits are exchanged between the parents to generate new children (strings). The idea behind the crossover operation is that good parents will mate to form better offspring. Let us take strings S-2 and S-4, which are randomly selected as parents from the mating pool (both of these strings are present in the selection procedure specified by roulette wheel and tournament selection). The next step is to generate the crossover site (position) randomly along the string length. Let the ninth position (from the left side or most significant bit side) is the crossover site. Then strings S-2 and S-4 after mating become

```
Parent S-2 100001010 111010          100001010 011101
Parent S-4 100000110 011101          100000110 111010
```

In this operation we have assumed a single crossover site. It is observed in nature that crossover can occur at one or more sites also. The number of crossovers follows a Poisson distribution (Hartl 1991) with mean as 2. Mutation is used to keep diversity in the population. The mutation operator changes the bit 1 to 0 and vice versa with a small probability.

Let us use the crossover (single-point) and mutation operation (with a probability of 0.02) for the test problem in which the mating pool is taken from tournament selection. The new population is given in Table 5.4.

This completes one generation (iteration) of the GA. The fitness of the new population is then computed and the cycle (reproduction, crossover, and mutation) is repeated. It is possible in GA that the objective function need not improve in a few successive generations.

The mathematical explanation of the GA is given as the *Schema* theorem

$$m(H, t+1) \geq m(H, t) \cdot \frac{f(H)}{\bar{f}} \left[1 - p_c \frac{\delta(H)}{l-1} - o(H) p_m \right] \tag{5.3}$$

where $m(H, t)$ represents m examples of a particular schema H at time t. $f(H)$ is the average fitness of the strings represented by the schema H; $\bar{f}$ is the average fitness of the entire population; p_c and p_m are the probabilities of occurrence of crossover and mutation; $\delta(H)$ is the defining length of the schema H, which is the distance between the first and last specific string

TABLE 5.4

New Population

Mating Pool	Crossover Site	Children	Mutation	New Population
100000110011101 (S-4)	9	100000110111010	No	100000110111010
100001010111010 (S-2)		100001010011101	No	100001010011101
010110100110001 (S-7)	6	010110100111001	No	010110100111001
001111100111001 (S-10)		001111100110001	No	001111100110001
100001010111010 (S-2)	11	100001010110001	Yes	101001010110001
010110100110001 (S-7)		010110100111010	No	010110100111010
100000110011101 (S-4)	5	100000101110011	No	100000101110011
100110101110011 (S-8)		100110110011101	Yes	100110110010101
100110101110011 (S-8)	13	100110101110001	No	100110101110001
001111100111001 (S-10)		001111100111011	No	001111100111011

TABLE 5.5

Different Optimization Runs with GA

No. of Generations	T	U
505	55.95	1225.58
501	53.99	1225.55
1029	55.08	1225.166
751	55.08	1225.166

position; $o(H)$ is the order of a schema defined by the number of fixed positions in a template; and l is the length of the string. The schema theorem states that the short, low-order, above average schemata receives increasing trials in subsequent generations.

The MATLAB code *prob.m* is the main program of the genetic algorithm. Its subroutines are given in *in.m, roulette.m, tournament.m*, and *func.m*. On executing the genetic algorithm code for the test problem, the output is displayed in Table 5.5. It is to be noted that each of the rows in the table corresponds to a separate optimization run. From Table 5.5 it is observed that minima are reached in most of the runs. However, the number of generations varies in each run. This is expected because each run of the genetic algorithm starts with a random set of the population. See Figure 5.5, which plots the minimum of the objective function achieved until that generation. The step region in the plot indicates that there is no reduction in the value of the objective function for a certain number of generations until in a particular generation, where there is a reduction in the value of the objective function.

5.2.5 Multimodal Test Functions

The main advantage of GA over gradient-based methods is that it does not get stuck at local optima. Let us take some multimodal test functions such

Guided Random Search Methods

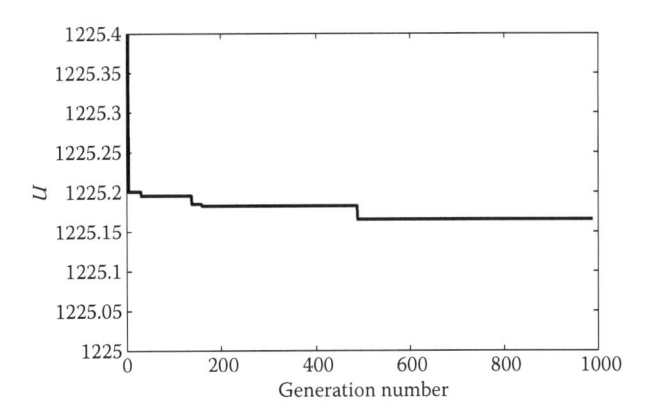

FIGURE 5.5
Variation of objective function value with increase in number of generations.

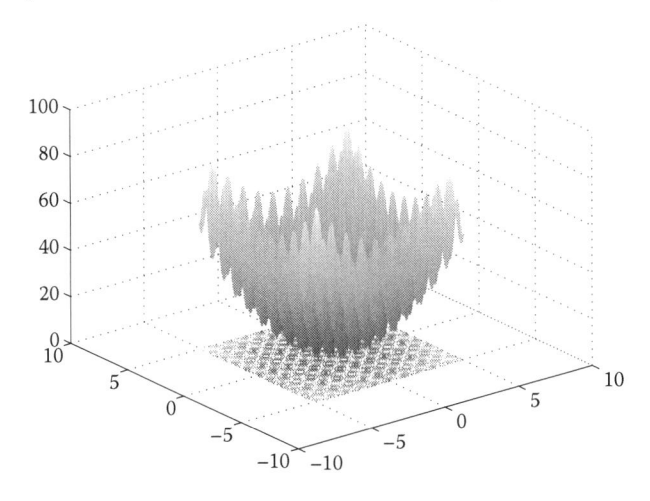

FIGURE 5.6
Rastrigin's function.

as Rastrigin and Schwefel's function to demonstrate that GA can locate the global minimum for these functions.

The two-variable Rastrigin's function (see Figure 5.6) is given by

$$f(x) = 20 + x_1^2 - 10\cos(2\pi x_1) + x_2^2 - 10\cos(2\pi x_2)$$

$$-5.12 \le x_1, x_2 \le 5.12$$

The function has a number of local optima and the global minimum value of the function is $f(x^*) = 0$ and occurs at $x^* = (0, 0)$. The function can be plotted in MATLAB using the following commands.

```
[X,Y] = meshgrid(-5.12:.1:5.12, -5.12:.1:5.12);
Z = 20 + (X.^2-10*cos(2*pi.*X) + Y.^2-10*cos(2*pi.*Y));
surfc(X,Y,Z)
shading interp
```

The input files for GA can be modified for the two-variable function as

```
%%%%%%%%%%%%%%%%%%%%%%%%%%%%%%%%%%%%%%%%%%%%%%%%%
% File name func.m
% Enter the function to be optimized
%%%%%%%%%%%%%%%%%%%%%%%%%%%%%%%%%%%%%%%%%%%%%%%%%
%
function [y,constr] = func(x)
y = 20 + (x(1)*x(1)-10*cos(2*pi*x(1))+ x(2)*x(2)-
  10*cos(2*pi*x(2)));
constr(1) = 10;% This is used with constraints.
% For unconstrained problems, define constr()with
% any positive value
%%%%%%%%%%%%%%%%%%%%%%%%%%%%%%%%%%%%%%%%%%%%%%%%%


%%%%%%%%%%%%%%%%%%%%%%%%%%%%%%%%%%%%%%%%%%%%%%%%%
% File name in.m
% Input parameters for Genetic algorithm
%%%%%%%%%%%%%%%%%%%%%%%%%%%%%%%%%%%%%%%%%%%%%%%%%
%
problem = 'min';      % used with roulette wheel
n_of_v = 2;           % number of variables
n_of_g = 10000;       % maximum number of generations
n_of_p = 40;          % population size
range(1,:) = [-5.12 5.12]; % variable bound
range(2,:) = [-5.12 5.12];
n_of_bits(1) = 20;    % number of bits
n_of_bits(2) = 20;
cross_prob = 0.9;     % crossover probability
multi_crossover = 0;% use multi-crossover
mut_prob = 0.1;       % mutation probability
tourni_flag = 0;      % use roulette wheel
epsilon = 1e-7;       % function tolerance
flag = 0;             % stall generations flag
flag1 = 0;            % scalin flag
stall_gen = 500;      % stall generations for termination
n_of_c = 0;           % for constraint handling
%
%%%%%%%%%%%%%%%%%%%%%%%%%%%%%%%%%%%%%%%%%%%%%%%%%
```

The above file uses a roulette wheel as the selection methodology. To change it to tournament selection simply change

```
tourni_flag = 1;
```

Guided Random Search Methods

Both selection criteria are able to locate the global minimum solution and the convergence to the optimum value by two selection methodologies given in Figures 5.7 and 5.8 respectively.

The two-variable Schwefel's function (see Figure 5.9) is given by

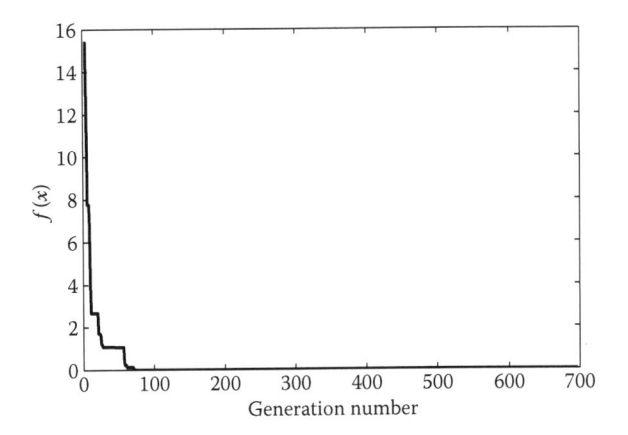

$$f(x) = -x_1 \sin\sqrt{|x_1|} - x_2 \sin\sqrt{|x_2|}$$

$$-500 \leq x_1, x_2 \leq 500$$

FIGURE 5.7
Convergence of genetic algorithm for Rastrigin function with roulette wheel selection.

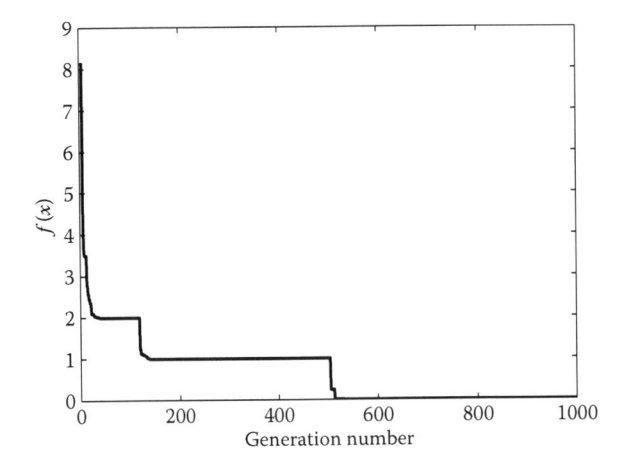

FIGURE 5.8
Convergence of genetic algorithm for Rastrigin function with tournament selection.

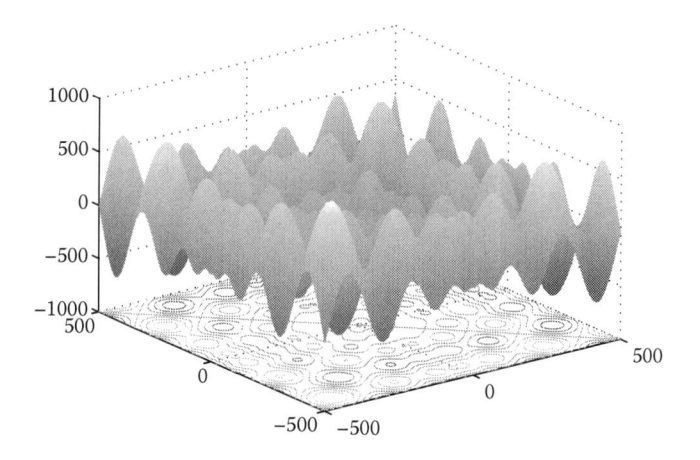

FIGURE 5.9
Schwefel's function.

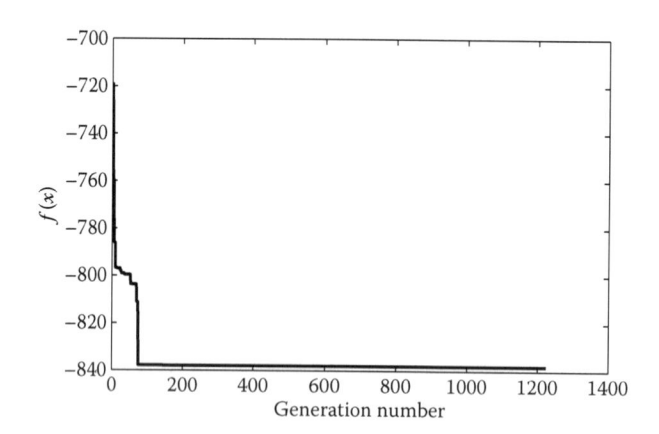

FIGURE 5.10
Convergence of genetic algorithm for Schwefel's function.

Guided Random Search Methods 153

The function has a number of local optima and the global minimum value of the function is $f(x^*) = -837.9658$ and occurs at $x^* = (420.9867, 420.9867)$.

The input files for GA can be modified for the two-variable function as

```
%%%%%%%%%%%%%%%%%%%%%%%%%%%%%%%%%%%%%%%%%%%%%
% File name func.m
% Enter the function to be optimized
%%%%%%%%%%%%%%%%%%%%%%%%%%%%%%%%%%%%%%%%%%%%%
%
function [y,constr] = func(x)
y = -x(1)*sin(sqrt(abs(x(1)))) -x(2)*sin(sqrt(abs(x(2))));
constr(1) = 10;
%%%%%%%%%%%%%%%%%%%%%%%%%%%%%%%%%%%%%%%%%%%%%

%%%%%%%%%%%%%%%%%%%%%%%%%%%%%%%%%%%%%%%%%%%%%
% File name in.m
% Input parameters for Genetic algorithm
%%%%%%%%%%%%%%%%%%%%%%%%%%%%%%%%%%%%%%%%%%%%%
%
problem = 'min';        % used with roulette wheel
n_of_v = 2;             % number of variables
n_of_g = 10000;         % maximum number of generations
n_of_p = 80;            % population size
range(1,:) = [-500 500]; % variable bound
range(2,:) = [-500 500];
n_of_bits(1) = 20;      % number of bits
n_of_bits(2) = 20;
cross_prob = 0.9;       % crossover probability
multi_crossover = 0;    % use multi-crossover
mut_prob = 0.1;         % mutation probability
tourni_flag = 1;        % use roulette wheel
epsilon = 1e-7;         % function tolerance
flag = 0;               % stall generations flag
flag1 = 0;              % scalin flag
stall_gen = 500;        % stall generations for termination
n_of_c = 0;             % for constraint handling
%
%%%%%%%%%%%%%%%%%%%%%%%%%%%%%%%%%%%%%%%%%%%%%
```

The GA code is able to achieve the global minimum and the convergence history is shown in Figure 5.10.

5.3 Simulated Annealing

Simulated annealing (SA) is an optimization technique that has derived its name from the process of annealing of solids where the solid is heated and then allowed to cool slowly until its molecules reach the minimum energy state. The solid in this state will be free from defects. In a similar manner, the optimization problem is transformed into an "ordered state" or a desired optimized state (solution). In the high-energy state, the molecules are free to move and their freedom gets restricted as the temperature is reduced (cooled). In a similar manner, SA methodology allows "hill climbing" when the temperature is high. That is, those points that are in the near vicinity of the search point, but have a higher objective function value can still be selected with certain probability. This allows the algorithm to escape from local optima. Thus, simulation methodology is a powerful technique in locating the global optimum solution.

The algorithm starts by picking any random value of the variable x_i using the equation

$$x_i = x_{i,\min} + (x_{i,\max} - x_{i,\min})u_i \tag{5.4}$$

where $x_{i,\min}$ and $x_{i,\max}$ are the bounds of the variable x_i and u_i is random number generated between 0 and 1 (uniform distribution). The energy (E_{old}) of this variable is given by its objective function value. That is,

$$E_{\text{old}} = f(x_i) \tag{5.5}$$

The next step in the algorithm is to perturb x_i in its neighborhood. The perturbation Δx_i can be computed as

$$\Delta x_i = \varepsilon x_i u_i \tag{5.6}$$

where ε is a small number fixed at the start of the simulation. The next search point is therefore given by

$$x_{i+1} = x_i + \Delta x_i \tag{5.7}$$

In case the variables x_{i+1} exceeds their bounds they are artificially brought back into the feasible design space using the equation

$$x_{i+1} = x_{i,\min} + (x_{i,\max} - x_{i,\min})u_i \tag{5.8}$$

The energy state for the new point is given by

$$E_{\text{new}} = f(x_{i+1}) \tag{5.9}$$

Guided Random Search Methods

If the new energy state E_{new} is lower than E_{old}, the objective function has improved and we replace the value of E_{old} with E_{new}. In case E_{new} is higher than E_{old} the following condition is checked

$$e^{\left(-\frac{E_{new}-E_{old}}{E_{old}}\right)} > u \quad (5.10)$$

If this condition is satisfied we allow for the "hill climbing" and replace E_{old} with E_{new}. If this condition is not satisfied previous value of x is restored. That is,

$$x_{i+1} = x_i - \Delta x_i \quad (5.11)$$

The iterations are repeated until there is no improvement in the objective function value for a fixed number of moves. The steps of simulation algorithm can thus be summarized in Table 5.6 and the MATLAB code *simann.m* is given subsequently.

The two-variable Rastrigin's function

$$f(x) = 20 + x_1^2 - 10\cos(2\pi x_1) + x_2^2 - 10\cos(2\pi x_2)$$

$$-5.12 \leq x_1, x_2 \leq 5.12$$

is optimized and it does not converge to the global minimum of $f(x^*) = 0$ at $x^* = (0, 0)$. The convergence history is given in Figure 5.11. By modifying the subroutine *func.m* other functions such as Schwefel's function and test

TABLE 5.6

Algorithm for Simulated Annealing

Step 1: Initialize ε and variable bounds $x_{i,min}$ and $x_{i,max}$
Step 2: Compute starting value of the variables as

$$x_i = x_{i,min} + (x_{i,max} - x_{i,min})u_i$$

Step 3: Compute $E_{old} = f(x_i)$
Step 4: Compute $\Delta x_{i+1} = \varepsilon x_i u_i$ and $x_{i+1} = x_i + \Delta x_i$
  If x_{i+1} exceeds bounds then $x_{i+1} = x_{i,min} + (x_{i,max} - x_{i,min})u_i$
Step 5: Compute $E_{new} = f(x_{i+1})$
Step 6: If $E_{new} < E_{old}$
  then $E_{old} = E_{new}$
  else if $e^{\left(-\frac{E_{new}-E_{old}}{E_{old}}\right)} > u$
  then $E_{old} = E_{new}$
  else $x_{i+1} = x_i - \Delta x_i$
Step 7: Go to Step 4 until termination criterion (function not improving for certain number of iterations) is satisfied.

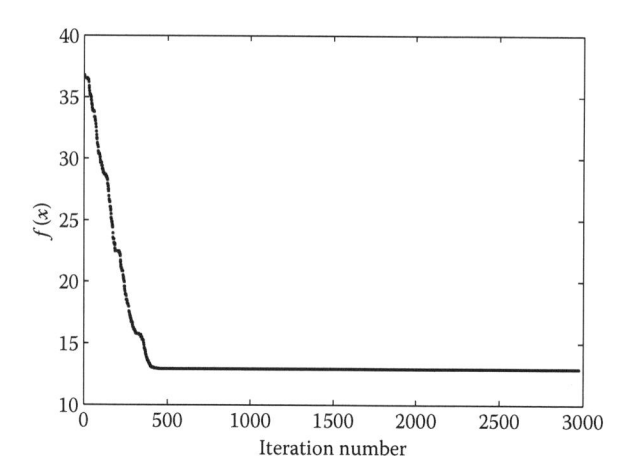

FIGURE 5.11
Convergence of simulated annealing for Rastrigin function.

problem on spring system (mentioned in Chapter 3) can also be optimized. See Figures 5.12 and 5.13, where the convergence history of these functions is shown. In all the plots, observe the hill-climbing region shown by the oscillatory nature of the curve. Because the algorithm starts from a random point, the convergence history will vary in each simulation run for each of the functions. The performance of the algorithm will also vary by choosing a different ε for a given function.

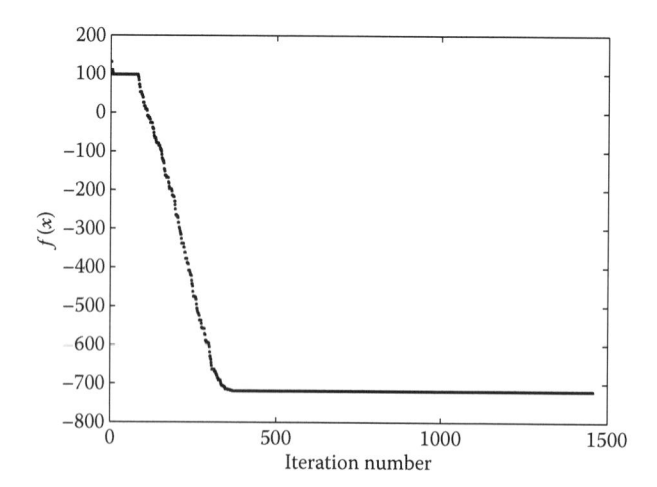

FIGURE 5.12
Convergence of simulated annealing for Schwefel's function.

Guided Random Search Methods

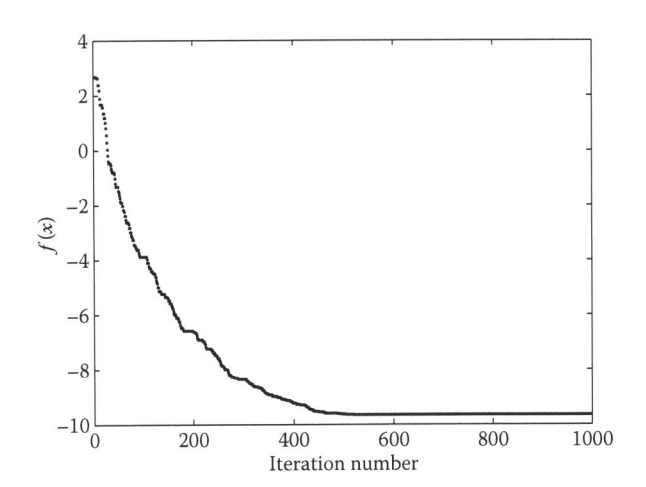

FIGURE 5.13
Convergence of simulated annealing for the test problem on the spring system (Chapter 3).

5.4 Particle Swarm Optimization

In the particle swarm optimization (PSO) technique, a number of search points are simultaneously explored in the iteration, similar to a search carried out by GA. The PSO technique is inspired by the collective wisdom of a group of individuals such as a flock of birds, animals moving in herds, or schools of fish moving together. The PSO algorithm keeps track of the best position of the individual as well as that of the population in terms of the objective function. The best objective function of the individual and that of the group is denoted by *pbest* and *gbest* respectively. Each individual in the group moves with a velocity that is a function of *pbest*, *gbest* and its initial velocity. The new position of the individual is updated based on its initial position and the velocity. The objective function value is again computed for the new positions and the PSO steps are repeated. Now, each step of the algorithm is described.

The initial position of the *k*th individual in the population is given by

$$x_{i,k} = x_{i,\min} + (x_{i,\max} - x_{i,\min})u_i \qquad (5.12)$$

where $x_{i,\min}$ and $x_{i,\max}$ are the bounds of the variable x_i and u_i is the random number generated between 0 and 1 (uniform distribution). Here *i* is the iteration number. Compute the fitness of the *k*th individual as

$$p_{i,k} = f(x_{i,k}) \qquad (5.13)$$

Since this is the initialization step, the best fitness of each individual is p_k itself. That is,

$$\text{pbest}_{i,k} = p_{i,k} \tag{5.14}$$

The global best fitness is computed as

$$\text{gbest}_i = \text{minimum}(\text{pbest}_{i,k}) \tag{5.15}$$

The location of pbest_k and gbest is given by p_{xik} and g_{ix}. Starting with an initial velocity of $v_{i,k}$, the velocity of the individual is updated using the equation

$$v_{i+1,k} = w_1 v_{i,k} + \phi_1 (p_{xik} - x_{i,k}) u_i + \phi_2 (g_{ix} - x_{i,k}) u_i \tag{5.16}$$

where w_1, ϕ_1, and ϕ_2 are the tuning factors of the algorithm. The position of each individual can be updated as

$$x_{i+1,k} = x_{i,k} + v_{i+1,k} \tag{5.17}$$

In case the variables x_{i+1} exceeds their bounds they are artificially brought back into the feasible design space using the equation

$$x_{i+1} = x_{i,\min} + (x_{i,\max} - x_{i,\min}) u_i \tag{5.18}$$

Based on the new position, the fitness of the kth individual is computed as

$$p_{i+1,k} = f(x_{i+1,k}) \tag{5.19}$$

If this fitness is lower than $\text{pbest}_{i,k}$, then replace $\text{pbest}_{i,k}$ with $p_{i+1,k}$. Compute the global best fitness as

$$\text{gbest}_{i+1} = \text{minimum}(\text{pbest}_{i+1,k}) \tag{5.20}$$

The steps are repeated for a finite number of iterations. The algorithm is given in Table 5.7 and the MATLAB code pso.m is given subsequently.

The two-variable Schwefel's function (see Figure 5.9)

$$f(x) = -x_1 \sin\sqrt{|x_1|} - x_2 \sin\sqrt{|x_2|}$$

$$-500 \leq x_1, x_2 \leq 500$$

is optimized and it converges to the global minimum of is $f(x^*) = -837.9658$ and occurs at $x^* = (420.9867, 420.9867)$. The convergence history of the PSO algorithm for Schwefel's function is shown in Figure 5.14.

Guided Random Search Methods

TABLE 5.7

Algorithm for Particle Swarm Optimization

Step 1: Initialize i_{max}, w_1, ϕ_1, ϕ_2, n (population size), $x_{i,min}$, and $x_{i,max}$.
Step 2: Initialize the starting position and velocities of the variables as

$$x_{i,k} = x_{i,min} + (x_{i,max} - x_{i,min})u_i \qquad k = 1 \cdots n$$

$$v_{i,k} = 0$$

Step 3: Compute $p_{i,k} = f(x_{i,k})$ $\qquad k = 1 \cdots n$
Step 4: Compute $pbest_{i,k} = p_{i,k}$ and $gbest_i = \text{minimum}(pbest_{i,k})$
  The location of $pbest_k$ and $gbest$ is given by p_{xik} and g_{ix}.
Step 5: Update velocity
$$v_{i+1,k} = w_1 v_{i,k} + \phi_1(p_{xik} - x_{i,k})u_i + \phi_2(g_{ix} - x_{i,k})u_i$$
Step 6: Update position $x_{i+1,k} = x_{i,k} + v_{i+1,k}$
Step 7: Update fitness $p_{i+1,k} = f(x_{i+1,k})$
Step 8: If $p_{i+1,k} < pbest_{i,k}$
  then $\qquad pbest_{i+1,k} = p_{i+1,k}$
Step 9: Update $gbest_{i+1} = \text{minimum}(pbest_{i+1,k})$
Step 10: If $i < i_{max}$ then increment i and go to Step 5, else stop.

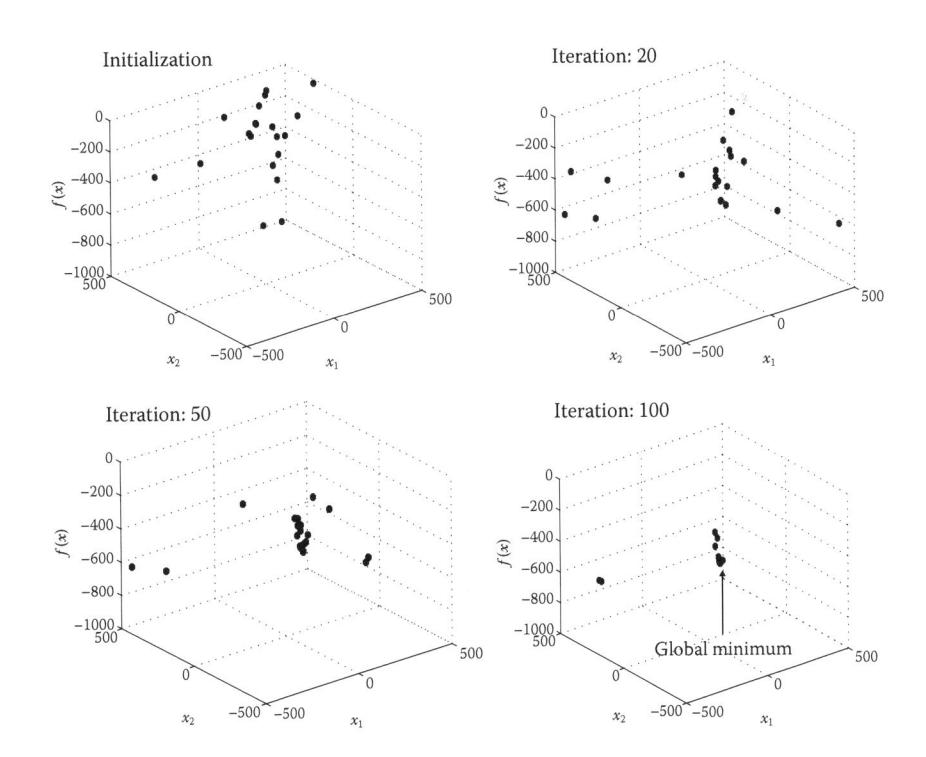

FIGURE 5.14
Convergence of particle swarm optimization for Schwefel's function.

5.5 Other Methods

In addition to the three methods (GA, SA, and PSO) that were discussed in the previous sections, there are numerous other optimization methods that have mimicked natural processes or some other physical analogies. Some of these methods are the bees algorithm, differential algorithm, evolutionary programming, tabu search, ant colony optimization (ACO), and so forth. Of these, ACO and tabu search are widely used for solving combinatorial problems (such as the traveling salesman problem or the job scheduling problem). As the name suggests, ACO mimics the behavior of ants in locating the minimum of a function. It may be noted that in a complex combinatorial problem, searching all the combinations is computationally expensive. Both ACO and tabu search provide a heuristic approach for such problems. These two techniques are briefly explained in this section.

5.5.1 Ant Colony Optimization

While on lookout for food, ants deposit a substance called a pheromone on the path. Other ants follow this favorable path to reach the food. The ant colony optimization (ACO) technique mimics the behavior of ants in solving the optimization problems. The ACO technique was proposed in the early 1990s (Dorigo 1992) and since then has been applied to solve a number of problems such as

- Protein folding problem (Shmygelska and Hoos 2005)
- Traveling salesman problem (Dorigo et al. 1996)
- Project scheduling (Merkle et al. 2002)
- Vehicle routing (Reimann et al. 2004)

In the ACO technique, the optimization problem is defined in terms of a number of layers and nodes. Each layer corresponds to the design variable and each node corresponds to the discrete values of the design variables. The ants have to pass through different "best" nodes to reach the destination, which is the minimum of the function. Let there be N ants in the colony. If the kth ant is at ith node, then the probability of choosing jth node is given by

$$p_{ij}^{(k)} = \frac{\tau_{ij}}{\sum_{j \in N_i^{(k)}} \tau_{ij}} \quad (5.21)$$

where $N_i^{(k)}$ indicates the set of neighborhood nodes of ant k at node i. Here τ_{ij} represents the pheromone trail and is given by the expression

$$\tau_{ij} = \tau_{ij} + \Delta \tau^{(k)} \quad (5.22)$$

Guided Random Search Methods

The pheromone content also evaporates according to the relation

$$\tau_{ij} = (1 - \rho)\tau_{ij} \tag{5.23}$$

where ρ is the evaporation rate. The typical value of ρ is 0.5. The pheromone content is updated using the relation

$$\tau_{ij} = (1 - \rho)\tau_{ij} + \sum_{k=1}^{N} \Delta\tau_{ij}^{(k)} \tag{5.24}$$

where the quantity $\Delta\tau_{ij}^{(k)}$ is given by

$$\Delta\tau_{ij}^{(k)} = \frac{\phi f_{best}}{f_{worst}} \tag{5.25}$$

where f_{best} and f_{worst} are the best and worst values of the objective function for the paths taken by the ants and ϕ is a scaling parameter. Let us explain the procedure of ACO to minimize the function

$$f(x) = 2(\sqrt{x} - 3) + \frac{\sqrt{x}}{100} \qquad 0.3 \leq x \leq 0.6$$

The problem has only one layer because there is there is only one design variable for the problem. Let there be seven nodes of this problem. Thus,

$$x_{11} = 0.30$$

$$x_{12} = 0.35$$

$$x_{13} = 0.40$$

$$x_{14} = 0.45$$

$$x_{15} = 0.50$$

$$x_{16} = 0.55$$

$$x_{17} = 0.60$$

Let us take the number of ants in the colony to be 5. To begin with, there is an equal probability of selection of any of the nodes. Using roulette wheel selection (as mentioned in Section 5.2), the following five nodes are selected:

$$x_{14} = 0.45$$
$$x_{16} = 0.55$$
$$x_{12} = 0.35$$
$$x_{17} = 0.60$$
$$x_{11} = 0.30$$

The corresponding function values are

$$f(x_{14}) = -4.6517$$
$$f(x_{16}) = -4.5093$$
$$f(x_{12}) = -4.8109$$
$$f(x_{17}) = -4.4310$$
$$f(x_{11}) = -4.8990$$

The best (fifth ant) and worst values of the objective function are

$$f_{best} = f(x_{11}) = -4.8990$$
$$f_{worst} = f(x_{17}) = -4.4310$$

Taking $\phi = 5$, the pheromone information is updated as

$$\Delta \tau^{(k=5)} = \frac{\phi f_{best}}{f_{worst}} = \frac{5 \times (-4.899)}{-4.431} = 5.528$$

Now,

$$\tau_{ij} = (1-\rho)\tau_{ij} + \sum_{k=1}^{N} \Delta \tau_{ij}^{(k)} = (1-0.5) \times 1 + 5.528 = 6.028 \quad \text{(for } j=1\text{)}$$

The probability of selecting this node in the next iteration is

$$p_{11} = \frac{6.028}{9.028} = 0.6677$$

Guided Random Search Methods

and the probability of selecting other nodes is

$$p_{1j} = \frac{0.5}{9.028} = 0.0554$$

Again using the roulette wheel selection, the following nodes are selected:

$$x_{11}(3 \text{ copies})$$

$$x_{15}(1 \text{ copy})$$

$$x_{17}(1 \text{ copy})$$

The iterations are repeated until all the ants follow the best path. The minimum value of the objective function is −4.899 and occurs at $x^* = 0.3$.

5.5.2 Tabu Search

The tabu search is a heuristic technique in which an approximate solution is used to tackle complex combinatorial problems such as job scheduling and traveling salesman problems. The method (Glover 1986) allows nonimproving moves whenever a local optimum is reached. However, the method prevents visiting earlier solutions by keeping a list of the search history. The list is called the tabu (or forbidden) list. To avoid stagnation of search process due to tabu, it is mandatory to modify tabu lists frequently. One such example is to allow a tabu move when the objective function value improves from the best value. Such moves are called as *aspiration criteria*. The following notation is used (Gendreau and Potvin 2010) in the algorithm to follow (Table 5.8).

TABLE 5.8

Algorithm for Tabu Search

Step 1: Start with an initial set S_0.
 $f^* = f(S_0)$
 $S^* = S_0$
 $T = \Phi$
Step 2: Select $S' \in \tilde{N}(S)$ and find $f(S')$.
 If $f(S') < f^*$
 $f^* = f(S')$
 $S^* = S'$
 Record current move in T.
Step 3: Go to Step 2 if termination criteria are not satisfied, else stop.
 (Termination criteria are set if the objective function does not show improvement for some fixed number of iterations.)

S current solution

S^* best solution

f^* best value of the objective function

$N(S)$ neighborhood of S

$\tilde{N}(S)$ admissible subset

T tabu list

Chapter Highlights

- GA and PSO work with a population of points in a search space whereas SA propagates through iterations with a single search point.
- GA mimics the genetic process in which hereditary characteristics are transmitted from a parent to an offspring.
- GA variables are coded into binary strings.
- In the reproduction step, the best individuals in the population are selected for mating.
- The diversity in the population is created using crossover and mutation operations.
- The mutation operator changes the bit 1 to 0 and vice versa with a small probability.
- SA is an optimization technique that has derived its name from the process of annealing of solids, in which the solid is heated and then allowed to cool slowly until its molecules reach a minimum energy state.
- SA allows points with higher objective functions to be selected with certain probability. It is often called a "hill-climbing" algorithm.
- The PSO technique is inspired by the collective wisdom of a group of individuals such as a flock of birds or animals moving in herds or schools of fish moving together.
- The PSO algorithm keeps track of the best position of the individual as well as that of the population in terms of the objective function.
- In the ACO technique, the optimization problem is defined in terms of a number of layers and nodes. Each layer corresponds to the design variable and each node corresponds to the discrete values of the design variables. The ants have to pass through different "best" nodes to reach the destination, which is the minimum of the function.
- The tabu search method allows nonimproving moves whenever a local optimum is reached. However, the method prevents visiting

Guided Random Search Methods

earlier solutions by keeping a list of the search history. The list called as the tabu (or forbidden) list.
- To avoid stagnation of a search process due to tabu, it is mandatory to modify tabu lists frequently. One such example is to allow a tabu move when the objective function value improves from the best value. Such moves are called aspiration criteria.

Formulae Chart

Decoding of string from binary to real value:

$$x_i = x_{i\min} + \frac{(x_{i\max} - x_{i\min})DV(s_i)}{(2^{l_i} - 1)}$$

Schema theorem:

$$m(H, t+1) \geq m(H, t) \cdot \frac{f(H)}{\bar{f}}\left[1 - p_c \frac{\delta(H)}{l-1} - o(H)p_m\right]$$

Velocity update in PSO:

$$v_{i+1,k} = w_1 v_{i,k} + \phi_1(p_{xik} - x_{i,k})u_i + \phi_2(g_{ix} - x_{i,k})u_i$$

ACO:

$$p_{ij}^{(k)} = \frac{\tau_{ij}}{\sum_{j \in N_i^{(k)}} \tau_{ij}}$$

$$\tau_{ij} = (1-\rho)\tau_{ij} + \sum_{k=1}^{N} \Delta\tau_{ij}^{(k)}$$

$$\Delta\tau_{ij}^{(k)} = \frac{\phi f_{\text{best}}}{f_{\text{worst}}}$$

Problems

1. What is the minimum string length required to code the variable range (−3, 5) with an accuracy of 0.0001?

2. An optimization problem has three design variables that are to be coded in binary strings. The range of variables is (−100, 50), (0, 1), and (3, 7) and the accuracy required is 0.01, 0.00001, and 0.001 respectively. Compute the minimum length of the string required.

3. By modifying the input (tuning) parameters in GA (population size, crossover, and mutation probabilities), SA (ε), and PSO (population size, w_1, ϕ_1, ϕ_2), rerun the codes for the test problem on spring system (mentioned in Chapter 3), Rastragin and Schwefel's functions.

4. In a given generation of GA, the following fitness values are obtained for ten strings (S-1 to S-10) for a maximization problem. Find the number of copies that will be generated for each string using Roulette wheel selection.

String	Fitness
S-1	25
S-2	16
S-3	74
S-4	8
S-5	99
S-6	45
S-7	12
S-8	65
S-9	22
S-10	19

5. If instead, tournament selection is used for reproduction, find strings (see previous problem) that get selected in the mating pool. Compare the results with those obtained from the roulette wheel selection.

6. Minimize the two variable Griewangk's function

$$f(x) = \frac{x_1^2 + x_2^2}{4000} - \cos x \cos \frac{x_2}{\sqrt{2}} + 1$$

$$-600 \le x_1, x_2 \le 600$$

using GA, SA, and PSO techniques.

7. Minimize the two-variable Ackley's function

$$f(x) = -ae^{-b\sqrt{\frac{1}{2}(x_1^2+x_2^2)}} - e^{\frac{1}{2}(\cos x_1 + \cos x_2)} + a + e^1$$

$$-32.768 \le x_1, x_2 \le 32.768$$

using GA, SA, and PSO techniques. Take $a = 20$, $b = 0.2$, and $c = 2\pi$.

8. Minimize the function

$$f(x) = (x_1^2 + x_2 - 11)^2 + (x_2^2 + x_1 - 7)^2$$

$$-5 \le x_1, x_2 \le 5$$

using GA, SA, and PSO techniques.

9. Minimize the function

$$f(x) = e^{x_1^2 + x_2^2} + x_1 + x_2 - 3 - \sin(3(x_1 + x_2))$$

using GA, SA, and PSO techniques.

10. Optimize the *minmax* function

Minimize $F(x)$

where

$$F(x) = \max\{f_i(x)\}$$

$$f_1(x) = x_1^2 + x_2^4$$

$$f_2(x) = (2 - x_1)^2 + (2 - x_2)^2$$

$$f_3(x) = 2e^{-x_1 + x_2}$$

$$-50 \le x_1, x_2 \le 50$$

using GA, SA, and PSO techniques.

11. Optimize the *minmax* function

$$\min.\max\{|x_1 + 2x_2 - 7|, |2x_1 + x_2 - 5|\}$$

$$-50 \le x_1, x_2 \le 50$$

using GA, SA, and PSO techniques.

12. Minimize the Eggcrate function

$$f(x) = x_1^2 + x_2^2 + 25\left(\sin^2 x_1 + \sin^2 x_2\right)$$

$$-5 \le x_1, x_2 \le 5$$

using GA, SA, and PSO techniques.

6

Constrained Optimization

6.1 Introduction

Invariably all optimization problems carry constraints, and examples can be given from any area one can think of. The supply of a product is constrained by the capacity of a machine. The trajectory of a rocket is constrained by the final target as well as the maximum aerodynamic load it can carry. The range of an aircraft is constrained by its payload, fuel capacity, and its aerodynamic characteristics. So how does a constrained optimization problem differ from an unconstrained problem? In constrained optimization problems, the feasible region gets restricted because of the presence of constraints. This is more challenging because for a multivariable problem with several nonlinear constraints, arriving at any feasible point itself is a daunting task.

The constrained optimization problem can be mathematically stated as

Minimize

$$f(x) \qquad (6.1)$$

subject to

$$g_i(x) \leq 0 \quad i = 1, 2, \ldots, m < n \qquad (6.2)$$

$$h_j(x) = 0 \quad j = 1, 2, \ldots, r < n \qquad (6.3)$$

$$x_l \leq x \leq x_u$$

where x is a vector of n design variables given by

$$x = \begin{bmatrix} x_1 \\ x_2 \\ \vdots \\ x_n \end{bmatrix}$$

The functions f, g_i, and h_j are all *differentiable*. The design variables are bounded by x_l and x_u. The constraints g_i are called as *inequality constraints* and h_j are called *equality constraints*.

Consider the following constrained optimization problem.

Minimize

$$(x_1 - 2)^2 + (x_2 - 3)^2$$

subject to

$$x_1 \geq 3$$

If we apply the first-order optimality condition on the objective function, the function minimum is obtained at (2, 3). However, in the presence of a constraint, the minimum occurs at (3, 3). See Figure 6.1, where the function contours are plotted along with the constraint. Note that the gradient of the function (∇f) and the gradient of the constraint (∇g) are parallel to each other at the optimum point. At other points on the constraint (say, point A), the gradients are not parallel to each other. More of the optimality conditions for the constrained optimization problems are discussed in the next section.

The road map of this chapter is shown in Figure 6.2. After a discussion on optimality conditions, different solution techniques such as penalty function, augmented Lagrangian, sequential quadratic programming (SQP), and method of feasible directions are discussed. In the penalty function method,

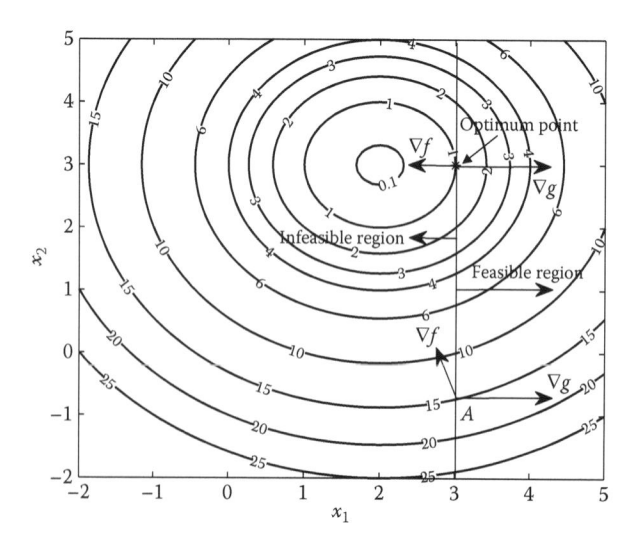

FIGURE 6.1
Constrained optimization problem.

Constrained Optimization

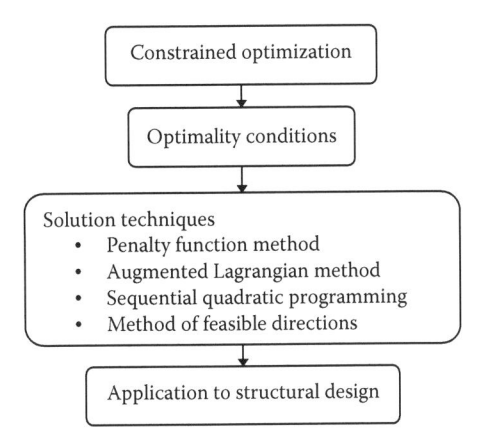

FIGURE 6.2
Road map of Chapter 6.

a constrained optimization problem is transformed into an unconstrained problem by penalizing the objective function for any violation of the constraints. The augmented Lagrangian method is a blend of both penalty function and Lagrangian multipliers methods. In the SQP method, the quadratic subproblem is solved in every iteration where the objective function is approximated by a quadratic function and the constraints are linearized. Some optimization problems require constraints to be satisfied in every iteration to ensure the meaningful value of the objective function. The method of feasible directions ensures meeting the constraints in every iteration.

6.2 Optimality Conditions

Let us define the Lagrange function for the constrained optimization problem with the equality and inequality constraints

$$L(x, \lambda, \mu) = f(x) + \sum_{j=1}^{r} \lambda_j h_j(x) + \sum_{i=1}^{m} \mu_i g_i(x) \tag{6.4}$$

The optimality conditions are given by

$$\nabla_x L = 0$$

$$\nabla_\lambda L = 0$$

$$\nabla_\mu L = 0$$

The first optimality condition results in the equation

$$\nabla_x L = \nabla f(x) + \sum_{j=1}^{r} \lambda_j \nabla h_j(x) + \sum_{i=1}^{m} \mu_i \nabla g_i(x) = 0 \quad (6.5)$$

If a particular inequality constraint is inactive ($g_i(x) \leq 0$), corresponding $\mu_i = 0$. This condition can also be written as

$$-\nabla f(x) = \sum_{j=1}^{r} \lambda_j \nabla h_j(x) + \sum_{i=1}^{m} \mu_i \nabla g_i(x) \quad (6.6)$$

That is, negative of the gradient of the objective function can be expressed as a linear combination of the gradient of the constraints.

For any feasible point x, the set of active inequality constraints is denoted by

$$A(x) = \{i \mid g_i(x) = 0\}$$

The second and third optimality conditions result in the constraints themselves. The multipliers λ_j and μ_i are called as *Lagrange multipliers* and these must be ≥ 0 at the optimum point. The optimality conditions of the constrained optimization problem are referred to as *Karush–Kuhn–Tucker* (KKT) conditions. These conditions are valid if x is a *regular point*. A point is regular if the gradient of active inequality and all equality constraints are linearly independent. It is important to note that KKT conditions are necessary but not sufficient for optimality. That is to say, there may be other local optima where KKT conditions are satisfied. The sufficient condition for $f(x)$ to be minimum is that $\nabla^2_{xx} L$ must be positive definite.

Let us take the example mentioned in the previous section and write the Lagrangian as

$$L(x, \mu) = (x_1 - 2)^2 + (x_2 - 3)^2 + \mu(-x_1 + 3)$$

The KKT conditions are given by the equations

$$2(x_1 - 2) - \mu = 0$$

$$2(x_2 - 3) = 0$$

$$-x_1 + 3 = 0$$

Constrained Optimization

Solving these equations gives the solution as $x_1 = 3$ and $x_2 = 3$, which is the optimum point with $\mu = 2$. The minimum value of the function is 1.

The Lagrange multipliers provide information on the sensitivity of the objective function with respect to sensitivity of the right-hand side of the constraint equation (say, b). Then,

$$\Delta f = \mu \Delta b = 2\Delta b$$

Therefore,

$$f \approx 1 + 2\Delta b$$

If the right-hand side of the constraint is changed by +1 unit, then a new value of the function minimum is 3 (approximately).

Example 6.1

Consider the optimization problem.

Minimize

$$f(x) = (x_1 - 1)^2 + (x_2 - 5)^2$$

subject to

$$g_1(x) = -x_1^2 + x_2 - 4 \leq 0$$

$$g_2(x) = -(x_1 - 2)^2 + x_2 - 3 \leq 0$$

Plot the function contours along with constraints. Check whether KKT conditions are satisfied at point A (0.75, 4.5625).

The function contours are given in Figure 6.3 along with the constraints. The gradient of the function and the constraints are given by

$$\nabla f(x) = \begin{bmatrix} 2(x_1 - 1) \\ 2(x_2 - 5) \end{bmatrix} = \begin{bmatrix} 2(0.75 - 1) \\ 2(4.5625 - 5) \end{bmatrix} = \begin{bmatrix} -0.5 \\ -0.875 \end{bmatrix}$$

$$\nabla g_1(x) = \begin{bmatrix} -2x_1 \\ 1 \end{bmatrix} = \begin{bmatrix} -1.5 \\ 1 \end{bmatrix}$$

$$\nabla g_2(x) = \begin{bmatrix} -2(x_1 - 2) \\ 1 \end{bmatrix} = \begin{bmatrix} 2.5 \\ 1 \end{bmatrix}$$

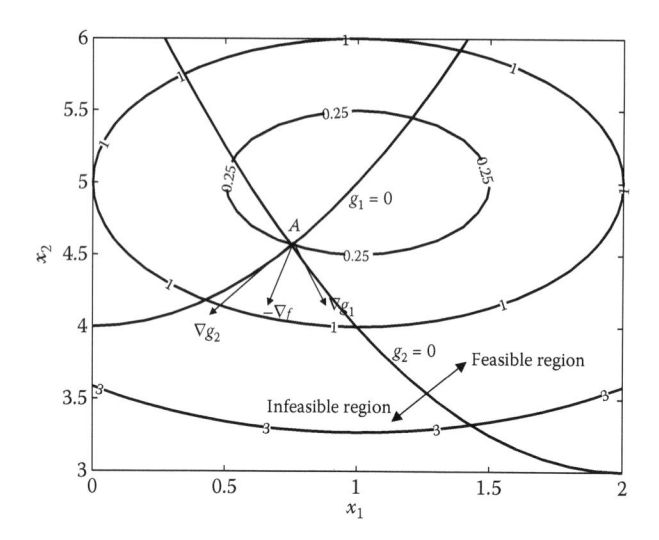

FIGURE 6.3
Function contours with the constraints for the test problem.

Let us check for the optimality condition

$$-\nabla f(x) = \mu_1 \nabla g_1(x) + \mu_2 \nabla g_2(x)$$

for some μ_1 and μ_2 which are ≥ 0.
It can be shown that at point A

$$\begin{bmatrix} 0.5 \\ 0.875 \end{bmatrix} \approx 0.4218 \begin{bmatrix} -1.5 \\ 1 \end{bmatrix} + 0.4531 \begin{bmatrix} 2.5 \\ 1 \end{bmatrix}$$

Thus, for positive value of multipliers ($\mu_1 = 0.4218$ and $\mu_2 = 0.4513$), the negative of the gradient of the objective function can be expressed as a linear combination of the gradient of the constraints. KKT conditions are satisfied at point A. Thus, point A is a candidate for the minimum of the function. Let us check the second-order condition as

$$\nabla^2 L = \begin{bmatrix} 2 - 2(\mu_1 + \mu_2) & 0 \\ 0 & 2 \end{bmatrix} = \begin{bmatrix} 0.25 & 0 \\ 0 & 2 \end{bmatrix}$$

As this matrix is positive definite, the minimum of the function occurs at point A.

6.3 Solution Techniques

For a simple optimization problem (say, with two variables) with one equality constraint, the simplest approach would be to use a *variable substitution method*. In this method, one variable is written in the form of another variable using the equality constraint. Then it is substituted in the objective function to make it an unconstrained optimization problem that is easier to solve. For instance, consider the optimization problem

Minimize

$$(x_1 - 2)^2 + (x_2 - 3)^2$$

subject to

$$-x_1 + x_2 = 4$$

Substituting $x_2 = 4 + x_1$ in the objective function, we can rewrite the optimization problem as

Minimize

$$(x_1 - 2)^2 + (x_1 + 1)^2$$

Using the first-order condition, it is easy to show that minimum of this function occurs at (1/2, 9/2). The main disadvantage of this method is that it is difficult to implement when there is a large number of variables and constraints are nonlinear.

Another way of converting a constrained optimization problem to an unconstrained problem is to penalize the objective function when constraints are violated. Such methods are termed are termed *penalty function methods* and are very easy to implement. Once the unconstrained problem is formed using the penalty functions, it can be solved using both gradient- and non–gradient-based methods described in previous chapters. The method, however, has one serious drawback. The original objective function gets distorted when modified with the penalty terms. The modified function may not be differentiable at all points. Non–gradient-based solution techniques for unconstrained problems (converted by penalty functions) are suggested for such cases.

The Lagrange function and multipliers were discussed in the previous section. In the augmented Lagrange multiplier (ALM) method, both the Lagrange multiplier and the penalty function methods are combined. Lagrange multipliers are updated on each iteration. One significant advantage of this method is that it provides an optimal value of the multipliers in addition to the solution of the optimization problem. This helps in generating a quick solution for the same optimization problem whose right-hand sides of the constraint equations are changed.

The most popular method to date is sequential quadratic programming (SQP) method for handling nonlinear objective function and constraints. In this method the objective function is approximated by a quadratic function and constraints are approximated by linear functions. The quadratic subproblem is then solved at each iteration. Hence, the method derives the name SQP.

In some optimization problems, the meaningful value of an objective function can be generated only if constraints are satisfied. The method of feasible directions ensures that design variables are always in the feasible region. Zoutendijk's method of feasible directions and Rosen's gradient projection method are discussed in this chapter.

6.3.1 Penalty Function Method

The motivation of the penalty function method is to solve the constrained optimization problem using algorithms for unconstrained problems. As the name suggests, the algorithm penalizes the objective function in case constraints are violated. The modified objective function with penalty terms is written as

$$F(x) = f(x) + r_k \sum_{j=1}^{r} h_j^2(x) + r_k \sum_{i=1}^{m} \langle g_i(x) \rangle^2 \qquad (6.7)$$

where r_k (>0) is a penalty parameter and the function

$$\langle g_i(x) \rangle = \max[0, g_i(x)] \qquad (6.8)$$

In case constraints are satisfied ($g_i(x) \leq 0$), $\langle g_i(x) \rangle$ will be zero and there will be no penalty on the objective function. In case constraints are violated ($g_i(x) \geq 0$), $\langle g_i(x) \rangle$ will be a positive value resulting in a penalty on the objective function. The penalty will be higher for higher infeasibility of the constraints. The function $F(x)$ can be optimized using the algorithms for unconstrained problems. The penalty function method of this form is called the *exterior penalty function* methods.

The parameter r_k has to be appropriately selected by the algorithm. If r_k is selected as a small value (say, 1), constraints may not be fully satisfied at the termination of the algorithm. If r_k is selected as a large value, there is a danger of ill-conditioning the objective function (see Figure 6.4). The correct approach would be to start the algorithm with a small r_k and increase it to

Constrained Optimization

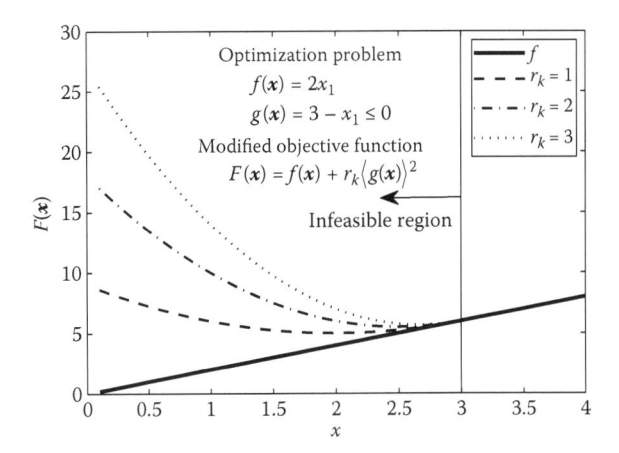

FIGURE 6.4
Exterior penalty function method.

a larger value for the purpose of tightening the constraints. The following strategy is suggested to take appropriate value of r_k during an iteration:

$$r_k = \max\left[1, \frac{1}{\left\|[\langle g_i(x)\rangle \quad h_j(x)]\right\|}\right] \quad (6.9)$$

Let us use the Davidon–Fletcher–Powell (DFP) method to solve the unconstrained problem. To account for varying penalty terms in each iteration, the MATLAB® code *DFP.m* is modified and reproduced at the end of the book. On executing the MATLAB code *DFP.m* for the optimization problem

Minimize

$$f(x) = (x_1 - 1)^2 + (x_2 - 5)^2$$

subject to

$$g_1(x) = -x_1^2 + x_2 - 4 \leq 0$$

$$g_2(x) = -(x_1 - 2)^2 + x_2 - 3 \leq 0$$

with a starting value of x of $(-1, 1)$, following output is displayed. Note from the output that penalty parameter becomes larger as constraints are tightened.

```
              Initial function value = 20.0000
No.      x-vector       f(x)    |Constr|     Penalty param.

 1    0.812   4.624    0.2219   11.70470          1
 2    0.751   4.643    0.2501    0.21589          5
 3    0.739   4.602    0.2544    0.11436          9
 4    0.742   4.572    0.2574    0.05665         18
 5    0.745   4.562    0.2589    0.02322         43
 6    0.751   4.562    0.2540    0.01459         69
 7    0.750   4.562    0.2543    0.00180        555
 8    0.750   4.562    0.2542    0.00109        915
 9    0.750   4.562    0.2541    0.00092       1082
10    0.750   4.562    0.2540    0.00052       1913
11    0.750   4.562    0.2540    0.00021       4720
12    0.750   4.562    0.2540    0.00018       5495
13    0.750   4.562    0.2542    0.00015       6526
14    0.750   4.563    0.2539    0.00056       1801
15    0.750   4.563    0.2541    0.00029       3435
16    0.750   4.562    0.2539    0.00030       3309
17    0.750   4.562    0.2539    0.00007      14692
18    0.750   4.562    0.2539    0.00006      15596
19    0.750   4.562    0.2539    0.00001      70933
```

The main advantages of the penalty function method are

- It can be started from an infeasible point.
- Unconstrained optimization methods can be directly used.

The main disadvantages of the penalty function method are

- The function becomes ill-conditioned as the value of the penalty terms is increased. Owing to abrupt changes in the function value, the gradient value may become large and the algorithm may show divergence.
- As this method does not satisfy the constraints exactly, it is not suitable for optimization problems where feasibility must be ensured in all iterations.

So far we have discussed the exterior penalty function method, which can be started even from an infeasible point. Some problems require feasibility to be maintained in all the iterations. In the *interior penalty function* method, a feasible point is first selected. The objective function is modified in such a way that it does not leave the feasible boundary. They are therefore frequently referred to as *barrier function* methods. The modified objective function in the interior penalty function approach would be

Constrained Optimization

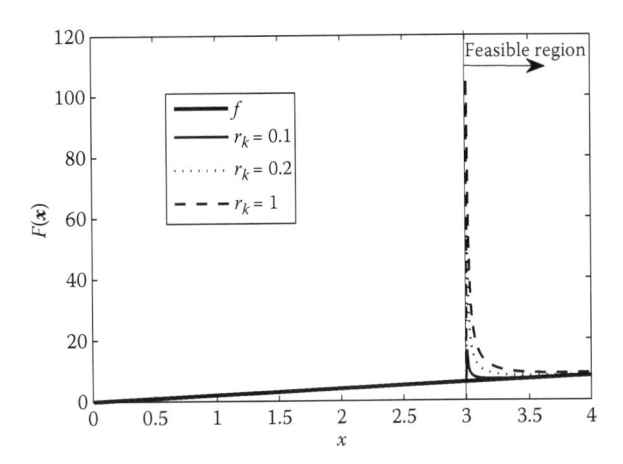

FIGURE 6.5
Interior penalty function method.

$$F(x) = f(x) - r_k \sum_{i=1}^{m} \frac{1}{g_i(x)} \qquad (6.10)$$

See Figure 6.5, where we observe that modified function remains feasible for different values of r_k.

Example 6.2

A welded beam (Ragsdell and Philips 1976) has to be designed at minimum cost whose constraints are shear stress in weld (τ), bending stress in the beam (σ), buckling load on the bar (P), and deflection of the beam (δ). The design variables (see Figure 6.6) are

$$\begin{bmatrix} x_1 \\ x_2 \\ x_3 \\ x_4 \end{bmatrix} = \begin{bmatrix} h \\ l \\ t \\ b \end{bmatrix}$$

The optimization problem is

Minimize

$$f(x) = 1.10471 x_1^2 x_2 + 0.04811 x_3 x_4 (14 + x_2)$$

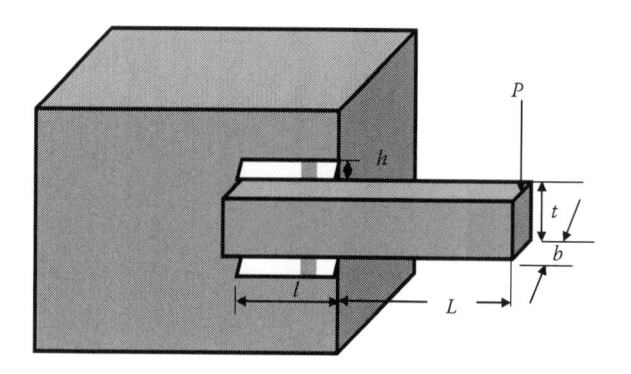

FIGURE 6.6
Welded beam.

subject to

$$g_1(x) = \tau(x) - \tau_{max} \leq 0$$

$$g_2(x) = \sigma(x) - \sigma_{max} \leq 0$$

$$g_3(x) = x_1 - x_4 \leq 0$$

$$g_4(x) = 0.10471x_1^2 + 0.04811x_3x_4(14 + x_2) - 5 \leq 0$$

$$g_5(x) = 0.125 - x_1 \leq 0$$

$$g_6(x) = \delta(x) - \delta_{max} \leq 0$$

$$g_7(x) = P - P_c(x) \leq 0$$

$$0.1 \leq x_1, x_4 \leq 2.0$$

$$0.1 \leq x_2, x_3 \leq 10.0$$

where

$$\tau(x) = \sqrt{\tau'^2 + 2\tau'\tau''\frac{x_2}{2R} + \tau''^2}$$

$$\tau' = \frac{P}{\sqrt{2}x_1x_2}$$

$$\tau'' = \frac{MR}{J}$$

Constrained Optimization

$$M = P\left(L + \frac{x_2}{2}\right)$$

$$R = \sqrt{\frac{x_2^2}{4} + \left(\frac{x_1 + x_3}{2}\right)^2}$$

$$J = \left\{\frac{x_1 x_2}{\sqrt{2}}\left[\frac{x_2^2}{12} + \left(\frac{x_1 + x_3}{2}\right)^2\right]\right\}$$

$$\tau(x) = \frac{6PL}{x_4 x_3^2}$$

$$\delta(x) = \frac{4PL^3}{E x_4 x_3^3}$$

$$P_c(x) = \frac{4.013\sqrt{EG x_3^6 x_4^6 / 36}}{L^2}\left(1 - \frac{x_3}{2L}\sqrt{\frac{E}{4G}}\right)$$

$P = 6000$ lb, $L = 14$ in., $E = 30 \times 10^6$ psi, $G = 12 \times 10^6$ psi, $\tau_{max} = 13{,}600$ psi, $\sigma_{max} = 30{,}000$ psi, $\delta_{max} = 0.25$ in.

To give equal weightage to all the constraints, the first step is to normalize all the constraints. For example, the constraint

$$\tau(x) - \tau_{max} \le 0$$

can be normalized as

$$\frac{\tau(x)}{\tau_{max}} - 1 \le 0$$

The penalty function method is used and the unconstrained optimization technique used is particle swarm optimization (PSO). The PSO code along with cost and constraint functions is given at the end of the book. On executing the code, the optimum value of objective function obtained is 2.381 and the corresponding variables are

$$\begin{bmatrix} x_1 \\ x_2 \\ x_3 \\ x_4 \end{bmatrix}^* = \begin{bmatrix} h \\ l \\ t \\ b \end{bmatrix} = \begin{bmatrix} 0.244 \\ 6.212 \\ 8.299 \\ 0.244 \end{bmatrix}$$

The termination criterion for the algorithm is the point at which the maximum number of iterations are completed. The output is reproduced below.

No.	x-vector			f(x)	
1	0.379	4.305	9.021	0.427	4.081
2	0.319	4.211	9.862	0.363	3.611
3	0.291	4.149	9.603	0.317	3.050
4	0.177	3.941	3.820	0.171	3.050
5	0.107	3.814	6.680	1.358	3.050
6	1.674	3.800	1.484	0.920	3.050
7	1.658	3.787	5.360	0.479	3.050
8	1.532	3.804	9.572	1.449	3.050
9	0.293	4.130	7.868	1.764	3.050
10	0.781	4.472	8.763	1.174	3.050
11	0.976	4.771	9.202	1.250	3.050
12	1.214	4.987	2.249	0.375	3.050
13	0.301	5.758	7.995	0.307	2.909
...					
2993	0.244	6.212	8.299	0.244	2.381
2994	0.244	6.212	8.299	0.244	2.381
2995	0.244	6.212	8.299	0.244	2.381
2996	0.244	6.212	8.299	0.244	2.381
2997	0.244	6.212	8.299	0.244	2.381
2998	0.244	6.212	8.299	0.244	2.381
2999	0.244	6.212	8.299	0.244	2.381
3000	0.244	6.212	8.299	0.244	2.381

6.4 Augmented Lagrange Multiplier Method

As the name suggests, the augmented Lagrange multipliers (ALM) method combines both Lagrange multipliers and penalty function methods. For an optimization problem with both equality and inequality constraints, the augmented Lagrangian function is given by

$$A(x,\lambda,\beta,r_k) = f(x) + \sum_{j=1}^{r} \lambda_j h_j(x) + \sum_{i=1}^{m} \beta_i \alpha_i + r_k \sum_{j=1}^{r} h_j^2(x) + r_k \sum_{i=1}^{m} \alpha_i^2 \quad (6.11)$$

where λ_j and β_i are the Lagrange multipliers, r_k is a penalty parameter fixed at the start of the iteration and

Constrained Optimization

$$\alpha_i = \max\left\{g_i(x), \frac{-\beta_i}{2r_k}\right\} \qquad (6.12)$$

The Lagrange multipliers are updated in each iteration (k) using the expressions

$$\lambda_j^{(k+1)} = \lambda_j^{(k)} + 2r_k h_j(x) \qquad (6.13)$$

$$\beta_i^{(k+1)} = \beta_i^{(k)} + 2r_k \max\left\{g_i(x), \frac{-\beta_i}{2r_k}\right\} \qquad (6.14)$$

The augmented Lagrange function can be minimized using algorithms for unconstrained optimization. Here the DFP method is used for unconstrained minimization. Consider again the optimization problem

Minimize

$$f(x) = (x_1 - 1)^2 + (x_2 - 5)^2$$

subject to

$$g_1(x) = -x_1^2 + x_2 - 4 \leq 0$$

$$g_2(x) = -(x_1 - 2)^2 + x_2 - 3 \leq 0$$

The MATLAB code *ALM.m* solves the constrained optimization problem using the ALM method with starting point as (0, 1). On executing the code the following output is displayed on the command window.

```
        Initial function value = 17.0000
No.     x-vector      rk        f(x)        |Cons.|

1       0.887  4.547  1.000     0.218       0.308
2       0.887  4.547  1.000     0.249       0.146
3       0.685  4.613  1.000     0.254       0.022
4       0.739  4.569  1.000     0.220       0.059
5       0.751  4.594  1.000     0.240       0.029
6       0.757  4.576  1.000     0.253       0.013
7       0.756  4.562  1.000     0.265       0.000
8       0.752  4.551  1.000     0.252       0.012
9       0.746  4.568  1.000     0.252       0.018
10      0.743  4.569  1.000     0.254       0.003
```

11	0.748	4.564	1.000	0.253	0.003
12	0.750	4.563	1.000	0.253	0.002
13	0.750	4.563	1.000	0.254	0.001
14	0.750	4.563	1.000	0.254	0.000
15	0.750	4.561	1.000	0.252	0.002
16	0.750	4.563	1.000	0.253	0.002

```
KKT conditions are satisfied

Lagrange multipliers
  0  0.4201  0.4500
```

Since there are no equality constraints, the first Lagrange multiplier is zero. The other two positive multipliers correspond to the inequality constraints. Since the multipliers are positive, both inequality constraints are active.

6.5 Sequential Quadratic Programming

We discussed in the earlier section that for a constrained optimization problem

Minimize

$$f(x)$$

subject to

$$h(x) = 0$$

the corresponding Lagrangian function would be

$$L(x, \lambda) = f(x) + \lambda h(x) \tag{6.15}$$

and the first-order optimality condition would be

$$\nabla_x L(x, \lambda) = 0 \tag{6.16}$$

The variables x and λ are updated using the equation

$$\begin{bmatrix} x^{k+1} \\ \lambda^{k+1} \end{bmatrix} = \begin{bmatrix} x^k \\ \lambda^k \end{bmatrix} + \begin{bmatrix} \Delta x \\ \Delta \lambda \end{bmatrix} \tag{6.17}$$

Constrained Optimization

where $\begin{bmatrix} \Delta x \\ \Delta \lambda \end{bmatrix}$ can be obtained by solving the linear system of equations

$$\begin{bmatrix} \nabla^2 L & \nabla h \\ \nabla h & 0 \end{bmatrix} \begin{bmatrix} \Delta x \\ \Delta \lambda \end{bmatrix} = - \begin{bmatrix} \nabla L \\ \nabla h \end{bmatrix} \tag{6.18}$$

This is equivalent to solving a quadratic problem with linear constraints. Thus a nonlinear optimization problem with both equality and inequality constraints can be written as a quadratic problem.

Minimize

$$Q = \Delta x^T \nabla f(x) + \frac{1}{2} \Delta x^T \nabla^2 L \Delta x \tag{6.19}$$

subject to

$$h_j(x) + \nabla h_j(x)^T \Delta x = 0 \tag{6.20}$$

$$g_i(x) + \nabla g_i(x)^T \Delta x = 0 \tag{6.21}$$

The SQP method approximates the objective function to a quadratic form and linearizes the constraints in each iteration. The quadratic programming problem is then solved to get Δx. The value of x is updated with Δx. Again the objective function is approximated with a quadratic function and constraints are linearized with new value of x. The iterations are repeated until there is no further improvement in the objective function.

Trust region approach is a useful technique for solving quadratic problems. In this approach, a region around x has to be evaluated (Δx) where a quadratic approximation of the function holds. The region is adjusted so that $f(x + \Delta x) < f(x)$. Refer to Byrd et al. (1988, 2000) and Moré and Sorensen (1983) for more details.

The Lagrangian function is often replaced by an augmented Lagrangian function in the SQP method. Let us show the steps of SQP for the constrained optimization problem

Minimize

$$f(x) = (x_1 - 1)^2 + (x_2 - 2)^2$$

subject to

$$h_1(x) = 2x_1 - x_2 = 0$$

$$g_1(x) = x_1 \leq 5$$

from a starting point (10, −5).

Iteration 1

$$f(x) = 130; \quad \nabla f(x) = \begin{bmatrix} 18 \\ -14 \end{bmatrix}; \quad \nabla h = \begin{bmatrix} 2 \\ -1 \end{bmatrix}; \quad \nabla g = \begin{bmatrix} 1 \\ 0 \end{bmatrix}; \quad \nabla^2 L = \begin{bmatrix} 12 & -4 \\ -4 & 4 \end{bmatrix}$$

The quadratic problem is

Minimize

$$Q = \Delta x^T \begin{bmatrix} 18 \\ -14 \end{bmatrix} + \frac{1}{2} \Delta x^T \begin{bmatrix} 12 & -4 \\ -4 & 4 \end{bmatrix} \Delta x$$

subject to

$$25 + [2 \quad -1]\Delta x = 0$$

$$5 + [1 \quad 0]\Delta x = 0$$

The solution of the quadratic problem is

$$\Delta x = \begin{bmatrix} -7.5 \\ 10 \end{bmatrix}$$

Now x is updated as

$$x = x + \Delta x = \begin{bmatrix} 10 \\ -5 \end{bmatrix} + \begin{bmatrix} -7.5 \\ 10 \end{bmatrix} = \begin{bmatrix} 2.5 \\ 5 \end{bmatrix}$$

Iteration 2

$$f(x) = 11.25; \quad \nabla f(x) = \begin{bmatrix} 3 \\ 6 \end{bmatrix}; \quad \nabla h = \begin{bmatrix} 2 \\ -1 \end{bmatrix}; \quad \nabla g = \begin{bmatrix} 1 \\ 0 \end{bmatrix}; \quad \nabla^2 L = \begin{bmatrix} 10 & -4 \\ -4 & 4 \end{bmatrix}$$

Constrained Optimization

The quadratic problem is

Minimize

$$Q = \Delta x^T \begin{bmatrix} 3 \\ 6 \end{bmatrix} + \frac{1}{2} \Delta x^T \begin{bmatrix} 10 & -4 \\ -4 & 4 \end{bmatrix} \Delta x$$

subject to

$$0 + [2 \quad -1]\Delta x = 0$$

$$-2.5 + [1 \quad 0]\Delta x = 0$$

The solution of the quadratic problem is

$$\Delta x = \begin{bmatrix} -1.5 \\ -3.0 \end{bmatrix}$$

Now x is updated as

$$x = x + \Delta x = \begin{bmatrix} 2.5 \\ 5 \end{bmatrix} + \begin{bmatrix} -1.5 \\ -3.0 \end{bmatrix} = \begin{bmatrix} 1 \\ 2 \end{bmatrix}$$

Iteration 3

$$f(x) = 0; \quad \nabla f(x) = \begin{bmatrix} 0 \\ 0 \end{bmatrix}; \quad \nabla h = \begin{bmatrix} 2 \\ -1 \end{bmatrix}; \quad \nabla g = \begin{bmatrix} 1 \\ 0 \end{bmatrix}; \quad \nabla^2 L = \begin{bmatrix} 10 & -4 \\ -4 & 4 \end{bmatrix}$$

Thus minimum of the function is at $\begin{bmatrix} 1 \\ 2 \end{bmatrix}$. The value of multiplier is zero for the inequality constraint. That is, the inequality constraint is inactive at the optimum point. The MATLAB code *sqp.m* solves the constrained optimization problem using SQP method.

Example 6.3

Solve the welded beam constrained optimization problem using the SQP method with an initial guess of (0.4, 6.0, 8.0, 0.5). Which constraints are active at the optimum point?

On executing the SQP code, the following output is displayed on the command screen.

No.	x-vector				f(x)	|Cons.|
1.0000	0.1250	10.0000	7.3810	0.1250	1.2379	7.2894
2.0000	0.1562	9.4551	8.6532	0.1562	1.7806	2.7531
3.0000	0.1920	7.1120	9.2150	0.1920	2.0868	0.9222
4.0000	0.2215	5.7879	9.4825	0.2215	2.3129	0.2302
5.0000	0.2352	5.5031	9.4865	0.2352	2.4297	0.0259
6.0000	0.2377	5.5289	9.3717	0.2377	2.4386	0.0006
7.0000	0.2385	5.6017	9.2434	0.2385	2.4308	0.0002
8.0000	0.2392	5.6751	9.1197	0.2392	2.4234	0.0002
9.0000	0.2399	5.7473	9.0011	0.2399	2.4166	0.0002
10.0000	0.2405	5.8180	8.8876	0.2405	2.4102	0.0002
11.0000	0.2412	5.8871	8.7793	0.2412	2.4044	0.0002
12.0000	0.2418	5.9543	8.6761	0.2418	2.3991	0.0001
13.0000	0.2425	6.0194	8.5781	0.2425	2.3942	0.0001
14.0000	0.2431	6.0822	8.4853	0.2431	2.3897	0.0001
15.0000	0.2436	6.1426	8.3977	0.2436	2.3856	0.0001
16.0000	0.2442	6.2004	8.3151	0.2442	2.3819	0.0001
17.0000	0.2444	6.2175	8.2914	0.2444	2.3809	0.0000
18.0000	0.2444	6.2175	8.2915	0.2444	2.3810	0.0000
19.0000	0.2444	6.2175	8.2915	0.2444	2.3810	0.0000

The minimum function value is 2.3810 and occurs at

$$\begin{bmatrix} x_1 \\ x_2 \\ x_3 \\ x_4 \end{bmatrix} = \begin{bmatrix} 0.2444 \\ 6.2175 \\ 8.2915 \\ 0.2444 \end{bmatrix}$$

On typing BETA at the command prompt, the following values are displayed.

```
1.4584    0.9876    0.0000    0    0    0  22.0248    0
0         0         0         0    0    0   0
```

The positive value of multipliers for the first, second, and seventh constraints indicate that these are active constraints at the optimum point.

Example 6.4

A cylindrical pressure vessel capped at both ends by hemispherical heads is to be designed for minimum cost (Sandgren 1990) whose design variables are the thickness of the shell (x_1), thickness of the head (x_2), inner radius (x_3), and length of the cylindrical section of the vessel (x_4).

Constrained Optimization

The optimization problem is

Minimize

$$f(x) = 0.6224 x_1 x_3 x_4 + 1.7781 x_2 x_3^2 + 3.1661 x_1^2 x_4 + 19.84 x_1^2 x_3$$

subject to

$$g_1(x) = -x_1 + 0.0193 x_3 \le 0$$

$$g_2(x) = -x_2 + 0.00954 x_3 \le 0$$

$$g_3(x) = -\pi x_3^2 x_4 - \frac{4}{3}\pi x_3^3 + 1{,}296{,}000 \le 0$$

$$g_4(x) = x_4 - 240 \le 0$$

where

$$0 \le x_1, x_2 \le 10, \quad 10 \le x_3, x_4 \le 200$$

Solve the constrained optimization problem using the SQP method with an initial guess of (4, 4, 100, 100).

On executing the SQP code, the function minimum obtained is 5885.3407 and occurs at (0.7782, 0.3848, 40.3196, 200). The convergence history is shown in the following table.

No.	x-vector				f(x)	|Cons.|
1.0000e+000	1.2397e+000	6.1277e-001	6.4231e+001	2.0000e+002	1.7338e+004	7.2861e-012
2.0000e+000	8.8933e-001	4.3960e-001	4.6079e+001	2.0000e+002	7.9846e+003	3.1070e-012
3.0000e+000	7.8712e-001	3.8908e-001	4.0784e+001	2.0000e+002	6.0404e+003	0
4.0000e+000	7.7823e-001	3.8468e-001	4.0323e+001	2.0000e+002	5.8865e+003	0
5.0000e+000	7.7817e-001	3.8465e-001	4.0320e+001	2.0000e+002	5.8853e+003	0
6.0000e+000	7.7817e-001	3.8465e-001	4.0320e+001	2.0000e+002	5.8853e+003	5.5511e-017
7.0000e+000	7.7817e-001	3.8465e-001	4.0320e+001	2.0000e+002	5.8853e+003	0
8.0000e+000	7.7817e-001	3.8465e-001	4.0320e+001	2.0000e+002	5.8853e+003	5.5511e-017
9.0000e+000	7.7817e-001	3.8465e-001	4.0320e+001	2.0000e+002	5.8853e+003	6.9849e-010

Example 6.5

The optimized production rate (Thygeson and Grossmann 1970) of a through-circulation system for drying catalyst pellets depends on the fluid velocity (x_1) and bed depth (x_2).

The optimization problem is

Minimize

$$f(x) = 0.0064 x_1 \left[\exp(-0.184 x_1^{0.3} x_2) - 1 \right]$$

subject to

$$g_1(x) = (3000 + x_1)x_1^2 x_2 - 1.2 \times 10^{13} \leq 0$$

$$g_2(x) = \exp(0.184 x_1^{0.3} x_2) - 4.1 \leq 0$$

where

$$0 \leq x_1 \leq 40{,}000, \quad 0 \leq x_2 \leq 1$$

Solve the constrained optimization problem using the PSO method.

On executing the PSO code, the function minimum obtained is −153.716 and occurs at (31,766, 0.342). The convergence history is shown in the following table.

No.	x-vector		f(x)
1	31475.978	0.340	-151.749
2	31532.201	0.341	-152.112
3	32464.351	0.321	-152.958
4	37080.697	0.262	-152.958
5	38808.857	0.240	-152.958
6	39261.784	0.234	-152.958
7	33844.159	0.303	-152.958
8	27509.668	0.384	-152.958
9	26009.433	0.404	-152.958
...			
992	31766.001	0.342	-153.716
993	31766.001	0.342	-153.716
994	31766.001	0.342	-153.716
995	31766.001	0.342	-153.716
996	31766.001	0.342	-153.716
997	31766.001	0.342	-153.716
998	31766.001	0.342	-153.716
999	31766.001	0.342	-153.716
1000	31766.001	0.342	-153.716

6.6 Method of Feasible Directions

Some optimization problems require constraints to be satisfied in every iteration. For example, consider the shape optimization problem of a body whose drag is to be minimized. The drag force is computed using computational fluid dynamics (CFD) analysis for a given shape of the body. It is obvious

Constrained Optimization

that CFD analysis will provide reliable results if only there is a meaningful shape of the body. This can be achieved by not only giving a proper definition of the constraints, but also satisfying them at each iteration. Consider a constrained optimization problem

Minimize

$$f(x)$$

subject to

$$g_i(x) \leq 0 \quad i = 1, 2, \ldots, m$$

A direction S is feasible at point x if

$$S^T \nabla g_i(x) < 0 \tag{6.22}$$

If the objective function also has to be reduced, then the following inequality must also be satisfied:

$$S^T \nabla f(x) \leq 0 \tag{6.23}$$

Zoutendijk's method of feasible directions and Rosen's gradient projection method are two popular methods of feasible directions that are explained in this section.

6.6.1 Zoutendijk's Method

The method starts with a feasible point x. That is, $g_i(x) \leq 0$ are satisfied. Set the search direction as the steepest descent direction. That is,

$$S = -\nabla f(x)$$

If at least one of the constraint is active $g_i(x) = 0$, then the following optimization subproblem is to be solved with respect to S.

Minimize

$$\beta \tag{6.24}$$

subject to

$$S^T \nabla g_i(x) + \beta \leq 0 \tag{6.25}$$

$$S^T \nabla f(x) + \beta \leq 0 \qquad (6.26)$$

$$-1 \leq s_k \leq 1 \quad k = 1, 2, \ldots, n \qquad (6.27)$$

where n denotes number of variables and s_k are the components of the search direction. A line search algorithm can be used to determine the next point $\bar{x}$ as

$$\bar{x} = x + \alpha S$$

such that

$$f(x + \alpha S) = f(x)$$

$$g_i(x + \alpha S) \leq 0$$

In case constraints are not met with $\bar{x}$, the optimization subproblem has to be solved again with $\bar{x}$ to obtain new S. The algorithm is terminated if any of the following criteria are met:

- The objective function value does not show improvement over successive iterations.
- Design variables do not change over successive iterations.
- β is close to zero.

6.6.2 Rosen's Gradient Projection Method

In this method, the search direction (negative of the gradient of the objective function) is projected into the subspace tangent of the active constraints. This condition of projection is sufficient for linear constraints. However, if the constraints are nonlinear, the projected search direction moves away from the search boundary (see Figure 6.7). A restoration move is carried out in case nonlinear constraints are present.

Let the matrix N denote gradient of active constraints. That is,

$$N = [\nabla g_1, \nabla g_2, \ldots, \nabla g_m] \qquad (6.28)$$

The projected matrix is given by

$$P = I - N(N^T N)^{-1} N^T \qquad (6.29)$$

Constrained Optimization

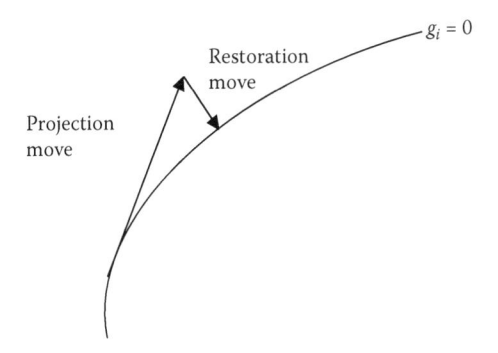

FIGURE 6.7
Rosen's gradient projection method with restoration move.

The search direction is given by

$$S = -P\nabla f(x) \tag{6.30}$$

The restoration move is given by

$$-N(N^T N)^{-1} g_i(x) \tag{6.31}$$

Combining the projection and restoration move, the design variable can be updated as

$$\bar{x} = x + \alpha S - N(N^T N)^{-1} g_i(x) \tag{6.32}$$

where

$$\alpha = -\frac{\gamma f(x)}{S^T \nabla f(x)} \tag{6.33}$$

and γ specifies the desired reduction in the objective function (Haug and Arora 1979).

Example 6.6

Show all the important variables in the first iteration of the Rosen's gradient projection method for the following optimization problem from a starting value of (2, 1).

Minimize

$$f(x) = (x_1 - 1)^2 + (x_2 - 2)^2$$

subject to

$$g_1(x) = 2x_1 - x_2 \leq 0$$

$$g_2(x) = x_1 \leq 5$$

At $x = (2, 1)$

$$f(x) = 2$$

$$\nabla f(x) = \begin{bmatrix} 2(x_1 - 1) \\ 2(x_2 - 2) \end{bmatrix} = \begin{bmatrix} 2 \\ -2 \end{bmatrix}$$

The constraint $g_1(x)$ is also violated. Therefore,

$$N = \begin{bmatrix} 2 \\ -1 \end{bmatrix}$$

The projection matrix is given by

$$P = I - N(N^T N)^{-1} N^T = \frac{1}{5} \begin{bmatrix} 1 & 2 \\ 2 & 4 \end{bmatrix}$$

Therefore, the search direction is given by

$$S = -P \nabla f(x) = \begin{bmatrix} 1 \\ 2 \end{bmatrix}$$

Taking $\gamma = 0.1$,

$$\alpha = -\frac{\gamma f(x)}{S^T \nabla f(x)} = 0.1$$

The value of x can now be updated as

$$x = x + \alpha S - N(N^T N)^{-1} g_i(x) = \begin{bmatrix} 2 \\ 1 \end{bmatrix} + 0.1 \begin{bmatrix} 1 \\ 2 \end{bmatrix} - \begin{bmatrix} 1.2 \\ -0.6 \end{bmatrix} = \begin{bmatrix} 0.9 \\ 1.8 \end{bmatrix}$$

At $x = (0.9, 1.8)$

$$f(x) = 0.05$$

Both the constraints are also feasible at this point.

6.7 Application to Structural Design

A structure is to be designed that has members with square cross sections (Figure 6.8). The design variables are the cross-sectional sizes of the columns (x_1) and beam (x_2). The objective function is to minimize the volume of the structure. The stresses are to be restricted at three critical sections: top of the column and end and midspan of the beam. The optimization problem (Horowitz et al. 2008) is stated as

Minimize

$$V = 2\alpha l x_1^2 + l x_2^2$$

subject to

$$g_1(x) \le \frac{ql}{2x_1^2} + \frac{3ql^2 x_1}{6x_1^4 + 4\alpha x_2^4} - \sigma_1 \le 0$$

$$g_2(x) \le \frac{x_1^4(x_2 + 6\alpha l)}{x_2^3(6x_1^4 + 4\alpha x_2^4)} \frac{ql}{2\alpha} - \sigma_2 \le 0$$

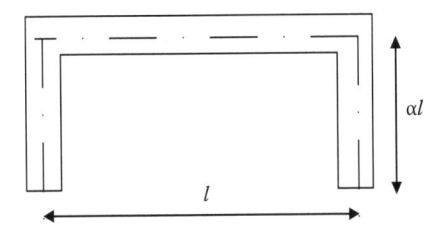

FIGURE 6.8
Structural frame.

$$g_3(x) \le \frac{x_1^4 x_2 + \alpha l \left(3x_1^4 + 6\alpha x_2^4\right)}{x_2^3 \left(6x_1^4 + 4\alpha x_2^4\right)} \frac{ql}{2\alpha} - \sigma_3 \le 0$$

where

$$0 \le x_1, x_2 \le 40$$

$$q = 15 \text{ kgf/cm}$$

$$l = 550 \text{ cm}$$

$$\sigma_1 = \sigma_2 = \sigma_3 = 103 \text{ kgf/cm}^2$$

The constrained optimization problem has two optima. It is desired to achieve a global optima for this problem. We use stochastic algorithm PSO for this purpose. On executing MATLAB code *pso.m* the following output is obtained.

No.	x1	x2	f(x)
1	10.881	31.486	597348.396
2	18.732	31.907	597348.396
3	25.175	32.252	597348.396
4	22.256	32.096	597348.396
5	11.958	31.131	595963.381
...			
96	9.294	31.083	569385.788
97	9.294	31.083	569385.788
98	9.294	31.083	569385.788
99	9.294	31.083	569385.788
100	9.294	31.083	569385.788

The global optimum value of the design variable is $(x_1, x_2) = (9.294, 31.083)$ and optimum value of the objective function is 569,385 cm^3.

Chapter Highlights

- A point is regular if the gradient of active inequality and all equality constraints are linearly independent.

Constrained Optimization

- The optimality conditions for constrained optimization problems are frequently referred to as *Karush–Kuhn–Tucker* (KKT) conditions. KKT conditions are necessary but not sufficient for optimality.
- The Lagrange multiplier provides information on the sensitivity of the objective function with respect to the sensitivity of the right-hand side of the constraint equation.
- A constrained optimization problem can be converted to an unconstrained problem by penalizing the objective function when constraints are violated. Such methods are termed *penalty function methods* and are very easy to implement.
- The motivation of using the penalty function method is to solve the constrained optimization problem using algorithms for unconstrained problems.
- The augmented Lagrange multiplier (ALM) method combines both Lagrange multiplier and penalty function methods.
- The sequential quadratic programming (SQP) method approximates the objective function to a quadratic form and linearizes the constraints in each iteration.
- The method of feasible directions ensures meeting the constraints in every iteration.
- In Rosen's gradient projection method, the search direction (negative of the gradient of the objective function) is projected into the subspace tangent of the active constraints.

Formulae Chart

Lagrange function:

$$L(x, \lambda, \mu) = f(x) + \sum_{j=1}^{r} \lambda_j h_j(x) + \sum_{i=1}^{m} \mu_i g_i(x)$$

Optimality condition:

$$-\nabla f(x) = \sum_{j=1}^{r} \lambda_j \nabla h_j(x) + \sum_{i=1}^{m} \mu_i \nabla g_i(x)$$

Penalty function:

$$f(x) = f(x) + r_k \sum_{j=1}^{r} h_j^2(x) + r_k \sum_{i=1}^{m} \langle g_i(x) \rangle^2$$

$$\langle g_i(x) \rangle = \max[0, g_i(x)]$$

Augmented Lagrangian function:

$$A(x, \lambda, \beta, r_k) = f(x) + \sum_{j=1}^{r} \lambda_j h_j(x) + \sum_{i=1}^{m} \beta_i \alpha_i + r_k \sum_{j=1}^{r} h_j^2(x) + r_k \sum_{i=1}^{m} \alpha_i^2 \langle g_i(x) \rangle$$

$$= \max[0, g_i(x)]$$

$$\alpha_i = \max\left\{ g_i(x), \frac{-\beta_i}{2r_k} \right\}$$

Quadratic problem:
  Minimize

$$Q = \Delta x^T \nabla f(x) + \frac{1}{2} \Delta x^T \nabla^2 L \Delta x$$

  subject to

$$h_j(x) + \nabla h_j(x)^T \Delta x = 0$$

$$g_i(x) + \nabla g_i(x)^T \Delta x = 0$$

Rosen's gradient projection method:

$$P = I - N(N^T N)^{-1} N^T$$

$$\bar{x} = x + \alpha S - N(N^T N)^{-1} g_i(x)$$

$$\alpha = -\frac{\gamma f(x)}{S^T \nabla f(x)}$$

Constrained Optimization

Problems

1. For the following optimization problem
   Minimize
   $$f(x) = 2x_1 + x_2$$
   subject to
   $$1 + x_1^2 - x_2 \leq 0$$
   check whether the following points are feasible
   i. (0, 0)
   ii. (1, 2)
   iii. (2, 1)
   iv. (1, 3)

2. For the following optimization problem
   Minimize
   $$f(x) = (x_1 - 3)^2 + (x_2 - 4)^2 + 2$$
   subject to
   $$1 + x_1^2 - x_2 \leq 0$$
   check which of the constraints are active at the following points
   i. (2, −1)
   ii. (1, 2)
   iii. (1, 1)
   iv. (13/5, 1/5)

3. Solve the following optimization problem using the *variable-elimination* method.
   Minimize
   $$f(x) = (3x_1 - 2x_2)^2 + (x_1 + 2)^2$$
   subject to
   $$x_1 + x_2 = 7$$

4. Write the Lagrangian for the problem
   Minimize
   $$f(x) = (3x_1 - 2x_2)^2 + (x_1 + 2)^2$$
   subject to
   $$x_1 + x_2 = 7$$
   and then write down the optimality conditions. Find the optimal value of x_1 and x_2. Also compute the value of multiplier and comment whether the constraint is active at the optimal point. What is the approximate change in the optimal value of $f(x)$ if the right-hand side of the constraint equation is changed to 6 from 7.

5. Solve the following optimization problem
   Minimize
   $$f(x) = \frac{5x_1}{x_2} + \frac{x_2}{x_1^2}$$
   subject to
   $$x_1 x_2 - 2 = 0$$
   $$x_1 + x_2 \geq 1$$
   using the SQP method with an initial guess of (1, 1). Define the quadratic sub-problem at each step.

6. Solve the previous optimization problem using the PSO method. Compare the results obtained from the SQP method.

7. The welded beam constrained optimization problem was solved using PSO and SQP methods in this chapter. For the SQP method, the initial guess for the design variables was taken as (0.4, 6.0, 8.0, 0.5), which was close to the optimum point. Using different initial guesses for the design variables, execute the SQP code and observe the sensitivity of the convergence.

8. Solve the pressure vessel problem using the PSO method and Rosen's gradient projection method.

9. Solve the through-circulation dryer problem using the SQP method with different initial guesses of the design variables.

10. Solve the spring design problem (Rao 2009), which minimizes the weight of a spring subject to constraints on deflection, shear stress, and frequency. The design variables are the mean coil diameter (x_1), the wire diameter (x_2), and the number of active coils (x_3).

Minimize

$$f(x) = (x_3 + 2)x_2 x_1^2$$

subject to

$$g_1(x) = 1 - \frac{x_2^3 x_3}{71,785 x_1^4} \leq 0$$

$$g_2(x) = \frac{4x_2^2 - x_1 x_2}{12,566(x_2 x_1^3 - x_1^4)} + \frac{1}{5108 x_1^2} - 1 \leq 0$$

$$g_3(x) = 1 - \frac{140.45 x_1}{x_2^2 x_3} \leq 0$$

$$g_3(x) = \frac{x_1 + x_2}{1.5} - 1 \leq 0$$

where

$$0.05 \leq x_1 \leq 2,\ 0.25 \leq x_2 \leq 1.3,\ 2 \leq x_3 \leq 15$$

7

Multiobjective Optimization

7.1 Introduction

In previous chapters, optimization problems with a single objective function were discussed and these problems were with or without constraints. Typical single-variable objective functions are cost minimization, efficiency maximization, weight minimization, and so on. The solution to single-variable optimization problems results in a single point in the design space and the corresponding objective function value at that point gives the minimum value of the function.

In the multiobjective optimization problem, two or more objective functions are to be simultaneously optimized. For example, the criteria in manufacturing a product could be cost minimization and efficiency maximization. The general form of a multiobjective optimization problem can be mathematically stated as

Minimize

$$f_k(x) \quad k = 1, 2, \ldots, K \tag{7.1}$$

subject to

$$g_i(x) \leq 0 \quad i = 1, 2, \ldots, m < n \tag{7.2}$$

$$h_j(x) = 0 \quad j = 1, 2, \ldots, r < n \tag{7.3}$$

$$x_l \leq x \leq x_u \tag{7.4}$$

where x is a vector of n design variables given by

$$x = \begin{bmatrix} x_1 \\ x_2 \\ \vdots \\ x_n \end{bmatrix}$$

The solution to a multiobjective problem results in a number of points in the objective function space referred to as *Pareto optimal solutions*. For a multiobjective problem with two objective functions (the first function is efficiency maximization and the second function is cost minimization), a typical Pareto optimal front is shown in Figure 7.1. The first objective (f_1) function "efficiency" is along the x-axis of this figure and the y-axis contains the second objective (f_2) function "cost." The Pareto optimal front is obtained using the principle of *domination*. In this concept, each solution is compared to check whether it dominates another solution or not.

A solution x^1 is said to dominate another solution x^2 if the following conditions are satisfied

- The solution x^1 is no worse than x^2 in all objectives.
- The solution x^1 is better than x^2 in at least one objective.

Consider points A and C for domination. Clearly, point C dominates point A in both the objective functions. However, point C is itself dominated by at least one of the points in the Pareto optimal front. The points along the Pareto optimal front are referred to as nondominated solutions. In Figure 7.1, the Pareto optimal front is convex. However, this front can be concave, partially convex/concave or discontinuous. The trade-off between the objective functions defines the shape of the Pareto front.

In this chapter, we discuss the methods for obtaining the nondominated solutions for a multiobjective optimization problem. These methods will be applied on some well-known test functions. The road map of this chapter is shown in Figure 7.2. The weighted sum approach, ε-constraint method, goal programming, and utility function method are explained as the techniques for solving multiobjective problems. In the weighted sum approach,

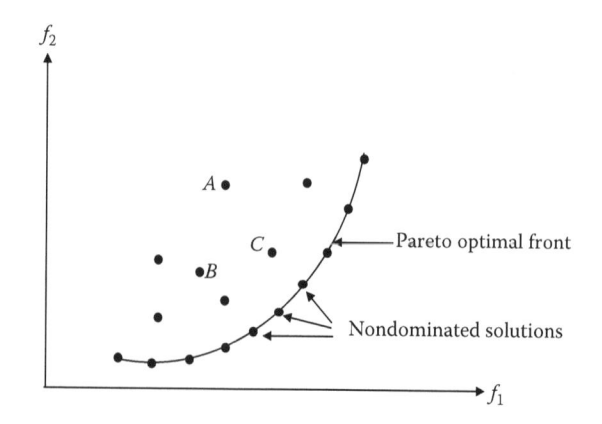

FIGURE 7.1
Pareto optimal front.

Multiobjective Optimization 205

```
Multiobjective optimization
          ↓
Weighted sum approach
          ↓
  ε-Constraints method
          ↓
   Goal programming
          ↓
 Utility function method
          ↓
      Application
```

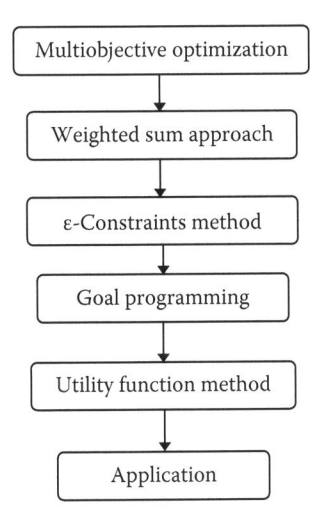

FIGURE 7.2
Road map of Chapter 7.

different objectives are combined into a single objective function using different weights. This method is simple and easy to implement. However, it can locate one Pareto point in one optimization run using the gradient-based method. The particle swarm optimization (PSO) technique, which works with a number of solution points, can locate the Pareto front on one single run. In the ε-constraint method, one objective function is minimized and remaining objective functions are transformed into constraints which are to be specified by the user. The transformed problem is then solved using the gradient-based method. The method can locate the Pareto fronts of the nonconvex problems. In goal programming, a target is set for each of the objective functions and the optimizer aims to minimize the deviations from the set goals. In the utility function method, all the objectives are combined into a single function which is then solved along with the constraints. In the last section, shape optimization of a reentry body is carried out that has two conflicting objectives: weight minimization and stability maximization, along with constraints.

7.2 Weighted Sum Approach

The simplest approach to solve a multiobjective optimization problem is to combine all the objective functions into a single objective function, which then can be solved using any of the methods described in previous chapters. Different objective functions can be combined into a single objective function using user-supplied weights and the optimization problem can be stated as

Minimize

$$\sum_{k=1}^{K} w_k f_k(x) \quad k = 1, 2, \ldots, K \tag{7.5}$$

subject to

$$g_i(x) \leq 0 \quad i = 1, 2, \ldots, m < n \tag{7.6}$$

$$h_j(x) = 0 \quad j = 1, 2, \ldots, r < n \tag{7.7}$$

$$x_l \leq x \leq x_u \tag{7.8}$$

where w_k is a nonnegative weight of the kth objective function such that

$$\sum_{k=1}^{K} w_k = 1 \tag{7.9}$$

The value of weight to be selected for an objective function depends on the relative importance of that objective function over the other objective functions. For example, in the cost-efficiency multiobjective problem discussed in the previous section, a weight of 0.2 for the objective function "cost" and a weight of 0.8 for the other objective function efficiency will result in an optimized solution given by a single point. To obtain the Pareto optimal front, the procedure has to be repeated with different weights.

Consider the following multiobjective optimization problem whose Pareto optimal front is to be obtained.

Minimize

$$f_1(x) = \frac{1}{2}\left(x_1^2 + x_2^2\right)$$

Minimize

$$f_2(x) = \frac{1}{2}\left[(x_1 - 1)^2 + (x_2 - 3)^2\right]$$

Multiobjective Optimization

Using the weighted sum approach, two objective functions can be combined to form a single objective function as given below

Minimize

$$F(x) = \frac{1}{2}w_1\left(x_1^2 + x_2^2\right) + \frac{1}{2}w_2\left[(x_1 - 1)^2 + (x_2 - 3)^2\right]$$

The above function is optimized by varying the weight w_1 from 0 to 1 in steps of 0.01. The other weight w_2 is selected using the following equality,

$$w_2 = 1 - w_1 \tag{7.10}$$

For each value of $[w_1, w_2]$, the above function will be optimized to obtain the optimal solution. Thus, different values of $[w_1, w_2]$ will result in a number of optimal solutions that result in the Pareto optimal front. Let us use the sequential quadratic programming (SQP) method, as discussed in the previous chapter, to solve the multi-objective problem. The MATLAB® code *sqp.m* is modified so that in a single execution, the Pareto optimal front can be obtained. Only the modified main program and functions are listed under the heading of this chapter in Appendix B. Other routines remain the same as previous chapters. On executing the code, the Pareto optimal front is obtained and is shown in Figure 7.3. The shape of the Pareto front is convex. The tangent line at point A represents the equal-cost line for the function $F(x)$ and its slope depends on the choice of weights w_1 and w_2.

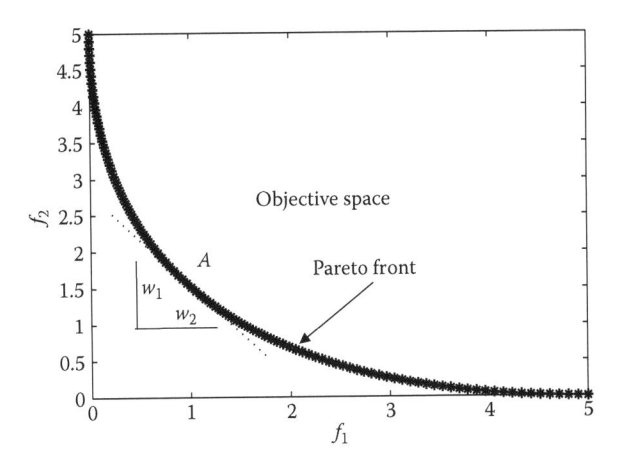

FIGURE 7.3
Pareto front.

Let us check the performance of the weighted sum method for multiobjective problems that have a nonconvex Pareto front. Consider the following multiobjective optimization problem:

Minimize
$$f_1(x) = x_1$$

Minimize
$$f_2(x) = 1 + x_2^2 - x_1 - 0.1\sin(3\pi x_1)$$

subject to
$$0 \le x_1 \le 1, \quad -2 \le x_2 \le 2$$

On executing the modified SQP code for these functions, an incomplete Pareto front is generated and is shown in Figure 7.4. The weighted sum approach, though simple to implement, has difficulty in locating the Pareto front of the nonconvex type. Another disadvantage of the weighted sum approach is that even if weights are uniformly distributed, it may not result in uniform distribution of Pareto optimal solutions (see Figure 7.4). The advantages and disadvantages of this method are

Advantages

- It is simple and easy to use.
- It ensures a solution for convex problems.

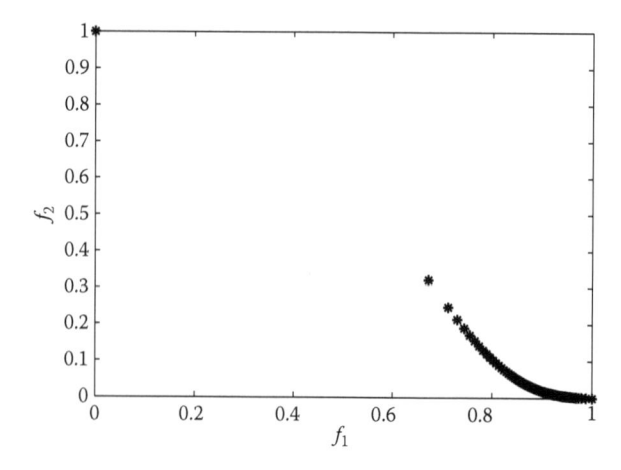

FIGURE 7.4
Incomplete Pareto front.

Multiobjective Optimization

Disadvantages

- The computational burden is higher.
- Different weights may lead to the same solution.
- A solution is not obtained for nonconvex problems.
- All problems have to be converted to the same type (min or max type).

An interesting alternative is to use evolutionary algorithms (such as genetic algorithm or PSO) to locate the Pareto optimal front because it works simultaneously on a number of points. In one such strategy using the PSO technique (Parsopoulos and Vrahatis 2002), the weights are updated on each iteration using the equation

$$w_1(t) = \left| \sin\left(\frac{2\pi t}{F}\right) \right|$$

where t is the iteration index and F is the weights' change frequency. The dynamic change of weights during the iterations forces the PSO algorithm to keep the solutions on the Pareto front. The PSO algorithm given in Chapter 5 is modified with the dynamic weight strategy and the revised MATLAB code is given in *pso.m*. On executing the code for the multiobjective problem, the Pareto optimal front is obtained and is shown in Figure 7.5. It is important to note that the modified PSO algorithm is able to locate the nonconvex Pareto front where weigthed sum approach failed to achieve the full Pareto front.

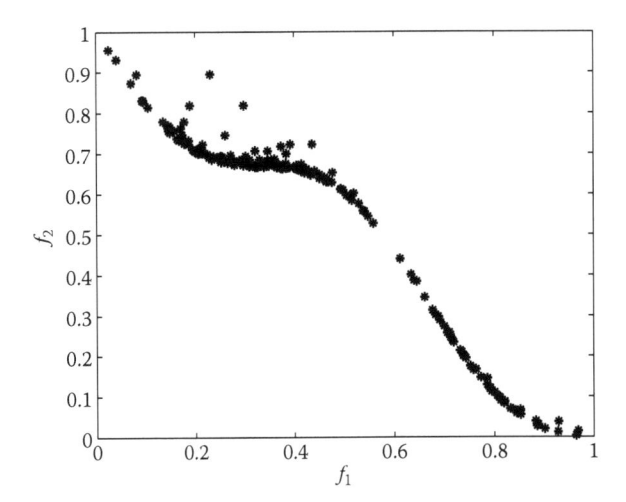

FIGURE 7.5
Nonconvex Pareto front generated with particle swarm optimization.

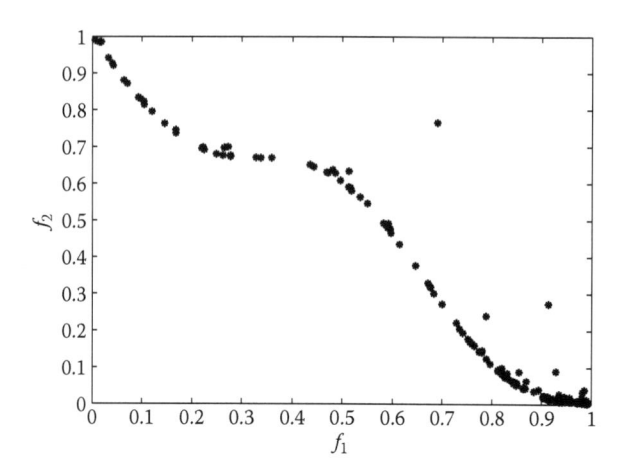

FIGURE 7.6
Pareto front for the test problem.

Example 7.1

Consider the multiobjective optimization problem:

Minimize
$$f_1(x) = x_1^3 + x_2^2$$

Minimize
$$f_2(x) = 2x_2^2 - 3x_1$$

subject to
$$0 \le x_1 \le 1, \quad -2 \le x_2 \le 2$$

Modify the PSO code for these functions and obtain the Pareto optimal front.

The tuning parameters for PSO algorithm are changed to different values. On executing the code with one such change in tuned parameters, the Pareto optimal front is obtained and given in Figure 7.6.

7.3 ε-Constraints Method

In this method, the decision-maker chooses one objective out of K objectives that needs to be minimized and the remaining objectives are put as constraints to some target values (which are to be defined by the decision-maker).

Multiobjective Optimization

If we select $f_3(x)$ as the objective function that needs to be minimized, then the ε-constraint problem is

Minimize

$$f_3(x) \tag{7.11}$$

subject to

$$f_k(x) \le \varepsilon_k \quad k = 1\ldots K, \quad k \ne 3 \tag{7.12}$$

For a simple multiobjective problem with two objectives, the concept of this method is explained through Figure 7.7. Using different values of ε, Pareto optimal solutions can be obtained. The method can also provide solutions for multiobjective problems with nonconvex Pareto fronts. The disadvantage of this method is that prior information on ε is required to obtain a proper solution.

Let us solve the following problem using this method.

Minimize

$$f_1(x) = x_1$$

Minimize

$$f_2(x) = 1 + x_2^2 - x_1 - 0.1\sin(3\pi x_1)$$

subject to

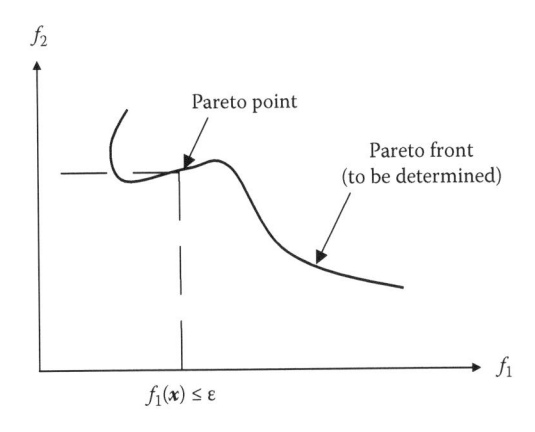

$$0 \le x_1 \le 1, \quad -2 \le x_2 \le 2$$

FIGURE 7.7
Concept of ε-constraints method.

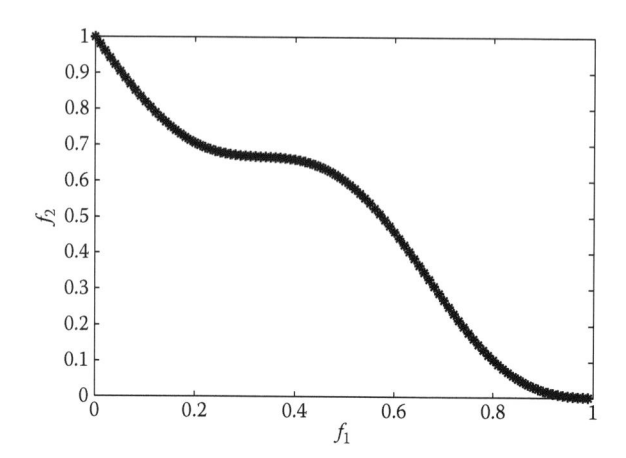

FIGURE 7.8
Nonconvex Pareto front generated using ε-constraints method.

The second objective function $f_2(x)$ is used as an objective function and the first objective function $f_1(x)$ is put as a constraint:

$$f_1(x) \leq \varepsilon$$

The optimization problem is solved for different ε using the SQP method. The ε is varied from 0.01 to 0.99, resulting each time in an optimization problem with different constraints. The solution of each of these problems results in a single Pareto point. The SQP code, mentioned in Chapter 6, is suitably modified and the Pareto front obtained is shown in Figure 7.8.

7.4 Goal Programming

In goal programing, a target or goal is set for each objective function. Then the optimization problem is to minimize the deviation from the set targets. For example, if the functions $f_k(x)$ are to be minimized and we set a goal for this function as τ_k, then the optimization problem becomes

Minimize

$$\sum_{k=1}^{K} w_{1,k} p_k + w_{2,k} n_k \quad k = 1, 2, \ldots, K \tag{7.13}$$

subject to

$$f_k(x) + p_k - n_k = \tau_k \tag{7.14}$$

$$p_k, n_k \geq 0 \tag{7.15}$$

where $w_{1,k}$ and $w_{2,k}$ are the weights of the kth goal and p_k and n_k are the underachievement and overachievement for the kth goal. The main advantage of goal programming is that multiobjectives are transformed into the constraints of a single-objective optimization problem.

Let us consider the following multiobjective optimization problem.

Minimize

$$f_1(x) = x_1^2 + x_2$$

Minimize

$$f_2(x) = x_2^2 - x_1$$

subject to

$$-5 \leq x_1 \leq 5, \quad -3 \leq x_2 \leq 3$$

Assuming goals for the two objective functions as 1 and 2, the goal programming problem can be written as

Minimize

$$f_1(x) = w_{11}p_1 + w_{12}p_2 + w_{21}n_1 + w_{22}n_2$$

subject to

$$x_1^2 + x_2 + p_1 - n_1 - 1 = 0$$

$$x_2^2 - x_1 + p_2 - n_2 - 2 = 0$$

$$-5 \leq x_1 \leq 5, \quad -3 \leq x_2 \leq 3$$

where the design variables for this problem are x_1, x_2, p_1, p_2, n_1, and n_2. For the user-supplied value of the weight variables, the optimization problem is solved. The optimal value of the design variables is substituted in the original multiobjective problem to obtain the values of f_1 and f_2. The Pareto front can be obtained by repeating the procedure with different weights.

In the *lexicographic goal programming* method, different objectives of the multiobjective problem are to be ranked in the order of importance or priority. The most important objective is selected first and it is solved to obtain x^*. The next objective function in the order of priority is then selected and solved with the additional constraint being the value of the objective function obtained from the first step. The process is repeated until all the objectives are covered. Let $f_1(x)$ be the most important objective function selected by the designer; then the first step is to solve the optimization problem:

Minimize

$$f_1(x)$$

subject to

$$g_i(x) \le 0 \quad i = 1, 2, \ldots, m$$

The optimal solution for this problem is denoted by x^*. In the next step of lexicographic goal programming, the second most important objective function $f_2(x)$ is selected for optimization and the problem can be stated as

Minimize

$$f_2(x)$$

subject to

$$g_i(x) \le 0 \quad i = 1, 2, \ldots, m$$

$$f_1(x) = f_1(x^*)$$

The process is repeated until all the objectives are covered and let the optimum solution obtained for the multiobjective problem be denoted by x^*. It is important to note that if the priorities of the objective functions are changed, the optimal solution obtained will be a different x^*.

7.5 Utility Function Method

In this method a utility function U is defined that combines all the objective functions of the multiobjective optimization problem. The utility function then becomes the objective function of the optimization problem that can be solved along with the constraints. Mathematically, the utility function method can be described as

Minimize

$$U(f_k(x)) \quad k = 1, 2, \ldots, K \qquad (7.16)$$

subject to

$$g_i(x) \leq 0 \quad i = 1, 2, \ldots, m < n \qquad (7.17)$$

$$h_j(x) = 0 \quad j = 1, 2, \ldots, r < n \qquad (7.18)$$

7.6 Application

Reentry bodies enter the Earth's atmosphere with high velocities. The large kinetic energy possessed by these bodies has to be dissipated by appropriately designing the shape of these bodies. The shape design of the reentry body is a typical multiobjective optimization problem with conflicting objectives (Adimurthy et al. 2012; Arora and Pradeep 2003). The weight of the reentry body is to be minimized and its stability is to be maximized. The weight minimization is the same as minimizing the surface area of the reentry body. The stability of the body is dictated by its location of center of pressure and is denoted by X_{cp}. An aerodynamic body is more stable if its X_{cp} is located as aft as possible. Thus, for aerodynamic stability X_{cp} is to be maximized.

The shape of a reentry body is typically a spherical nose-cone-flare type (Figure 7.9). The design parameters for the multiobjective optimization are the

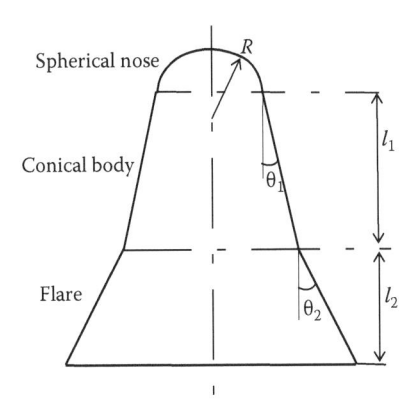

FIGURE 7.9
Design variables of the reentry body.

nose radius (R_n), first conical flare angle (θ_1), and its length (l_1) and second conical flare angle (θ_2) and its length (l_2). During the launch phase of the reentry body, it has to be accommodated inside the payload fairing of the rocket. So there are restrictions on the dimensions of the nose radius and flare lengths. In addition, to avoid flow separation, variable θ_2 is linked to θ_1. Further, electronic packages and other equipment have to be housed inside the reentry body, leading to volume (V) constraint in the optimization problem.

The constraints of the problem are

$$V \geq 1$$

$$0.4 < R_n < 0.6$$

$$22 < \theta_1 < 27$$

$$\theta_1 + 5 < \theta_2 < \theta_1 + 10$$

$$0.4 < l_1 < 0.8$$

$$0.4 < l_2 < 0.8$$

The surface area and volume are computed using the expressions

$$A = 2\pi R_n^2 (1 - \sin\theta_1) + \pi(R_1 + R_2)\sqrt{(R_2 - R_1)^2 + l_1^2}$$
$$+ \pi(R_2 + R_B)\sqrt{(R_B - R_2)^2 + l_2^2} + \pi R_B^2$$

$$V = \frac{\pi R_n (1 - \sin\theta_1)}{6}\left(3R_1^2 + R_1^2(1 - \sin\theta_1)^2\right) + \frac{1}{3}\pi l_1\left(R_1^2 + R_2^2 + R_1 R_2\right)$$
$$+ \frac{1}{3}\pi l_2\left(R_B^2 + R_2^2 + R_B R_2\right)$$

where

$$R_1 = R_n \cos\theta_1$$
$$R_2 = R_n \cos\theta_1 + l_1 \tan\theta_1$$
$$R_B = R_2 + l_2 \tan\theta_2$$

The other objective function X_{cp} is computed using the expression

$$X_{cp} = \frac{C_m}{C_n}$$

Multiobjective Optimization 217

where C_m and C_n are pitching moment coefficient and the normal force coefficient respectively. The value of X_{cp} is computed for a unit reference length. The value of aerodynamic coefficients can be computed using the flow field analysis for different geometrical shapes. As an alternate, one can build a response surface for C_m and C_n as a function of input parameters R_n, θ_1, l_1, θ_2 and l_2. The responses are generated using modified Newtonian flow (Chernyi 1961), which is valid for hypersonic (Mach>5) flows. The analysis is valid for a small angle of attack. The response matrix is generated for an angle of attack of 5 degrees and is shown in Table 7.1.

The response surface model is then generated with MATLAB using the *regstats* function. For example,

```
>> regstats(cm,inputparam,'purequadratic')
>> beta
```

will generate the polynomial coefficients (beta) of the respective input parameters. The response surface of the aerodynamic coefficients is thus given as

TABLE 7.1

Response Surface Matrix

	Input Parameters				Responses	
R_n m	θ_1 degrees	θ_2 degrees	l_1 m	l_2 m	C_m	C_n
0.5	20	25	0.65	0.65	0.07954	0.12198
0.5	20	25	1.00	0.65	0.12409	0.15979
0.5	20	25	0.20	0.65	0.03992	0.08037
0.5	20	25	0.65	1.00	0.13359	0.16864
0.5	20	25	0.65	0.20	0.03420	0.072529
0.2	20	25	0.65	0.65	0.03848	0.06818
0.2	20	25	1.00	0.65	0.06647	0.09546
0.2	20	25	0.20	0.65	0.01552	0.03951
0.2	20	25	0.65	1.00	0.07245	0.10236
1.0	20	25	0.65	0.65	0.01974	0.23620
1.0	20	25	1.00	0.65	0.27559	0.29093
1.0	20	25	0.20	0.65	0.12147	0.17304
1.0	20	25	0.65	1.00	0.29286	0.30292
1.0	20	25	0.65	0.20	0.10818	0.16111
0.5	15	20	0.65	0.65	0.05917	0.09790
1.2	15	25	0.20	0.65	0.12767	0.17713
0.2	15	20	1.00	0.65	0.04339	0.06695
0.5	25	35	0.65	0.65	0.11970	0.15591
0.5	25	30	0.65	1.00	0.17626	0.20450

$$C_m = \beta_0 + \beta_1 R_n + \beta_2 \theta_1 + \beta_3 l_1 + \beta_4 \theta_2 + \beta_5 l_2 + \beta_6 R_n^2 + \beta_7 \theta_1^2 + \beta_8 l_1^2 + \beta_9 \theta_2^2 + \beta_{10} l_2^2$$

$$C_n = \beta_{0'} + \beta_{1'} R_n + \beta_{2'} \theta_1 + \beta_{3'} l_1 + \beta_{4'} \theta_2 + \beta_{5'} l_2 + \beta_{6'} R_n^2 + \beta_{7'} \theta_1^2 + \beta_{8'} l_1^2 + \beta_{9'} \theta_2^2 + \beta_{10'} l_2^2$$

where the polynomial coefficients are given by

$$\begin{bmatrix} \beta_0 \\ \beta_1 \\ \beta_2 \\ \beta_3 \\ \beta_4 \\ \beta_5 \\ \beta_6 \\ \beta_7 \\ \beta_8 \\ \beta_9 \\ \beta_{10} \end{bmatrix} = \begin{bmatrix} -0.2785277 \\ 0.07575931 \\ 0.00138183 \\ 0.00582562 \\ 0.08788085 \\ 0.07978807 \\ 0.10309911 \\ 0.00009141 \\ -0.0000837 \\ 0.02437810 \\ 0.05287244 \end{bmatrix} \text{ and } \begin{bmatrix} \beta_{0'} \\ \beta_{1'} \\ \beta_{2'} \\ \beta_{3'} \\ \beta_{4'} \\ \beta_{5'} \\ \beta_{6'} \\ \beta_{7'} \\ \beta_{8'} \\ \beta_{9'} \\ \beta_{10'} \end{bmatrix} = \begin{bmatrix} -0.3142861 \\ 0.15013042 \\ -0.0039655 \\ 0.01401952 \\ 0.10363958 \\ 0.10800186 \\ 0.04790623 \\ 0.00018685 \\ -0.0002199 \\ -0.0004853 \\ 0.01629809 \end{bmatrix}$$

The response surface model is validated by using some arbitrary values of input parameters, but within their constraint bounds and generating the responses for these parameters. The accuracy of responses is checked by comparing them with those generated by flow field analysis for the same input parameters. The accuracy of the response surface model is within 5% of those generated by the flow field analysis. Let us summarize the multiobjective problem:

Minimize

$$A = 2\pi R_n^2 (1 - \sin\theta_1) + \pi(R_1 + R_2)\sqrt{(R_2 - R_1)^2 + l_1^2} + \pi(R_2 + R_B)\sqrt{(R_B - R_2)^2 + l_2^2} + \pi R_B^2$$

Maximize

$$X_{cp}$$

subject to

$$V \geq 1$$

$$0.4 < R_n < 0.6$$

$$22 < \theta_1 < 27$$

Multiobjective Optimization

$$\theta_1 + 5 < \theta_2 < \theta_1 + 10$$

$$0.4 < l_1 < 0.8$$

$$0.4 < l_2 < 0.8$$

The multiobjective optimization problem is solved using the ε-constraints method. The objective function X_{cp} is taken as the function to be optimized and the other objective function A becomes a constraint. Since X_{cp} is to be maximized, it is put as $-X_{cp}$ in the SQP method which is written for minimization of a function. The area is varied from 6.4 m² to 16.3 m². This is put as a constraint in the optimization problem:

$$A < \varepsilon$$

where ε is varied from 6.4 to 16.3 in steps of 0.2 resulting in different constraint optimization problems. Each of these problems is then solved using the SQP method. The MATLAB code for the objective functions is given in *func.m* and *func1.m*. On executing the MATLAB code *sqp.m* the Pareto front is obtained and is given in Figure 7.10. The shape of the reentry body for the extreme cases of Pareto front is also shown in this figure. For achieving higher stability, the flare lengths are higher and for achieving lower surface area, flare lengths are smaller. Along the Pareto front, the maximum X_{cp} that can be obtained is 0.87 where A is 16.3 m². For area of 6.4 m², the X_{cp} achieved will be 0.58.

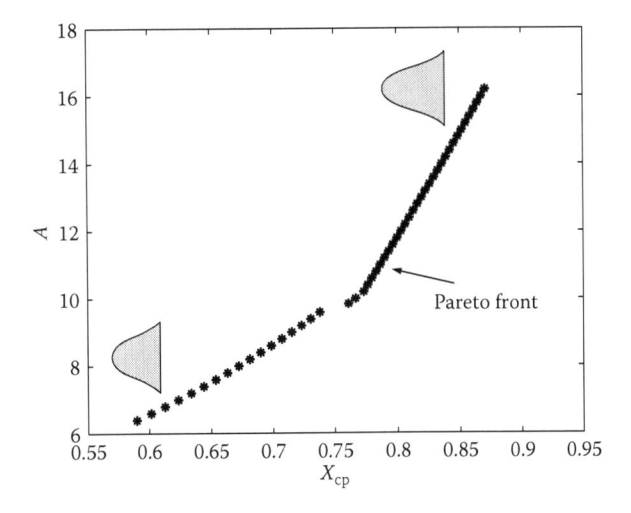

FIGURE 7.10
Pareto front of reentry test body.

Chapter Highlights

- In the multiobjective optimization problem, two or more objective functions are to be simultaneously optimized.
- The solution to a multiobjective problem results in a number of points in the objective function space referred to as Pareto optimal solutions.
- The Pareto front can be concave, partially convex/concave or discontinuous.
- The points along the Pareto optimal front are referred to as non-dominated solutions.
- In the weighted-sum approach, different objective functions are combined into a single objective function using user-supplied weights.
- The weighted sum approach, though simple to implement, has difficulty in locating the Pareto front of the nonconvex type.
- In the ε-constraint method, the decision-maker chooses one objective out of K objectives that needs to be minimized and the remaining objectives are put as constraints to some target values.
- Evolutionary algorithms (such as genetic algorithm or particle swarm optimization) are often used to locate the Pareto optimal front since they work simultaneously on a number of points.
- In goal programing, a target or goal is set for each objective function. Then the optimization problem is to minimize the deviation from the set targets.

Formulae Chart

Multiobjective problem:
Minimize/maximize

$$f_k(x) \quad k = 1, 2, \ldots, K$$

Weighted sum approach:
Minimize

$$\sum_{k=1}^{K} w_k f_k(x) \quad k = 1, 2, \ldots, K$$

$$\sum_{k=1}^{K} w_k = 1$$

Goal programming:
Minimize
$$f_3(x)$$
subject to
$$f_k(x) \le \varepsilon_k \quad k = 1\ldots K, \quad k \ne 3$$

Goal programming:
Minimize
$$\sum_{k=1}^{K} w_{1,k} p_k + w_{2,k} n_k \quad k = 1, 2, \ldots, K$$
subject to
$$f_k(x) + p_k - n_k = \tau_k$$

Utility function method:
Minimize
$$U(f_k(x))$$

Problems

1. Find the convex Pareto front for the multiobjective optimization problem (Parsopoulos and Vrahatis 2002):
   Minimize $f_1 = x^2$
   Minimize $f_2 = (x-2)^2$
   where $x \in [-10^5, 10^5]$.

2. Find the concave Pareto front for the multiobjective optimization problem (Zitzler et al. 2000).
   Minimize $f_1 = x_1$
   Minimize $f_2 = g\left(1 - \left(\dfrac{f_1}{g}\right)^2\right)$

where $g = 1 + \dfrac{9}{n-1}\sum_{i=2}^{n} x_i$ and $x_i \in [0,1]$. Take $n = 30$.

3. Find the convex Pareto front for the multiobjective optimization problem (Zitzler et al. 2000).

   Minimize $f_1 = x_1$

   Minimize $f_2 = g\left(1 - \sqrt{\dfrac{f_1}{g}}\right)$

   where $g = 1 + \dfrac{9}{n-1}\sum_{i=2}^{n} x_i$ and $x_i \in [0,1]$. Take $n = 30$.

4. Find the convex/concave Pareto front for the multiobjective optimization problem (Zitzler et al. 2000).

   Minimize $f_1 = x_1$

   Minimize $f_2 = g\left(1 - \left(\dfrac{f_1}{g}\right)^4 - \sqrt[4]{\dfrac{f_1}{g}}\right)$

   where $g = 1 + \dfrac{9}{n-1}\sum_{i=2}^{n} x_i$ and $x_i \in [0,1]$. Take $n = 30$.

5. Find the concave Pareto front for the multiobjective optimization problem (Deb 2002):

   Minimize $f_1 = 1 - \exp\left(-\sum_{i=1}^{n}\left(x_i - 1/\sqrt{n}\right)^2\right)$

   Minimize $f_2 = 1 - \exp\left(-\sum_{i=1}^{n}\left(x_i + 1/\sqrt{n}\right)^2\right)$

   where $x \in [-4, 4]$. Take $n = 2$.

6. Find the convex Pareto front for the constrained multiobjective optimization problem (Binh and Korn 1997):

   Minimize $f_1 = 4x_1^2 + 4x_2^2$
   Minimize $f_2 = (x_1 - 5)^2 + (x_2 - 5)^2$
   subject to $(x_1 - 5)^2 + x_2^2 \leq 25$
   $(x_1 - 8)^2 + (x_2 + 3)^2 \geq 7.7$

   where $x_1 \in [0, 5]$, $x_2 \in [0, 3]$.

8

Geometric Programming

8.1 Introduction

Geometric programming can be applied to optimization problems in which the objective function and constraints have a special structure. The conventional format of the objective function and constraints can be converted into the format required for geometric programming. Once the problem is written in the required format, it is much easier to solve the optimization problem using geometric programming than using nonlinear programming (NLP) methods described in previous chapters. The geometric programming technique proposed by Zener, Duffin, and Peterson can solve large-scale optimization problems with high reliability and efficiency. Geometric programming is applied to various disciplines such as inventory model (Abuo-El-Ata et al. 2003), structural optimization (Hajela 1986), communication systems (Chiang 2005), very-large-scale integration (VLSI) design (Chu and Wong 2001), and so on.

In geometric programming, the objective function is written in *posynomial* form:

$$f(x) = c x_1^{a_1} x_2^{a_2} x_3^{a_3} \ldots x_n^{a_n} \tag{8.1}$$

where c is a positive constant, the exponents a_i are real numbers, and x_i are the design variables that can take positive values. It is important to note that in polynomials, c can take both positive and negative values. For example,

$$f(x) = 5x_1^2 - 2x_2^2 - 3x_1 x_2$$

is a polynomial, while

$$f(x) = 2x_1^2 + 5x_2^2 + 4x_1^{-2} x_2^{-1}$$

is a posynomial.

If the objective function is obtained in polynomial form, then it has to be transformed into a posynomial before geometric programming techniques can be used. For example, the maximization function $f(x) = x_1^2 x_2$ can be

```
┌─────────────────────────┐
│ Geometric programming   │
└─────────────────────────┘
             │
             ▼
┌─────────────────────────┐
│ Unconstrained problems  │
└─────────────────────────┘
             │
             ▼
┌─────────────────────────┐
│     Dual problem        │
└─────────────────────────┘
             │
             ▼
┌─────────────────────────┐
│ Constrained optimization│
└─────────────────────────┘
             │
             ▼
┌─────────────────────────┐
│      Application        │
└─────────────────────────┘
```

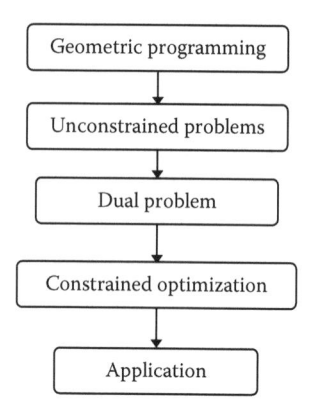

FIGURE 8.1
Road map of Chapter 8.

transformed into a posynomial form minimization function $f(x) = x_1^{-2} x_2^{-1}$. It is very interesting to note that in geometric programming, the objective function is evaluated first and then optimal design variables are obtained. That is, the optimized value of the objective function can be obtained without knowing the optimal value of design variables. Thus, the solution to geometric programming problems does not depend on the initial guess. In this chapter, both unconstrained and constrained optimization problems are solved using geometric programming. The chapter concludes with a practical application of geometric programming. The road map for this chapter is given in Figure 8.1.

8.2 Unconstrained Problem

Consider minimization of the function

$$f(x) = \sum_{j=1}^{N} U_j(x) = \sum_{j=1}^{N} \left(c_j \prod_{i=1}^{n} x_i^{a_{ij}} \right) \qquad (8.2)$$

where $x_i, c_j > 0$. The minimum or maximum of the function can be obtained using the first-order condition

$$\frac{\partial f}{\partial x_i} = 0 \qquad (8.3)$$

Geometric Programming

The solution of this equation leads to the orthogonality condition

$$\sum_{j=1}^{N} w_j^* a_{ij} = 0 \tag{8.4}$$

and the normality condition

$$\sum_{j=1}^{N} w_j^* = 1 \tag{8.5}$$

where

$$w_j^* = \frac{U_j(x^*)}{f^*} \tag{8.6}$$

The procedure for obtaining the optimal value of the objective function is to write the function as

$$f^* = \left(\frac{U_1^*}{w_1^*}\right)^{w_1^*} \left(\frac{U_2^*}{w_2^*}\right)^{w_2^*} \cdots \left(\frac{U_n^*}{w_n^*}\right)^{w_n^*} \tag{8.7}$$

where the values w_j^* are obtained by solving the orthogonality and normal equations.

The quantity $N - (n + 1)$ is called as the *degree of difficulty* in geometric programming, where n is the number of design variables and N is the number of posynomial terms in the objective function. If the degree of difficulty is zero, then the problem has a unique solution. If the degree of difficulty is positive (number of equations obtained through orthogonality and normality condition being less than the number of variables), some variables have to be expressed in terms of other variables to obtain the solution. In geometric programming, we do not have the negative degree of difficulty.

Using f^* and U_j^*, optimal values of the design variables can be evaluated using the expression

$$U_j^* = w_1^* f^* = c_j \prod_{i=1}^{n} (x_i^*)^{a_{ij}} \tag{8.8}$$

For a zero degree of difficulty problem, the above equation can be reduced to a set of simultaneous equations, which are easier to solve. This can be done by taking logarithms on both the sides, that is,

$$\ln \frac{w_1^* f^*}{c_j} = a_{1j} \ln x_1^* + a_{2j} \ln x_2^* + \ldots + a_{nj} \ln x_n^* \tag{8.9}$$

and then letting

$$k_i = \ln x_i^* \tag{8.10}$$

The design variables can be obtained as

$$x_i^* = e^{k_i} \tag{8.11}$$

The main advantage of using log summation terms is that the transformed function becomes a convex one.

The above procedure is explained by following examples that are of zero degree of difficulty.

Example 8.1

Solve the optimization problem using geometric programming:

Minimize

$$f(x) = 3x_1^{-1}x_2^{-3} + 4x_1^2 x_2 x_3^{-2} + 5x_1 x_2^4 x_3^{-1} + 6x_3$$

The degree of difficulty of this problem is 4 − (3 + 1) = 0. Also given are

$$\begin{bmatrix} a_{11} & a_{12} & a_{13} & a_{14} \\ a_{21} & a_{22} & a_{23} & a_{24} \\ a_{31} & a_{32} & a_{33} & a_{34} \\ a_{41} & a_{42} & a_{43} & a_{44} \end{bmatrix} = \begin{bmatrix} -1 & 2 & 1 & 0 \\ -3 & 1 & 4 & 0 \\ 0 & -2 & -1 & 1 \\ 1 & 1 & 1 & 1 \end{bmatrix} \text{ and } \begin{bmatrix} c_1 \\ c_2 \\ c_3 \\ c_4 \end{bmatrix} = \begin{bmatrix} 3 \\ 4 \\ 5 \\ 6 \end{bmatrix}$$

Writing the normality and orthogonality conditions in matrix form

$$\begin{bmatrix} -1 & 2 & 1 & 0 \\ -3 & 1 & 4 & 0 \\ 0 & -2 & -1 & 1 \\ 1 & 1 & 1 & 1 \end{bmatrix} \begin{bmatrix} w_1 \\ w_2 \\ w_3 \\ w_4 \end{bmatrix} = \begin{bmatrix} 0 \\ 0 \\ 0 \\ 1 \end{bmatrix}$$

Geometric Programming

Solving the above equation gives

$$\begin{bmatrix} w_1 \\ w_2 \\ w_3 \\ w_4 \end{bmatrix} = \begin{bmatrix} 7/20 \\ 1/20 \\ 1/4 \\ 7/20 \end{bmatrix}$$

Substituting these values in the following equation gives the optimal value of the objective function.

$$f^* = \left(\frac{3}{w_1}\right)^{w_1} \left(\frac{4}{w_2}\right)^{w_2} \left(\frac{5}{w_3}\right)^{w_3} \left(\frac{6}{w_4}\right)^{w_4} = \frac{3865}{256} = 15.1$$

The next step is to obtain the value of the design variables. The following equations are solved simultaneously to obtain k_i.

$$\begin{bmatrix} -1 & -3 & 0 \\ 2 & 1 & -2 \\ 1 & 4 & -1 \\ 0 & 0 & 1 \end{bmatrix} \begin{bmatrix} k_1 \\ k_2 \\ k_3 \end{bmatrix} = \begin{bmatrix} \ln \dfrac{15.1 \times \tfrac{7}{20}}{3} \\ \ln \dfrac{15.1 \times \tfrac{1}{20}}{4} \\ \ln \dfrac{15.1 \times \tfrac{1}{4}}{5} \\ \ln \dfrac{15.1 \times \tfrac{7}{20}}{6} \end{bmatrix}$$

These values of k_i are substituted in the equation $x_i^* = e^{k_i}$ to obtain the design variables as

$$\begin{bmatrix} x_1^* \\ x_2^* \\ x_3^* \end{bmatrix} = \begin{bmatrix} 0.4201 \\ 1.1407 \\ 0.8995 \end{bmatrix}$$

Example 8.2

The treatment of waste is accomplished by chemicals and dilation to meet effluent requirements (Stoecker 1971). The total cost is the sum of the treatment plant, pumping power requirements, and piping costs. This cost is given by the equation

228 Optimization: Algorithms and Applications

$$C = 150D + \frac{972{,}000Q^2}{D^5} + \frac{432}{Q}$$

where C is in dollars, D is in inches, and Q is in cubic feet per second. Find the minimum cost and best values of D and Q by geometric programming.

The degree of difficulty of this problem is $3 - (2+1) = 0$. Also given are

$$\begin{bmatrix} a_{11} & a_{12} & a_{13} \\ a_{21} & a_{22} & a_{23} \\ a_{31} & a_{32} & a_{33} \end{bmatrix} = \begin{bmatrix} 1 & -5 & 0 \\ 0 & 2 & -1 \\ 1 & 1 & 1 \end{bmatrix} \text{ and } \begin{bmatrix} c_1 \\ c_2 \\ c_3 \end{bmatrix} = \begin{bmatrix} 150 \\ 972{,}000 \\ 432 \end{bmatrix}$$

Writing the normality and orthogonality conditions in matrix form

$$\begin{bmatrix} 1 & -5 & 0 \\ 0 & 2 & -1 \\ 1 & 1 & 1 \end{bmatrix} \begin{bmatrix} w_1 \\ w_2 \\ w_3 \end{bmatrix} = \begin{bmatrix} 0 \\ 0 \\ 1 \end{bmatrix}$$

Solving the above equation gives

$$\begin{bmatrix} w_1 \\ w_2 \\ w_3 \end{bmatrix} = \begin{bmatrix} 5/8 \\ 1/8 \\ 1/4 \end{bmatrix}$$

Substituting these values in the following expression gives the optimal value of the objective function.

$$C^* = \left(\frac{150}{w_1}\right)^{w_1} \left(\frac{972{,}000}{w_2}\right)^{w_2} \left(\frac{432}{w_3}\right)^{w_3} = 1440$$

The next step is to obtain the value of the design variables. The following equations are solved simultaneously to obtain k_i.

$$\begin{bmatrix} 1 & 0 \\ -5 & 2 \\ 0 & -1 \end{bmatrix} \begin{bmatrix} k_1 \\ k_2 \end{bmatrix} = \begin{bmatrix} \ln \dfrac{1440 \times \frac{5}{8}}{150} \\ \ln \dfrac{1440 \times \frac{1}{8}}{972{,}000} \\ \ln \dfrac{1440 \times \frac{1}{4}}{432} \end{bmatrix}$$

Geometric Programming

This values of k_i are substituted in the equation $x_i^* = e^{k_i}$ to obtain the design variables as

$$\begin{bmatrix} D^* \\ Q^* \end{bmatrix} = \begin{bmatrix} 6 \\ 1.2 \end{bmatrix}$$

8.3 Dual Problem

Similar to linear programming, there is a dual problem in geometric programming. The minimization problem discussed in the previous section in this chapter is referred to as the *primal problem*. The corresponding maximum of the primal problem is referred to as the *dual problem*. The dual problem structure is helpful in solving geometric programming problems that have a degree of difficulty greater than zero. In the primal problem, the minimization of the function

$$f(x) = \sum_{j=1}^{N} \left(c_j \prod_{i=1}^{n} x_i^{a_{ij}} \right) \tag{8.12}$$

is replaced by maximization of the function

$$F(w) = \prod_{j=1}^{N} \left(\frac{c_j}{w_j} \right)^{w_j} \tag{8.13}$$

in the dual problem. Because it is easy to solve an objective function that has summation terms rather than product terms, the logarithm is taken on both sides of Equation 8.13.

$$\ln F(w) = -\sum_{j=1}^{N} w_j \ln\left(\frac{w_j}{c_j}\right) \tag{8.14}$$

This function is maximized subject to normality and orthogonality conditions, mentioned in the previous section. It is significant to note that the solution obtained from the dual problem (maximization) is the same as the solution of the primal problem (minimization).

Example 8.3

Solve the optimization problem:

Minimize
$$f(x) = x_1^2 + 2x_2^2 + 3x_1^{-1}x_2^{-1} + 2x_1 x_2$$

The degree of difficulty of this problem is 4 − (2 + 1) = 1. Writing the minimization problem in dual form as

Maximize
$$f(w) = \left(\frac{1}{w_1}\right)^{w_1} \left(\frac{2}{w_2}\right)^{w_2} \left(\frac{3}{w_3}\right)^{w_3} \left(\frac{2}{w_4}\right)^{w_4}$$

subject to orthogonality and normality conditions

$$\begin{bmatrix} 2 & 0 & -1 & 1 \\ 0 & 2 & -1 & 1 \\ 1 & 1 & 1 & 1 \end{bmatrix} \begin{bmatrix} w_1 \\ w_2 \\ w_3 \\ w_4 \end{bmatrix} = \begin{bmatrix} 0 \\ 0 \\ 0 \\ 1 \end{bmatrix}$$

In the above matrix notation, four unknowns are to be determined from three equations. One can write the three variables in the form of a fourth variable as

$$w_1 = \frac{1 - 2w_4}{4}$$

$$w_2 = \frac{1 - 2w_4}{4}$$

$$w_3 = \frac{1}{2}$$

Substituting these values in the dual objective function and taking the logarithm on both sides:

$$\ln F(w_4) = -\left[\left(\frac{1-2w_4}{4}\right)\ln\left(\frac{1-2w_4}{4}\right) + \left(\frac{1-2w_4}{4}\right)\ln\left(\frac{1-2w_4}{8}\right) + \frac{1}{3}\ln\frac{1}{6} + w_4 \ln\frac{w_4}{4}\right]$$

Geometric Programming

Differentiating the above equation with respect to w_4:

$$\ln w_4 - \ln 2 - \frac{1}{2}\left[\ln(1-2w_4)^2 - \ln 32\right] = 0$$

Solving the above equation gives $w_4 = 0.20711$. Substituting the value of w_4 gives

$$w_1 = w_2 = \frac{1-2w_4}{4} = 0.146445$$

This optimum value of the objective function can now be obtained as

$$f^* = \left(\frac{1}{w_1}\right)^{w_1}\left(\frac{2}{w_2}\right)^{w_2}\left(\frac{3}{w_3}\right)^{w_3}\left(\frac{2}{w_4}\right)^{w_4} = 7.6119$$

The next step is to determine the design variables x_1 and x_2. Now,

$$U_1^* = w_1 f^* = 0.146445 \times 7.6119 = x_1^2$$

$$U_2^* = w_2 f^* = 0.146445 \times 7.6119 = 2x_2^2$$

This gives

$$x_1^* = 1.0558$$

$$x_2^* = 0.7466$$

8.4 Constrained Optimization

In the constrained optimization problems, both the objective function and the constraints are given as posynomials. Consider minimization of the function

$$f(x) = g_0(x) = \sum_{j=1}^{N}\left(c_j \prod_{i=1}^{n} x_i^{a_{ij}}\right) \tag{8.15}$$

subject to k constraints

$$g_k(x) = \sum_{j=1}^{N}\left(c_{kj}\prod_{i=1}^{n} x_i^{a_{kij}}\right) \leq 1 \qquad (8.16)$$

where x_i, c_j, $c'_j > 0$. This the primal problem in standard form and its dual (maximization function) is given by

$$F(w) = \prod_{k=0}^{m}\prod_{j=1}^{N}\left(\frac{c_{kj}}{w_{kj}}\sum_{l=1}^{N} w_{kl}\right)^{w_{kl}} \qquad (8.17)$$

subject to orthogonality and normality conditions

$$\sum_{k=0}^{m}\sum_{j=1}^{N} a_{kij} w_{kl} = 0 \qquad (8.18)$$

$$\sum_{j=1}^{N} w_{kj} = 0, \quad k = 0 \qquad (8.19)$$

The problem is then solved in a manner similar to the unconstrained optimization problem. If the right-hand side of the constraints are given as posynomials such as

$$g_k(x) \leq v(x) \qquad (8.20)$$

the same can be transformed into the standard form as

$$\frac{g_k(x)}{v(x)} \leq 1 \qquad (8.21)$$

Example 8.4

Solve the optimization problem:

Minimize

$$f(x) = g_0(x) = 30 x_1^{-1} x_2^{-1} x_3^{-1} + 30 x_2 x_3$$

Geometric Programming

subject to

$$g_1(x) = 0.5x_1x_3 + 0.25x_1x_2 \leq 1$$

The degree of difficulty of this problem is 3 − (2 + 1) = 0. Writing the minimization problem in dual form as

Maximize

$$f(w) = \left(\frac{30}{w_1}\right)^{w_1} \left(\frac{30}{w_2}\right)^{w_2} \left(\frac{0.5}{w_3}\right)^{w_3} \left(\frac{0.25}{w_4}\right)^{w_4} (w_3 + w_4)^{w_3+w_4}$$

subject to orthogonality and normality conditions

$$\begin{bmatrix} -1 & 0 & 1 & 1 \\ -1 & 1 & 0 & 1 \\ -1 & 1 & 1 & 0 \\ 1 & 1 & 0 & 0 \end{bmatrix} \begin{bmatrix} w_1 \\ w_2 \\ w_3 \\ w_4 \end{bmatrix} = \begin{bmatrix} 0 \\ 0 \\ 0 \\ 1 \end{bmatrix}$$

Solving the above equation gives

$$\begin{bmatrix} -1 & 0 & 1 & 1 \\ -1 & 1 & 0 & 1 \\ -1 & 1 & 1 & 0 \\ 1 & 1 & 0 & 0 \end{bmatrix} \begin{bmatrix} w_1 \\ w_2 \\ w_3 \\ w_4 \end{bmatrix} = \begin{bmatrix} -1 & 0 & 1 & 1 \\ -1 & 1 & 0 & 1 \\ -1 & 1 & 1 & 0 \\ 1 & 1 & 0 & 0 \end{bmatrix}^{-1} \begin{bmatrix} 0 \\ 0 \\ 0 \\ 1 \end{bmatrix} = \begin{bmatrix} 2/3 \\ 1/3 \\ 1/3 \\ 1/3 \end{bmatrix}$$

This optimum value of the objective function can now be obtained as

$$f^* = f(w) = \left(\frac{30}{w_1}\right)^{w_1} \left(\frac{30}{w_2}\right)^{w_2} \left(\frac{0.5}{w_3}\right)^{w_3} \left(\frac{0.25}{w_4}\right)^{w_4} (w_3 + w_4)^{w_3+w_4} = 45$$

The next step is to determine the design variables x_1 and x_2. Now,

$$U_1^* = w_1 f^* = 30 = 30x_1^{-1}x_2^{-1}x_3^{-1}$$

$$U_2^* = w_2 f^* = 15 = 30x_2 x_3$$

This gives

$$x_1^* = 2$$

$$x_2^* = 1$$

$$x_3^* = \frac{1}{2}$$

Example 8.5

Solve the optimization problem (Dembo 1976):

Minimize

$$f(x) = g_0(x) = x_1 x_2 + x_1^{-1} x_2^{-1}$$

subject to

$$g_1(x) = 0.25 x_1^{1/2} + x_2 \leq 1$$

The degree of difficulty of this problem is 4 − (2 + 1) = 1. Writing the minimization problem in dual form as

Maximize

$$f(w) = \left(\frac{1}{w_1}\right)^{w_1} \left(\frac{1}{w_2}\right)^{w_2} \left(\frac{0.25}{w_3}\right)^{w_3} \left(\frac{1}{w_4}\right)^{w_4} (w_3 + w_4)^{w_3 + w_4}$$

subject to orthogonality and normality conditions

$$\begin{bmatrix} 1 & -1 & 0.5 & 0 \\ 1 & -1 & 0 & 1 \\ 1 & 1 & 0 & 0 \end{bmatrix} \begin{bmatrix} w_1 \\ w_2 \\ w_3 \\ w_4 \end{bmatrix} = \begin{bmatrix} 0 \\ 0 \\ 0 \\ 1 \end{bmatrix}$$

In the above matrix notation, four unknowns are to be determined from three equations. One can write the three variables in the form of a fourth variable as

$$w_1 = \frac{1 - w_4}{2}$$

$$w_2 = \frac{1 + w_4}{2}$$

$$w_3 = 2 w_4$$

Geometric Programming

Substituting these values in the dual function

$$f(w) = \left(\frac{2}{1-w_4}\right)^{\frac{1-w_4}{2}} \left(\frac{2}{1+w_4}\right)^{\frac{1+w_4}{2}} \left(\frac{0.25}{2w_4}\right)^{w_4} \left(\frac{1}{w_4}\right)^{w_4} (2w_4+w_4)^{2w_4+w_4}$$

Taking the logarithm on both sides of the above equation and then differentiating it with respect to w_4 and equating it to zero one gets $w_4 = 0$. Therefore,

$$w_1 = \frac{1-w_4}{2} = \frac{1}{2}$$

$$w_2 = \frac{1+w_4}{2} = \frac{1}{2}$$

$$w_3 = 2w_4 = 0$$

This optimum value of the objective function can now be obtained as

$$f^* = f(w) = \left(\frac{1}{w_1}\right)^{w_1} \left(\frac{1}{w_2}\right)^{w_2} \left(\frac{0.25}{w_3}\right)^{w_3} \left(\frac{1}{w_4}\right)^{w_4} (w_3+w_4)^{w_3+w_4} = 2$$

Now,

$$U_1^* = w_1 f^* = 1 = x_1 x_2$$

$$U_2^* = w_2 f^* = 1 = x_1^{-1} x_2^{-1}$$

The above equation is satisfied for a number of combinations of x_1 and x_2.

8.5 Application

A two-bar structure (Figure 8.2) is to be designed so as to minimize its weight (Dey and Roy 2013) while tolerating certain maximum tensile and compressive stresses. The optimization problem is written as

Minimize

$$W = \rho\left(A_1\sqrt{x_B^2 + (l-y_B)^2} + A_2\sqrt{x_B^2 + y_B^2}\right)$$

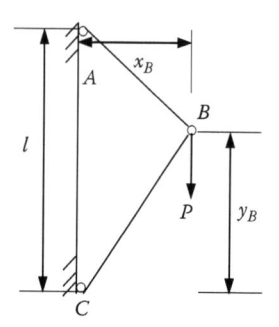

FIGURE 8.2
Two-bar truss.

subject to

$$\frac{P\sqrt{x_B^2 + (l-y_B)^2}}{lA_1} \leq \sigma_{t,max}$$

$$\frac{P\sqrt{x_B^2 + y_B^2}}{lA_2} \leq \sigma_{c,max}$$

$$0.5 \leq y_B \leq 1.5$$

$$A_1, A_2 \geq 0$$

where

Load = $P = 10^5$ N
Density = $\rho = 77$ kN/m^3
Length = $l = 2$ m
Width = $x_B = 1$ m
Maximum tensile stress = $\sigma_{t,max} = 150$ Mpa
Maximum compressive stress = $\sigma_{c,max} = 100$ Mpa

The design variables are A_1, A_2, and y_B.
The nonlinear optimization problem can be transformed into geometric programming problem with the following substitutions:

$$x_1 = A_1$$

$$x_2 = A_2$$

$$\sqrt{1+(2-y_B)^2} \leq x_3$$

$$\sqrt{1+y_B^2} \leq x_4$$

$$x_5 = y_B$$

$$x_6 = 1 + 4x_5 x_3^{-2}$$

The geometric programming problem becomes

Minimize

$$W = 77(x_1 x_3 + x_2 x_4)$$

subject to

$$\frac{1}{3} x_3 x_1^{-1} \leq 1$$

$$\frac{1}{2} x_4 x_2^{-1} \leq 1$$

$$x_4^{-2} + x_4^{-2} x_5^2 \leq 1$$

$$5 x_3^{-2} x_6^{-1} + x_3^{-2} x_3^2 x_6^{-1} \leq 1$$

$$\frac{1}{2} x_5^{-1} \leq 1$$

$$\frac{2}{3} x_5 \leq 1$$

$$x_6^{-1} + 4 x_3^{-2} x_5 x_6^{-1} = 1$$

The degree of difficulty is 12 − (6 + 1) = 5. The problem can be converted into a dual problem and then solved. The solution is given by

$$x_1^* = 0.52068$$

$$x_2^* = 0.640312$$

$$x_3^* = 1.56205$$

$$x_4^* = 1.280625$$

$$x_5^* = 0.8$$

$$x_6^* = 2.31147$$

$$W = 125.7667$$

Chapter Highlights

- In geometric programming, the objective function and constraints are written in posynomial form.
- In geometric programming, the objective function is determined first and then design variables are evaluated. The initial guess of the variables has no role in geometric programming.
- The degree of difficulty refers to the number of unknowns minus the number of equations (orthogonality and normal conditions).
- The dual problem structure is helpful in solving geometric programming problems that have a degree of difficulty greater than zero.
- The solution obtained from the dual problem (maximization) is the same as the solution of the primal problem (minimization).

Formulae Chart

Posynomial:

$$f(x) = c x_1^{a_1} x_2^{a_2} x_3^{a_3} \ldots x_n^{a_n}$$

Geometric Programming

Unconstrained minimization:
Minimize

$$f(x) = \sum_{j=1}^{N} U_j(x) = \sum_{j=1}^{N} \left(c_j \prod_{i=1}^{n} x_i^{a_{ij}} \right)$$

Normality and orthogonality conditions:

$$\sum_{j=1}^{N} w_j^* a_{ij} = 0$$

$$\sum_{j=1}^{N} w_j^* = 1$$

Optimal function value:

$$f^* = \left(\frac{U_1^*}{w_1^*} \right)^{w_1^*} \left(\frac{U_2^*}{w_2^*} \right)^{w_2^*} \cdots \left(\frac{U_n^*}{w_n^*} \right)^{w_n^*}$$

where

$$w_j^* = \frac{U_j(x^*)}{f^*}$$

Dual problem (unconstrained):

$$F(w) = \prod_{j=1}^{N} \left(\frac{c_j}{w_j} \right)^{w_j}$$

Dual problem (constrained):

$$F(w) = \prod_{k=0}^{m} \prod_{j=1}^{N} \left(\frac{c_{kj}}{w_{kj}} \sum_{l=1}^{N} w_{kl} \right)^{w_{kl}}$$

Problems

1. Minimize (Ojha and Biswal 2010)

$$10x_1^{-3}x_2^2x_3^{-1} + 40x_1x_2 + 40x_1x_2x_3$$

subject to

$$2x_1^{-2}x_2^{-2} + x_2^{-5}x_3^{-1} \leq 1$$

$$x_1, x_2, x_3 \geq 0$$

2. Minimize (Ojha and Biswal 2010)

$$x_1^{-4}x_2^{-1}x_3x_4^{-1} + 3x_1^{-2}x_2^{-3}x_3^{-2}$$

subject to

$$2x_1^3 x_3 + x_1^{-1}x_3^{-1} \leq 3$$

$$x_2^{-1}x_3^{-1}x_4^{-2} + 3x_1^2 x_2 x_4 \leq 1$$

$$x_1, x_2, x_3, x_4 \geq 0$$

3. Minimize (Rao 2009)

$$x_1$$

subject to

$$-4x_1^2 + 4x_2 \leq 1$$

$$x_1 + x_2 \geq 1$$

$$x_1, x_2 \geq 0$$

4. In a certain reservoir pump installation (Rao 2009), the first cost of the pipe is given by $(100D + 50D^2)$, where D is the diameter of the pipe in centimeters. The cost of the reservoir decreases with an increase in the quantity of fluid handled and is given by $20/Q$, where Q is the rate at which the fluid is handled (m^3/s). The pumping cost is given by $(300Q^2/D^5)$. Find the optimal size of the pipe and the amount of fluid handled for minimum overall cost.

Geometric Programming

5. A hydraulic power system (Stoecker 1971) must provide 300 W of power, where the power is the product of volume flow rate Q m³/s and the pressure build up Δp Pa. The cost of the hydraulic pump is a function of both the flow rate and pressure buildup.

$$\text{Cost} = 1200 Q^{0.4} \sqrt{10 + (\Delta p \times 10^{-4})} \text{ dollars}$$

Convert to a single-variable unconstrained problem and use geometric programming to determine the minimum cost of the pump and the optimum values of Q and Δp.

6. A newly harvested grain system (Stoecker 1971) has a high moisture content and must be dried to prevent spoilage. The drying can be achieved by blowing it with air. The seasonal operating cost in dollars per square meter of the grain bed for such a dryer consists of the cost of heating of the air.

$$\text{Heating cost} = 0.002 Q \Delta t$$

and

$$\text{Blower cost} = 2.6 \times 10^{-9} Q^3$$

where Q is air quantity delivered through the bed, m³/m² of bed area and Δt is the rise in temperature through heater in °C. The values of Q and Δt also influence the time required for adequate drying of the grain according to the equation

$$\text{Drying time} = \frac{80 \times 10^6}{Q^2 \Delta t} \text{ days}$$

Using the geometric programming method, compute the minimum operating cost and optimum values of Q and Δt that will achieve adequate drying in 60 days.

7. The torque T (Nm) developed by an internal combustion engine is represented by

$$T = 23.6 \omega^{0.7} - 3.17 \omega$$

where ω is the rotational speed in rad/s. Determine the maximum power of this engine and the ω at which this occurs (Stoecker 1971).

9
Multidisciplinary Design Optimization

9.1 Introduction

In Chapter 7 on multiobjective optimization, a number of objective functions were simultaneously handled along with constraints for a given discipline. In multidisciplinary design optimization (MDO), two or more disciplines are simultaneously optimized. For example, in rocket design, the disciplines could be structures, aerodynamics, propulsion, control, and mission. Each of these disciplines can have separate optimal requirements. For example, the propulsion discipline can have a constraint on chamber pressure, the structural discipline can have constraints on stresses on the members, and the mission can have trajectory constraints such as on dynamic pressure and heat loads. Further, in MDO there are interactions among the disciplines. For example, the variable dynamic pressure in the trajectory discipline has an effect on load computation in the structural discipline. The idea of MDO is to optimize the design in a global sense. This has the following advantages:

- The time required in the design cycle can be significantly reduced. For example, a given aerodynamic shape will give a certain higher load distribution of certain structural members that may require reworking of these members, which in turn can change the aerodynamic shape. The cycle is iterative and time consuming if the individual disciplines are optimized sequentially. In addition, sequential optimization of disciplines may lead to a suboptimal solution for the whole system. For example, lift distribution along the wing span changes if the aerodynamic and structure disciplines are optimized together instead of considering the aerodynamic discipline alone (Figure 9.1).
- Disciplines with conflicting objectives can be resolved. For example, to minimize wave drag on a supersonic aircraft, optimizing the aerodynamics discipline alone will result in thin wings. This, on the other hand, could result in aero-elastic problems (structure discipline).

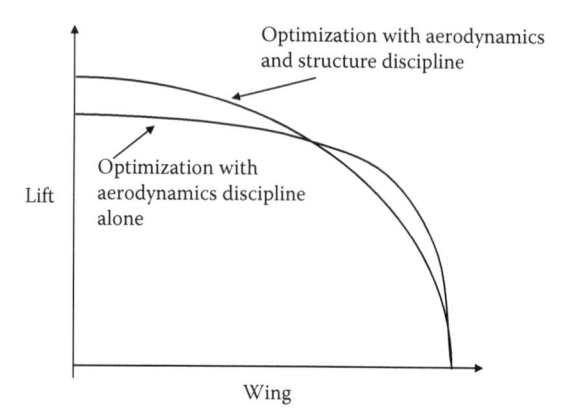

FIGURE 9.1
Optimization of a single versus two disciplines.

Here one may argue that when optimizing even a single discipline is highly time consuming, how can so many disciplines be optimized together? For example, the aerodynamic discipline has to generate lift and drag coefficients for a number of configurations through computational fluid dynamics (CFD) that requires a large computational time. Similarly, the structural discipline has to make finite element models and compute stresses on different members, which again are computationally intensive. In MDO, this problem can be alleviated by considering simplified mathematical models for each discipline. One such technique is response surface methodology (RSM). In RSM, one generates a response surface to variation in design variables by carrying out a limited number of tests. For example, aerodynamic response surface models can be generated with a limited number of CFD or wind tunnel tests carried out at certain Mach numbers and certain angles of attack only. The response surface model can then generate aerodynamic coefficients at any Mach number and angle of attack.

MDO is often used for aerospace problems (Balesdent et al. 2010; Manokaran et al. 2009; Xiaoqian et al. 2006; Yushin et al. 2006) as they are highly complex in nature owing to the presence of a large number of constraints in various disciplines, and even if optimization results in increasing only a few kilograms of payload, revenue can be increased by a few thousand dollars. However, MDO can also be applied to other areas (Geyer 2009; He and McPhee 2005) such as automobiles, where one can simultaneously optimize different disciplines such as body, engine, hydraulics, and so on. The road map of this chapter is shown in Figure 9.2. Through MDO architecture, the MDO problem is transformed into a series of optimization problems. A number of such architectures are discussed in this section along with their advantages and disadvantages. A very brief introduction is given about MDO framework that provides a platform for comparing different architectures. As MDO requires working with a number of

Multidisciplinary Design Optimization 245

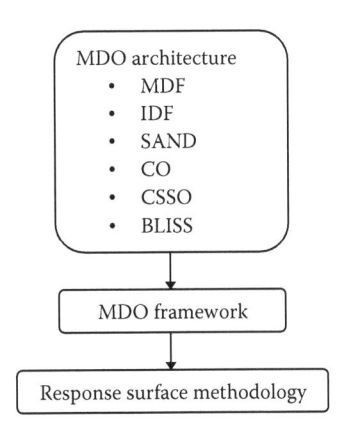

FIGURE 9.2
Road map of Chapter 9.

disciplines simultaneously, simplified but accurate models are required for each discipline. This is done through response surface methodology, which is explained in the last section.

9.2 MDO Architecture

Through MDO architecture, the multidisciplinary problem is transformed into a series of standard optimization problems that can be solved through either a gradient-based solver (Fletcher 1981) such as sequential quadratic programming (SQP) or through a non–gradient-based solver such as genetic algorithm (GA; Goldberg 1989) or particle swarm optimization (PSO). In the literature, different architectures are reported that transform the problems differently. It is quite obvious that each MDO architecture has certain advantages and disadvantages. The efficiency of different architectures can be measured in terms of number of disciplines or number of global/local variables. Some well-known MDO architectures are multidisciplinary feasible (MDF), individual discipline feasible (IDF), simultaneous analysis and design (SAND), collaborative optimization (CO), current subspace optimization (CSSO), and bilevel integrated system synthesis (BLISS). Excellent details of these architectures are mentioned in Martins and Lambe (2013) and Tedford and Martins (2006). Important highlights of these architectures are presented in this section. Let us define an MDO problem with two disciplines with x_i as the local variables, z_i as the global variables, and y_i as the coupling variables. See Figure 9.3 for more clarity. Each discipline solves the governing equations and provides feasible states and outputs in the form of coupling variables to the other discipline. The variables that belong to

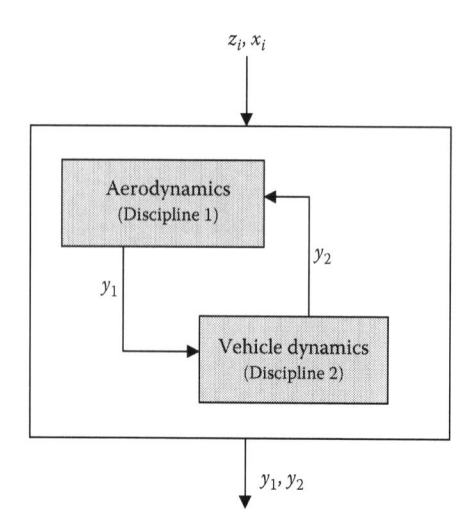

FIGURE 9.3
Multidisciplinary design analysis (MDA).

a single discipline are called local variables. The variables that affect more than one discipline are called global variables.

For example, discipline 1 could be aerodynamics, which feeds aerodynamic forces to discipline 2, where vehicle dynamics is simulated. Altitude and velocity information from discipline 2 is then fed to discipline 1 for computing aerodynamic forces. Here, angle of attack and bank angle are the global variables. The set of discipline analyses is repeated until a change in values of coupling variables becomes negligible. Mathematically, this can be stated as

$$y_i^{n+1} = y_i^n \qquad (9.1)$$

where y_i^n represents the value of ith discipline coupling variables after n iterations. The optimization problem can be stated as

Minimize

$$f(z, x) \qquad (9.2)$$

subject to

$$g_j(z, x) \leq 0 \quad j = 1, 2, \ldots, m < n \qquad (9.3)$$

$$h_k(z, x) = 0 \quad k = 1, 2, \ldots, r < n \qquad (9.4)$$

$$y_i^{n+1} - y_i^n = 0 \qquad (9.5)$$

9.2.1 Multidisciplinary Design Feasible

In this architecture, the design variables are iterated until coupling variables become consistent. The objective function and constraints are then computed and supplied to the optimizer (Figure 9.4). The procedure is said to converge if the coupling variables remain constant over successive iterations.

The main advantage of MDF is that it ensures feasible solution at each iterative step. By this we mean that constraints are satisfied with every iteration, but the optimum solution is not yet reached. The disadvantage of MDF is that it cannot be parallelized and computation of gradients for the coupled system is difficult. The MDF problem can be mathematically stated as

Minimize
$$f(z, x, y(x, z)) \tag{9.6}$$

subject to
$$g(z, x, y(x, z)) \leq 0 \tag{9.7}$$

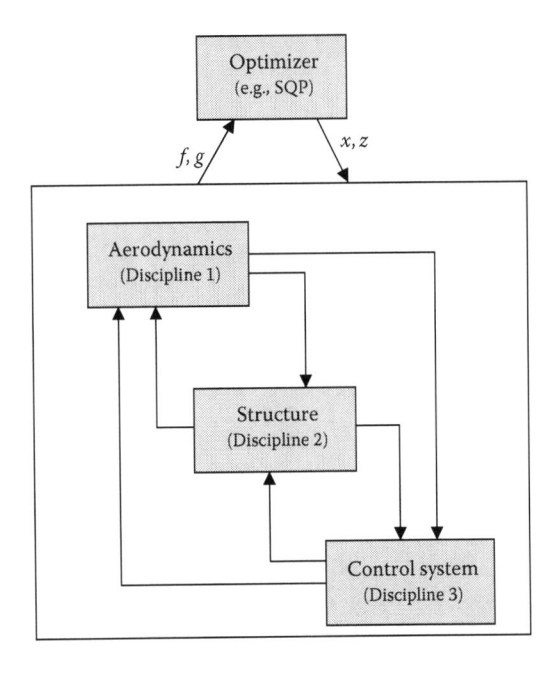

$$y_i^{n+1} - y_i^n = 0 \tag{9.8}$$

FIGURE 9.4
Multidisciplinary feasible (MDF) analysis.

9.2.2 Individual Discipline Feasible

In IDF architecture, discipline feasibility is ensured whereas a multidisciplinary feasible solution may not be present. The advantage of the IDF is that different disciplines can be evaluated in parallel. Further discipline computations are fewer as compared to MDF, and this can be a significant advantage because discipline evaluations are often time consuming. In IDF architecture, coupling variables are handled by the optimizer (Figure 9.5), which in turn provides design and coupling variables to different disciplines. This architecture is recommended for those MDO problems that have a small number of coupling variables.

The IDF problem can be mathematically stated as

Minimize

$$f(z, x, y^t) \tag{9.9}$$

with respect to z, x, y^t
subject to

$$g(z, x, y(x, y^t, z)) \leq 0 \tag{9.10}$$

$$y_i^t - y_i(x, y^t, z) = 0 \tag{9.11}$$

where y^t are the estimates of coupling variables by the optimizer, y_i are the coupling variable output of the discipline i, and y_i^t are the estimates of the nonlocal coupling variables. The last constraint ensures that at the optimum, the coupling variables computed by the discipline and the optimizer are matched.

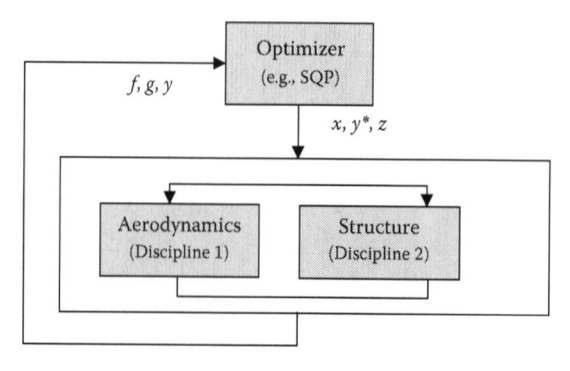

FIGURE 9.5
Individual discipline feasible (IDF) architecture.

9.2.3 Simultaneous Analysis and Design

In SAND architecture, the optimizer is given freedom to design (optimize) the system and solve the governing equations simultaneously. The residuals obtained from the discipline analyses are treated as equality constraints in the optimization problem. As compared to MDF and IDF, SAND architecture (Figure 9.6) does not maintain even discipline feasibility at different iterations.

The SAND problem can be mathematically stated as

Minimize

$$f(z, x, y(x, z, u)) \tag{9.12}$$

with respect to z, x, u

subject to

$$g(z, x, y(x, u, z)) \leq 0 \tag{9.13}$$

$$R(z, x, y(x, z, u), u) \leq 0 \tag{9.14}$$

where u is the state variable of the discipline and R represents the residuals of the discipline equations.

Example 9.1

Consider the MDO problem with two disciplines.

Minimize

$$x_1^2 + x_2 + y_1 + e^{-y_2}$$

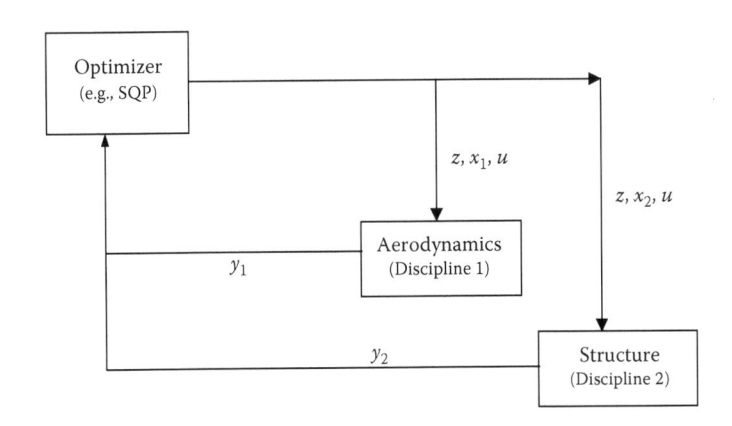

FIGURE 9.6
Simultaneous analysis and design (SAND) architecture.

with respect to x_1, x_2, z_1
subject to

$$1 - y_1/3.16 \leq 0$$

$$y_2/24 - 1 \leq 0$$

$$0 \leq x_1 \leq 10$$

$$0 \leq x_2 \leq 10$$

$$-10 \leq z_1 \leq 10$$

Discipline 1

$$y_1 = z_1^2 + x_1 + x_2 - 0.2y_2$$

Discipline 2

$$y_2 = \sqrt{y_1} + z_1 + x_2$$

Solve the MDO problem (Tedford and Martins 2010) using SAND architecture.

The MDO problem is solved using the SQP method. The MATLAB® codes from Chapter 6 are suitably modified to solve this problem. The starting value of the design variables is taken as $(x_1, x_2, z_1, y_1, y_2) = (1, 2, 5, 1, 0)$. The design variables y_1, y_2 are to be matched with the discipline outputs. The optimizer carries out this task by defining two additional equality constraints. On executing the *sqp.m* code, the following output is obtained.

No.	x-vector					f(x)	|Cons.|
1.0000	13.0989	0.000	5.3084	36.3832	24.0000	207.9656	12.6601
2.0000	0.0000	0	4.6986	19.8331	9.3585	19.8332	0.4254
3.0000	0.0000	0	3.7071	11.3189	7.2047	11.3197	0.9920
4.0000	0.0000	0.0000	3.2572	9.1468	6.2987	9.1487	0.2032
5.0000	0.0000	0.0000	2.8631	6.9373	5.5221	6.9413	0.1573
6.0000	0.0000	0.0000	2.5514	5.4328	4.8997	5.4403	0.0987

```
7.0000   0.0000   0.0000   2.2985   4.3400   4.3949   4.3523   0.0653

8.0000   0.0000   0.0000   2.0899   3.5283   3.9783   3.5471   0.0447

9.0000   0.0000   0.0000   1.9809   3.1600   3.7613   3.1833   0.0122

10.0000  0.0000   0.0000   1.9776   3.1600   3.7553   3.1834   0.0000

11.0000  0.0000   0.0000   1.9776   3.1600   3.7553   3.1834   0.0000

12.0000  0.0000   0.0000   1.9776   3.1600   3.7553   3.1834   0.0000
```

The optimal values of the design variables are $\left(x_1^*, x_2^*, z_1^*\right) = (0, 0, 1.9776)$. The converged values of the coupling variables are $(y_1, y_2) = (3.16, 3.7553)$. The minimum value of the objective function is 3.1834.

9.2.4 Collaborative Optimization

In CO architecture, optimization is carried out at discipline and system levels. Thus, discipline feasibility is guaranteed throughout the optimization process. The MDO problem is decomposed into a number of subproblems corresponding to each discipline (Figure 9.7). The discipline optimization is carried out in a conventional way in which local constraints to that discipline are satisfied. The system level is optimized with respect to global, coupling, and local variables. The constraints at the system level consist of global constraints as well as compatibility constraints of the discipline. The discipline

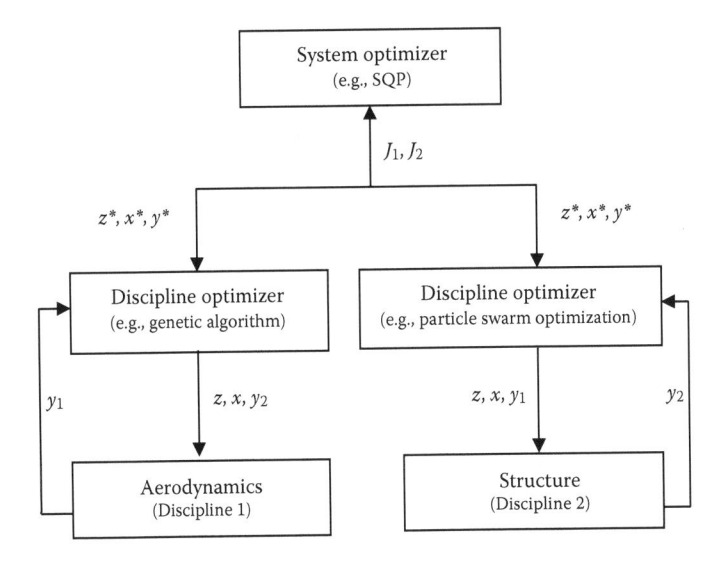

FIGURE 9.7
Collaborative optimization (CO) architecture.

optimizer, on the other hand, reduces the discrepancy between the system level variables and the discipline variables. One significant advantage of CO architecture is that each discipline can be optimized in parallel. Further, different optimization techniques (gradient- or non–gradient-based) can be used by different disciplines. The disadvantage of CO architecture is that the dimensionality of the system-level optimization problem increases significantly with increase in coupling variables.

The CO architecture at system level can be mathematically stated as

Minimize

$$f(z, y, x_{obj}) \tag{9.15}$$

with respect to z, y, x_{obj}
subject to

$$J\left(z_i, z^*, x_{obj}, x_{obj}^*, y_i, y^*\right) = 0 \tag{9.16}$$

where x_{obj} is the local variable affecting the objective function.

9.2.5 Concurrent Subspace Optimization

So far in MDO architecture, we have assumed discipline computations are easy to evaluate. This is far from true. As explained in the introduction, aerodynamic analysis through CFD and structural analysis through the finite-element method are time consuming. In this particular architecture, the problem of extensive computing is alleviated by making simplified mathematical models for each discipline. One such technique is RSM. In RSM, one generates a response surface to variation in design variables by carrying out a limited number of tests.

In CSSO architecture, RSM is used to provide information for the discipline subspace optimization. The response surface is constructed by carrying out a discipline analysis at a few design points. Thus, response surfaces provides state variables of each discipline for the given design variables. The CSSO architecture is depicted in Figure 9.8.

9.2.6 Bilevel Integrated System Synthesis

BLISS architecture is designed to suit a parallel computing environment. It is a bilevel architecture where each discipline optimization is fully autonomous and coordination is done at the system level to ensure multidisciplinary feasibility. In this architecture, discipline response levels are used

Multidisciplinary Design Optimization 253

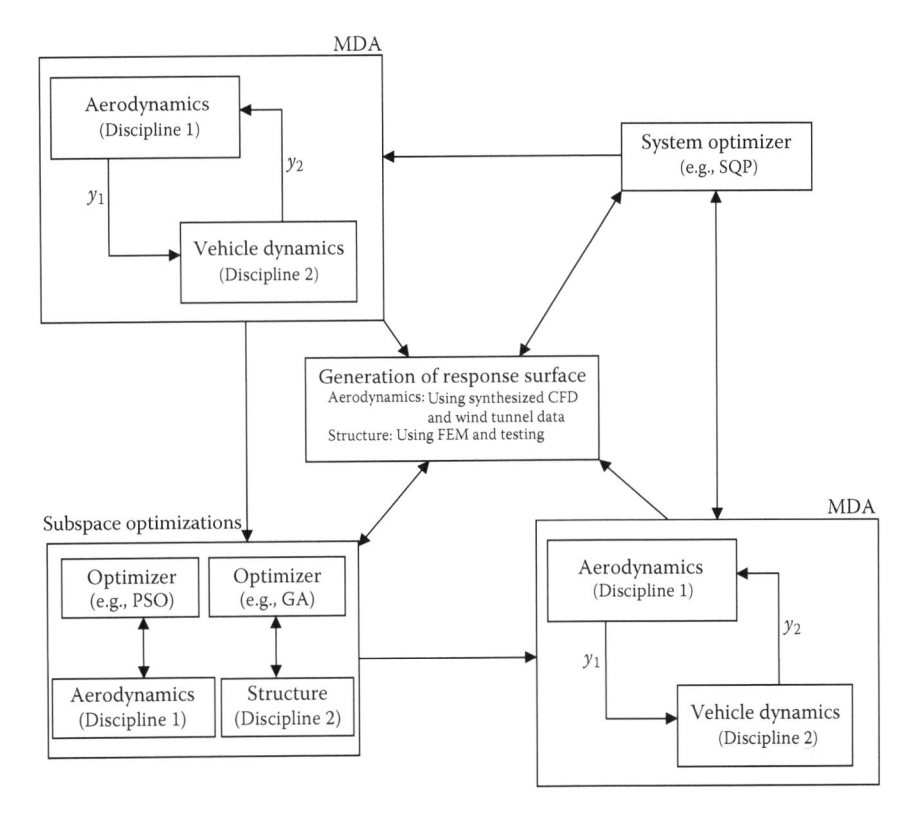

FIGURE 9.8
Current subspace optimization (CSSO) architecture.

by the system optimizer (Figure 9.9). A major difference between BLISS and CSSO is that subspace optimization is not carried out in BLISS.

9.3 MDO Framework

Different MDO architectures were presented in the previous section. One question that arises here is, how do we know which architecture is more efficient? To answer this, one needs a platform in which different MDO architectures can be compared. The MDO framework provides this platform. The MDO problem is first input in standard form. The user then has to select the architecture through which the problem needs to be solved. The framework then casts the MDO problem into the specified architecture form, which is then solved to get the solution.

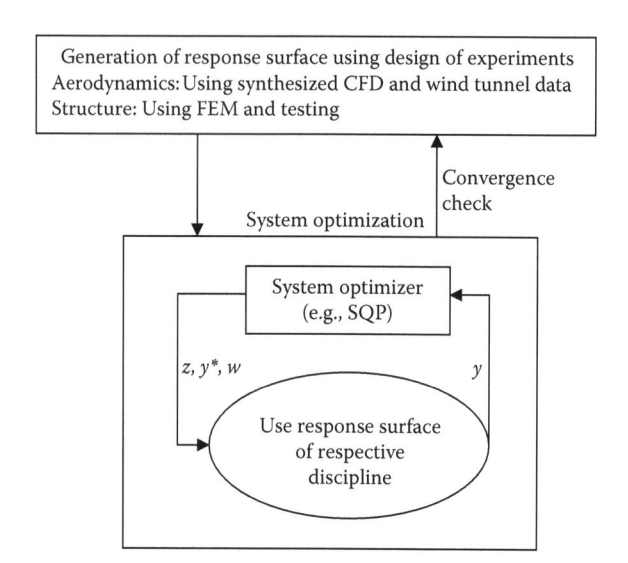

FIGURE 9.9
Bilevel integrated system synthesis (BLISS) architecture.

The standard MDO form is to define the objective function, constraints, design variables, coupled variables, state variables, and analysis functions. The following are some typical requirements of an ideal framework.

- It should be able to handle large problems.
- A majority of operations should be handled by the graphical user interface (GUI).
- It should support a collaborative design.
- It should support different optimizers on different disciplines.
- It should provide debugging support.
- It should offer feasibility of parallel processing.
- Data exchange between different modules should be possible.
- Visualization of intermediate and final results should be possible.

9.4 Response Surface Methodology

The motivation behind the use of RSM is that expensive computational procedures such as finite element methods, CFD, or experimentation are minimized. RSM is an empirical model building technique in which the objective is to generate one or more outputs (responses) from a system that has several input parameters (Cornell and Khuri 1996). The empirical model is built using simulations and experiments in which the output is computed or measured by changing the inputs, which are also known. For example, lift coefficient (C_L) for a wing varies with angle of attack (α) and Mach number (M). Here, response surface of parameter lift coefficient can be generated as a function of angle of attack and Mach number. That is,

$$C_L = f(\alpha, M) \tag{9.17}$$

A typical response surface plot is shown in Figure 9.10. One should have some idea of the relationship between the responses and input parameters. Typically, one can use the first- or second-order polynomial approximation between the output and input variables. For example, for a single input x, the estimated output $\hat{y}$ is given by the relationship

$$\hat{y} = a_0 + a_1 x + a_2 x^2 \tag{9.18}$$

The aim is to estimate the coefficients a_0, a_1, and a_2 so that output $\hat{y}$ can be estimated for any given x. These coefficients can be estimated by minimizing the function

$$\sum_{j=1}^{N} (\hat{y} - y)^2 \tag{9.19}$$

where y is the actual measurement made through experiments or through high-fidelity simulations such as CFD and finite element analysis. These

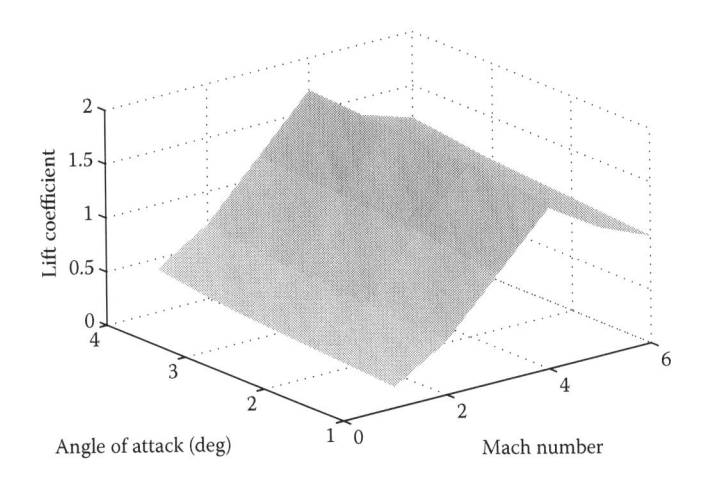

FIGURE 9.10
Response surface of lift coefficient as a function of α and M.

measurements are made for N cases. A question that needs to be asked here is, at what values of x should the experiments be performed? The answer to this is given by design of experiments (DoE). The accuracy and computational cost of constructing the response surface is given by DoE (Anderson and McLean 1974).

A number of design models are available that can capture the interactions among variables, with each having an advantage in either having a lower number of points or having higher accuracy. In one such model, called full factorial design, 3^N points are selected at which experiments need to be carried out. For a two-variable problem, nine experimental points are required (Figure 9.11). In this figure, the subscripts l and u stand for lower and upper bound of the input variable.

Let us assume the input variable angle of attack (α) varies from 1 to 10 degrees and Mach number (M) varies from 0.3 to 2.0. Then the design matrix at which experiments need to be carried out is given in Table 9.1.

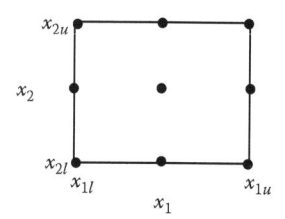

FIGURE 9.11
Full factorial design.

TABLE 9.1

Design Matrix

Experiment Number	α	M
1	1	0.3
2	1	1.15
3	1	2.0
4	5.5	0.3
5	5.5	1.15
6	5.5	2.0
7	10	0.3
8	10	1.15
9	10	2.0

Multidisciplinary Design Optimization

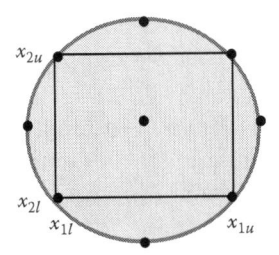

FIGURE 9.12
Central composite design.

In a central composite design, the corner points are augmented with central and axial points (Figure 9.12). This design is preferred for a second-order model. For more such design models refer to Cornell and Khuri (1996).

Chapter Highlights

- In multidisciplinary design optimization, two or more disciplines are simultaneously optimized with interaction among them.
- The main advantage of MDO is that the time required in the design cycle can be significantly reduced. In addition, disciplines with conflicting objectives can be resolved.
- Through MDO architecture, the multidisciplinary problem is transformed into a series of standard optimization problems that can be solved through either a gradient-based solver such as SQP or through a non–gradient-based solver such as GA or PSO.
- The main advantage of MDF architecture is that it ensures a feasible solution at each iterative step.
- The main advantage of the IDF is that different disciplines can be evaluated in parallel.
- In CO architecture, optimization is carried out at a discipline and system level. Thus, discipline feasibility is guaranteed throughout the optimization process. The MDO problem is decomposed into a number of subproblems corresponding to each discipline.
- BLISS architecture is designed to suit a parallel computing environment. It is a bilevel architecture in which each discipline optimization is fully autonomous and coordination is done at the system level

to ensure multidisciplinary feasibility. In this architecture, discipline response levels are used by the system optimizer.
- RSM is an empirical model building technique in which the objective is to generate one or more outputs (responses) from a system that has several input parameters. The empirical model is built using simulations and experiments in which the output is computed or measured by changing the inputs, which are also known.
- The accuracy and computational cost of constructing the response surface is given by design of experiments (DoE).
- Full factorial and central composite design are two such DoE techniques.

Formulae Chart

Multidisciplinary analysis:

Minimize

$$f(z, x)$$

subject to

$$g_j(z, x) \leq 0 \quad j = 1, 2, \ldots, m < n$$

$$h_k(z, x) = 0 \quad k = 1, 2, \ldots, r < n$$

$$y_i^{n+1} - y_i^n = 0$$

Multidisciplinary feasible:

Minimize

$$f(z, x, y(x, z))$$

subject to

$$g(z, x, y(x, z)) \leq 0$$

$$y_i^{n+1} - y_i^n = 0$$

IDF:

Minimize

$$f(z, x, y^t)$$

subject to

$$g(z, x, y(x, y^t, z)) \le 0$$

$$y_i^t - y_i(x, y^t, z) = 0$$

SAND:

Minimize

$$f(z, x, y(x, z, u))$$

subject to

$$g(z, x, y(x, z, u)) \le 0$$

$$R(z, x, y(x, z, u), u) \le 0$$

CSSO:

Minimize

$$f(z, y, x_{obj})$$

subject to

$$J\left(z_i, z^*, x_{obj}, x_{obj}^*, y_i, y^*\right) = 0$$

Problems

1. A DoE has to be carried out for a process that has three inputs and one output. The lower and upper bounds for the three inputs are [0.5, 2.0], [5, 10], and [0.01, 0.1] respectively. How many experiments are to be carried out using a full factorial design, and at what values of input variables?

2. A DoE has to be carried out for a process that has two inputs and one output. The lower and upper bounds for the two inputs are [0.5, 2.0] and [5, 10]. How many experiments are to be carried out using central composite design, and at what values of input variables?

3. A linear model

$$\hat{y} = a_0 + a_1 x$$

is to be used for a system where the following measurements are made.

x	y
0.1	0.3
0.5	0.4
1.0	0.7
2.0	1.2
3.0	2.0
5.0	3.0

Estimate the coefficients a_0 and a_1 by minimizing the function

$$\sum_{j=1}^{N=6} (\hat{y} - y)^2$$

4. The response variable y in a chemical process is a function of four variables: temperature (x_1), pressure (x_2), time (x_3), and stoichiometric ratio (x_4). The lower and upper limits of input variables are [350, 450], [5, 10], [10, 40], and [0.1, 0.5]. A full factorial design is used to fit a first-order model. The input variable combinations and corresponding response values are given in Table 9.2.

Fit a first-order model for this problem.

TABLE 9.2

Responses for Different Inputs

x_1	x_2	x_3	x_4	y
−1	−1	−1	−1	47.5
1	−1	−1	−1	73.2
−1	1	−1	−1	59.4
1	1	−1	−1	75.1
−1	−1	1	−1	74.0
−1	1	1	−1	72.0
1	1	1	−1	73.2
−1	−1	−1	1	82.3
1	−1	−1	1	61.9
−1	1	−1	1	63.8
1	1	−1	1	70.5
−1	−1	1	1	83.2
1	−1	1	1	69.7
−1	1	1	1	80.5
1	1	1	1	81.7

5. The speed reducer optimization problem is written in MDO form (Tedford and Martins 2010) as

Minimize
$$C_1 y_1 z_1^2 \left(C_2 z_2^2 + C_3 z_2 - C_4\right) - C_5 \left(y_2^2 + y_3^2\right) y_1 + C_6 \left(y_2^2 + y_3^2\right) + C_1 \left(x_1 y_2^2 + x_2 y_3^2\right)$$

with respect to x_1, x_2, z_1, z_2

subject to $1 - z_1 x_2 / C_7 \geq 0$
$0.7 \leq z_1 \leq 0.8$
$17 \leq z_2 \leq 28$
$7.3 \leq x_1 \leq 8.3$
$7.3 \leq x_2 \leq 8.3$

Discipline 1

$y_1 = \max(g_1, g_2, g_3, g_4)$

subject to $1 - y_1 / (C_8 z_1) \geq 0$
$1 - y_1 / C_9 \geq 0$

where $g_1 = C_{10} / z_1^2 z_2$
$g_2 = C_{11} / z_1^2 z_2^2$
$g_3 = C_{12} z_1$
$g_4 = C_{13}$

Discipline 2

$y_2 = \max(g_5, g_6, g_7)$

subject to $1 - y_2 / (C_{14}) \geq 0$
$1 - y_2 C_{15} C_{16} / x_1 \geq 0$

where $g_5 = \left(C_{17} x_1^3 / z_1 z_2\right)^{1/4}$

$g_6 = \left(1 / C_{18} C_{19} \sqrt{C_{20}^2 x_1^2 / \left(z_1^2 z_2^2\right) + C_{21}}\right)^{1/3}$

$g_7 = C_{22}$

Discipline 3

$y_3 = \max(g_8, g_9, g_{10})$

subject to $1 - y_3 / (C_{23}) \geq 0$
$1 - y_3 C_{24} C_{16} / x_2 \geq 0$

where $g_8 = \left(C_{25} x_2^3 / z_1 z_2\right)^{1/4}$

$$g_9 = \left(1/C_{26}C_{19}\sqrt{C_{20}^2 x_2^2/\left(z_1^2 z_2^2\right)} + C_{27}\right)^{1/3}$$

$$g_{10} = C_{28}$$

The values of constants for this problem are

C_1	0.7854	C_{15}	1.5
C_2	3.3333	C_{16}	1.9
C_3	14.9334	C_{17}	1.93
C_4	43.0934	C_{18}	1100
C_5	1.5079	C_{19}	0.1
C_6	7.477	C_{20}	1.69×10^9
C_7	40	C_{21}	745
C_8	12	C_{22}	2.9
C_9	3.6	C_{23}	5.5
C_{10}	27	C_{24}	1.1
C_{11}	397.5	C_{25}	1.93
C_{12}	5	C_{26}	850
C_{13}	2.6	C_{27}	1.575×10^8
C_{14}	3.9	C_{28}	5

10
Integer Programming

10.1 Introduction

In the previous chapters, decision variables in optimization problems were considered to be continuous and they could take any fractional values such as 10.5, 5.64, etc. Some optimization problems require design variables to be integers. For example, the number of cars produced in a day, number of maneuvers required by a spacecraft in an orbit, number of rivets required, amount of manpower required, and so forth, all have to be integers. It does not make much sense to get a solution such as 8.4 rivets for butting two plates. It is important to note that rounding off the decision variable to the nearest integer may not yield the optimum solution or may violate some of the constraints. Therefore, it is desirable to give a special formulation to integer programming problems.

Integer programming can be of different types. An *all-integer programming* problem contains design variables that can take integer values only. In *mixed-integer programming* problems, some decision variables are of an integer type and some can take fractional values or are of a continuous type. Optimization problems in which design variables can take only discrete values are referred to as *discrete programming* problems. For example, pipe sizes come in standard sizes such as 0.5, 0.8, 1.0, 1.4, 1.8, . . . inch. If pipe size is a decision variable, then it can take these discrete values only. There is also a special type of integer programming called a zero-one programming problem in which design variables can take a value of 0 or 1. For example, suppose we want to set up two plants from five candidate locations. If variable S_i corresponds to the setup of plant at ith location, then $S_i = 1$, else it takes the value 0.

Cutting plane and branch-and-bound methods are two popular techniques for solving integer programming problems. Gomory's cutting plane method is well suited for linear integer programming problems. The linear programming problem is first solved using the simplex method. If integer solutions are not obtained, additional constraints called "cuts" are added to the problem. The modified linear programming is then solved using the dual method. The procedure is repeated until the integer solutions are obtained. In the branch-and-bound method, the nonlinear integer optimization problem is

```
┌─────────────────────────────┐
│    Integer programming      │
└─────────────────────────────┘
              ↓
┌─────────────────────────────────────┐
│ Integer linear programming          │
│  • Gomory's cutting plane method    │
│  • Balas' method                    │
└─────────────────────────────────────┘
              ↓
┌─────────────────────────────────────┐
│ Integer nonlinear programming       │
│  • Branch-and-bound method          │
│  • Evolutionary method              │
└─────────────────────────────────────┘
```

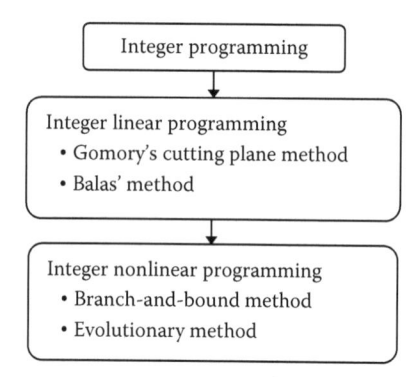

FIGURE 10.1
Road map of Chapter 10.

first solved as a continuous variable problem. Then the method branches into subproblems in which additional constraints are added to the problem to get integer solutions. The subproblems are again solved as continuous variable problems and the procedure is repeated until a feasible integer solution is obtained. The Balas algorithm is popular in solving the zero-one integer programming problems. This method selects few solutions from the possible 2^n enumerated solutions, where n is the number of binary variables in the problem. In this chapter, we also explore a particle swarm optimization (PSO) method for solving integer programming problems. The road map of this chapter is shown in Figure 10.1.

10.2 Integer Linear Programming

Consider the following integer programming problem.

Minimize

$$f(x) = -3x_1 - 2x_2$$

subject to

$$x_1 - x_2 \le 5$$

$$4x_1 + 7x_2 \le 50$$

$$x_1, x_2 \ge 0$$

where x_1 and x_2 are integers.

Integer Programming

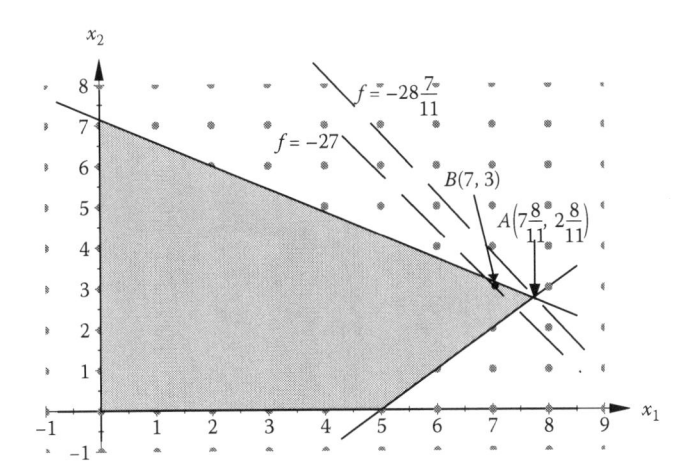

FIGURE 10.2
Continuous/integer variable solution.

Let us plot the objective function and the constraints (Figure 10.2) and for the time being ignore the integer aspects of the problem. The minimum of the optimization problem is $-28\frac{7}{11}$ and occurs at $x_1 = 7\frac{8}{11}$ and $x_2 = 2\frac{8}{11}$, which is shown by point A in Figure 10.2. Let us round off the values of x_1 and x_2 to obtain an integer solution. The truncated point (8,3) becomes an infeasible point where the constraint $4x_1 + 7x_2 \leq 50$ is not satisfied. The optimal point of this integral programming problem is $B(7, 3)$ and the value of the objective function at this point is -27. It is important to note that rounded off values of the decision variables obtained by solving the optimization problem as continuous variables may or may not lead to an optimal solution.

10.2.1 Gomory's Cutting Plane Method

To start with, the linear integer programming problem is solved using the simplex method described in Chapter 4 by ignoring the integer requirement of the variables. If the variables at the optimal solution happen to be integers, the algorithm is terminated. Otherwise, some additional constraints are imposed on the problem. The modified problem is then solved to obtain an integer solution (Gomory 1960).

Let us explain the procedure for this problem. The matrices A, b, and c are modified as follows.

```
A = [1 -1 1 0;
     4  7 0 1];
b = [5;50];
```

```
c = [-3;-2;0;0];
basic_set = [3 4];
nonbasic_set = [1 2];
```

The MATLAB® code (*simplex.m*) is executed with these initializations and the following output is obtained.

```
basic_set =
     3     4
nonbasic_set =
     1     2
Initial_Table =
     1     0     1    -1     5
     0     1     4     7    50
Cost =
     0     0    -3    -2     0

basic_set =
     1     4
nonbasic_set =
     2     3
Table =
     1     0    -1     1     5
     0     1    11    -4    30
Cost =
     0     0    -5     3    15

basic_set =
     1     2
nonbasic_set =
     3     4
Table =
     1     0   7/11   1/11   85/11
     0     1  -4/11   1/11   30/11
Cost =
     0     0  13/11   5/11  315/11
- - -SOLUTION- - -
basic_set =
     1     2
xb =
     85/11
     30/11
zz =
     -315/11
```

The minimum of the optimization problem is $-28\frac{7}{11}$ and occurs at $x_1 = 7\frac{8}{11}$ and $x_2 = 2\frac{8}{11}$. Because the decision variables are nonintegers, a Gomory

Integer Programming 267

constraint is to be added. This requires the addition of another slack variable, x_5. We have to select a variable from x_1 or x_2 that is to be made an integer. The one with the largest fractional value is selected. As both x_1 and x_2 have the same fractional value $\left(\dfrac{8}{11}\right)$, we select x_1 randomly as the variable that has to be made an integer. The Gomory constraint is written as

$$x_5 - \frac{7}{11}x_3 - \frac{1}{11}x_4 = -\frac{8}{11}$$

The Gomory constraint is written in the following manner. First, consider the row corresponding to the variable that is to be made an integer. Because it is x_1 for this problem, the final row from the simplex table is selected as

1   0   7/11   1/11   85/11

Take the negative for the nonbasic variables and add it to the new slack variable x_5, which then becomes the left-hand side of the Gomory constraint. The right-hand side of the Gomory constraint is given by the negative of the fractional value corresponding to $\left(\dfrac{85}{11}\right)$, which is $\left(-\dfrac{8}{11}\right)$. When this constraint is added to the primal problem, it becomes infeasible because one of b_i is negative. The problem can be solved using the dual simplex method. The MATLAB code (*dual.m*) is executed with following initialization.

```
A = [1 0  7/11  1/11 0;
     0 1 -4/11  1/11 0;
     0 0 -7/11 -1/11 1];
b = [85/11;30/11;-8/11];
c = [0;0;13/11;5/11;0];
basic_set = [1 2 5];
nonbasic_set = [3 4];
zz = -315/11;
```

On executing the code the following output is obtained.

```
basic_set =
         1    2    5
nonbasic_set =
         3    4
Initial_Table =
         1    0    0     7/11    1/11    85/11
         0    1    0    -4/11    1/11    30/11
         0    0    1    -7/11   -1/11    -8/11
Cost =
         0    0    0    13/11    5/11   -315/11
```

```
basic_set =
       1      2      3
nonbasic_set =
       4      5
Table =
       1      0      0      0      1       7
       0      1      0      1/7   -4/7    22/7
       0      0      1      1/7   -11/7    8/7
Cost =
       0      0      0      2/7    13/7   191/7
- - -FINAL SOLUTION- - -
basic_set =
       1      2      3
xb =
       7
       22/7
       8/7
zz = -191/7
```

The variable x_1 has taken an integer value ($x_1 = 7$). The variables x_2 and x_3 are still not integers. A Gomory constraint is to be added. This requires the addition of another slack variable x_6. Picking the row

```
0      1      0      1/7   -4/7    22/7
```

The Gomory constraint is given by

$$x_6 - \frac{1}{7}x_3 + \frac{4}{7}x_4 = -\frac{1}{7}$$

The MATLAB code (*dual.m*) is again executed with following initialization.

```
A = [1 0 0 0 1 0;
     0 1 0 1/7 -4/7 0;
     0 0 1 1/7 -11/7 0;
     0 0 0 -1/7 4/7 1];
b = [7;22/7;8/7;-1/7];
c = [0;0;0;2/7;13/7;0];
basic_set = [1 2 3 6];
nonbasic_set = [4 5];
zz = -191/7
```

On executing the code the following output is obtained.

```
basic_set =
       1      2      3      6
nonbasic_set =
       4      5
```

Integer Programming

```
Initial_Table =
    1      0      0      0      0      1      7
    0      1      0      0     1/7   -4/7    22/7
    0      0      1      0     1/7  -11/7    8/7
    0      0      0      1    -1/7    4/7   -1/7
Cost =
    0      0      0      0     2/7   13/7  -191/7

basic_set =
    1      2      3      4
nonbasic_set =
    5      6
Table =
    1      0      0      0      1      0      7
    0      1      0      0      0      1      3
    0      0      1      0     -1      1      1
    0      0      0      1     -4     -7      1
Cost =
    0      0      0      0      3      2     27
- - -FINAL SOLUTION- - -
basic_set =
    1      2      3      4
xb =
    7
    3
    1
    1
zz =
    -27
```

The minimum of the optimization problem is −27 and occurs at $x_1 = 7$ and $x_2 = 3$. Observe that other basic variables x_3 and x_4 have also achieved integer values at the optimum point for an all-integer problem.

Consider the following mixed-integer programming problem.

Minimize
$$f(x) = -3x_1 - 2x_2$$

subject to
$$x_1 + x_2 \le 6$$

$$5x_1 + 2x_2 \le 20$$

$$x_1, x_2 \ge 0$$

where x_2 is an integer.

The first step is to solve the linear programming problem by neglecting the integer constraint. The matrices A, b, and c are modified as below.

```
A = [1 1 1 0;
     5 2 0 1];
b = [6;20];
c = [-3;-2;0;0];
basic_set = [3 4];
nonbasic_set = [1 2];
```

The MATLAB code (*simplex.m*) is executed with these initializations and the following output is obtained.

```
basic_set =
        3       4
nonbasic_set =
        1       2
Initial_Table =
        1       0       1       1       6
        0       1       5       2       20
Cost =
        0       0       -3      -2      0

basic_set =
        3       1
nonbasic_set =
        2       4
Table =
        1       0       3/5     -1/5    2
        0       1       2/5     1/5     4
Cost =
        0       0       -4/5    3/5     12

basic_set =
        2       1
nonbasic_set =
        3       4
Table =
        1       0       5/3     -1/3    10/3
        0       1       -2/3    1/3     8/3
Cost =
        0       0       4/3     1/3     44/3
- - -SOLUTION- - -
basic_set =
        2       1
xb =
        10/3
        8/3
zz =
        -44/3
```

Integer Programming

Because the variable x_2 is noninteger $\left(\dfrac{10}{3}\right)$, we have to add the Gomory constraint. The Gomory constraint is written as

$$x_5 - \frac{5}{3}x_3 + \frac{1}{3}x_4 = -\frac{1}{3}$$

The MATLAB code (*dual.m*) is executed with the following initialization.

```
A = [1 0 -2/3 1/3 0;
     0 1 5/3 -1/3 0;
     0 0 -5/3 1/3 1];
b = [8/3;10/3;-1/3];
c = [0;0;4/3;1/3;0];
basic_set = [1 2 5];
nonbasic_set = [3 4];
zz = -44/3
```

On executing the code the following output is obtained.

```
basic_set =
         1     2     5
nonbasic_set =
         3     4
Initial_Table =
         1     0     0     -2/3    1/3     8/3
         0     1     0      5/3   -1/3    10/3
         0     0     1     -5/3    1/3    -1/3
Cost =
         0     0     0      4/3    1/3   -44/3

basic_set =
         1     2     3
nonbasic_set =
         4     5
Table =
         1     0     0      1/5   -2/5    14/5
         0     1     0      0      1      3
         0     0     1     -1/5   -3/5    1/5
Cost =
         0     0     0      3/5    4/5    72/5
- - -FINAL SOLUTION- - -
basic_set =
         1     2     3
xb =
         14/5
         3
         1/5
```

$$zz = -72/5$$

The minimum of the optimization problem is $-\dfrac{72}{5}$ and occurs at $x_2 = 3$ and $x_1 = \dfrac{14}{5}$.

10.2.2 Zero-One Problems

In these problems, the decision variables can only take the values 0 or 1. For example, if a plant is to be located at a particular site, the variable takes a value 1, else it takes the value 0. If there are n integer variables to be evaluated, an enumerated search would require 2^n evaluations of the objective function and constraints. For a problem with a few variables, an explicit enumerated search should be good enough. However, for a problem with a large number of variables, an enumerated search will be computationally expensive. For example, for a 20-variable problem, the number of function (and constraints) evaluations would be 1,048,576. Balas' method uses an implicit enumeration (Balas 1965) technique to find the optimal solution.

The standard form of linear programming problem where Balas' method can be applied is given by

Minimize

$$z = c^T x \tag{10.1}$$

subject to

$$Ax = b, \quad x \in \{0, 1\} \tag{10.2}$$

$$c \geq 0 \tag{10.3}$$

where A is $m \times n$ constraint matrix given by

$$A = \begin{bmatrix} a_{11} & a_{12} & \cdots & a_{1n} \\ a_{21} & a_{22} & \cdots & a_{2n} \\ \vdots & & & \\ a_{m1} & a_{m2} & \cdots & a_{mn} \end{bmatrix}$$

and b, c, and x are column vectors given by

Integer Programming

$$b = \begin{bmatrix} b_1 \\ b_2 \\ \vdots \\ b_m \end{bmatrix}, \quad c = \begin{bmatrix} c_1 \\ c_2 \\ \vdots \\ c_n \end{bmatrix}, \quad x = \begin{bmatrix} x_1 \\ x_2 \\ \vdots \\ x_n \end{bmatrix}$$

If some of the cost coefficients (x_i) are negative, they can be put in the standard form by the substitution

$$x_i = 1 - y_i, \quad y_i \in \{0, 1\} \tag{10.4}$$

For example, the following problem

Minimize

$$f(x) = x_1 - x_2$$

subject to

$$-2x_1 - 3x_2 \le -5$$

is written in standard form as

Minimize

$$f(x) = x_1 + y_2$$

subject to

$$-2x_1 + 3y_2 \le -2$$

$$x_1, y_2 \in \{0, 1\}$$

Let us explain Balas' method through an example. Consider the following zero-one integer programming problem (Bricker 1999).

Minimize

$$f(x) = 4x_1 + 8x_2 + 9x_3 + 3x_4 + 4x_5 + 10x_6$$

subject to

$$4x_1 - 5x_2 - 3x_3 - 2x_4 - x_5 + 8x_6 \leq -8$$

$$-5x_1 + 2x_2 + 9x_3 + 8x_4 - 3x_5 + 8x_6 \leq 7$$

$$8x_1 + 5x_2 - 4x_3 + x_5 + 6x_6 \leq 6$$

$$x \in \{0, 1\}$$

Start with the solution

$$x_1 = x_2 = x_3 = x_4 = x_5 = x_6 = 0$$

This is an initial solution and no variable is fixed. Thus the solution vector is a null set and is given by

$$S = \{\ \}$$

On substituting these values of variables, second and third constraints are satisfied, whereas the first constraint is infeasible. The violated constraint is denoted as

$$V = \{1\}$$

To check the sensitivity of different variables on the feasibility of the first constraint, we observe that if variables x_1 and x_6 become 1, they only increase the infeasibility of the first constraint. These two variables are not helpful. The helpful variables are therefore given by

$$H = \{2, 3, 4, 5\}$$

At the end of the first iteration we can write

$$S_1 = \{\ \}$$

$$V_1 = \{1\}$$

$$H_1 = \{2, 3, 4, 5\}$$

Integer Programming

We must select a helpful variable for branching. The variable x_2 will reduce infeasibility in the first constraint and therefore can be selected for branching. The solution vector is therefore written as

$$S_2 = \{2\}$$

This means that variable x_2 is now fixed at 1. The first constraint is, however, still violated. That is,

$$V_2 = \{1\}$$

Again, we observe that variables x_3, x_4, and x_5 are helpful. Therefore, at the end of second iteration we can write

$$S_2 = \{2\}$$

$$V_2 = \{1\}$$

$$H_2 = \{3, 4, 5\}$$

The variable x_5 will reduce more infeasibility in the constraints as compared to the variables x_3 and x_4. Therefore the variable x_5 is also fixed at 1. Hence,

$$S_3 = \{2, 5\}$$

The first constraint is still violated. Therefore, at the end of third iteration, we can write

$$S_3 = \{2, 5\}$$

$$V_3 = \{1\}$$

$$H_3 = \{3, 4\}$$

Similarly, at the end of the fourth iteration, we can write

$$S_4 = \{2, 5, 4\}$$

$$V_4 = \{\,\}$$

$$H_4 = \{\,\}$$

Thus, if the variables x_2, x_5, and x_4 are fixed at 1, no constraints are violated and the value of the objective function is given by 15. Because a violated constraint set is a null set, we backtrack and fix x_4 to zero. This is written as

$$S_5 = \{2, 5, \bar{4}\}$$

Therefore, at the end of the fifth iteration, we can write

$$S_5 = \{2, 5, \bar{4}\}$$

$$V_5 = \{1\}$$

$$H_5 = \{3\}$$

In the next iteration, x_5 is fixed at 0 and variable x_4 is removed from the solution set. Therefore,

$$S_6 = \{2, \bar{5}\}$$

$$V_6 = \{1\}$$

$$H_6 = \{3, 4\}$$

Similarly, the last node is written as

$$S_7 = \{\bar{2}\}$$

$$V_7 = \{1\}$$

$$H_7 = \{3, 4, 5\}$$

Thus the optimal value of variables, as obtained in the fourth iteration, is given by

$$x_1 = x_3 = x_5 = 0 \quad \text{and} \quad x_2 = x_4 = x_5 = 1$$

Integer Programming

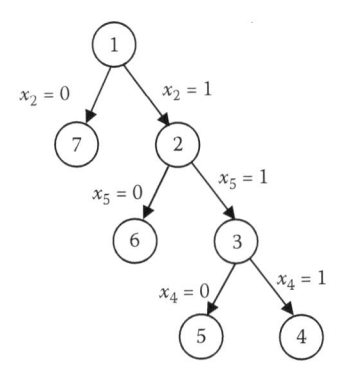

FIGURE 10.3
Tree diagram for the test problem.

The optimal value of the objective function is 15. The different steps of the Balas method can be understood with the tree diagram (Figure 10.3).

10.3 Integer Nonlinear Programming

The branch-and-bound method is one of the popular methods of solving both integer linear and nonlinear programming. The technique was developed by Land and Doig and can also be used for mixed-integer programming. We will also explore the utility of the PSO technique in solving mixed-integer nonlinear problems. The constrained mixed-integer optimization problem can be mathematically stated as

Minimize

$$f(x) \qquad (10.5)$$

subject to

$$g_i(x) \le 0 \quad i = 1, 2, \ldots, m < n \qquad (10.6)$$

$$h_j(x) = 0 \quad j = 1, 2, \ldots, r < n \qquad (10.7)$$

$$x_k = \text{integers} \quad k = 1, 2, \ldots, p < n \qquad (10.8)$$

where there are n variables to be determined out of which p are integers and the remaining variables are continuous.

10.3.1 Branch-and-Bound Method

In this method (Land and Doig 1960), the optimization problem is solved with continuous variables, and the integer constraints are relaxed. If the solution obtained is integers, the algorithm is terminated as it represents the optimal solution of the integer problem. If one of the integer variables x_k is continuous, then one has to solve two additional subproblems with the upper bound constraint

$$x_k \leq [x_k] \qquad (10.9)$$

and lower bound constraint

$$x_k \geq [x_k] + 1 \qquad (10.10)$$

This process of the branching ensures that feasible integer solutions are not eliminated. The branching problem is again solved (as continuous variables) with these additional constraints. The process is continued until an integer solution is obtained. This solution corresponds to the upper bound of the objective function for a minimization problem. During the course of further branchings, if any of the branches have the value of the objective function greater than this upper bound value then that node is terminated or fathomed. If a lower value of the objective function is reached than the upper bound value, then the upper bound value is updated. The method continues to branch until all the nodes have been evaluated or fathomed. The lowest value of the objective function corresponding to the integer feasible solution gives the optimal value of the objective function.

Consider the following integer programming problem that is solved using the branch-and-bound method.

Minimize

$$f(x) = -4x_1 - 5x_2$$

subject to

$$2x_1 + 5x_2 \leq 16$$

$$2x_1 - 3x_2 \leq 7$$

$$x_1, x_2 \geq 0$$

where x_1 and x_2 are integers.

Integer Programming

As a first step, integer constraints are relaxed and the linear programming problem is solved with continuous variables. The MATLAB code (*simplex.m*) is executed with the initializations

```
A = [2  5  1  0;
     2 -3  0  1];
b = [16;7];
c = [-4;-5;0;0];
basic_set = [3 4];
nonbasic_set = [1 2];
```

and following output is obtained:

```
basic_set =
         3       4
nonbasic_set =
         1       2
Initial_Table =
         1       0       2       5      16
         0       1       2      -3       7
Cost =
         0       0      -4      -5       0

basic_set =
         2       4
nonbasic_set =
         1       3
Table =
         1       0      2/5     1/5    16/5
         0       1     16/5     3/5    83/5
Cost =
         0       0      -2       1      16

basic_set =
         2       1
nonbasic_set =
         3       4
Table =
         1       0      1/8    -1/8     9/8
         0       1     3/16    5/16   83/16
Cost =
         0       0     11/8     5/8   211/8
- - -SOLUTION- - -
basic_set =
         2       1
xb =
        9/8
       83/16
zz =
      -211/8
```

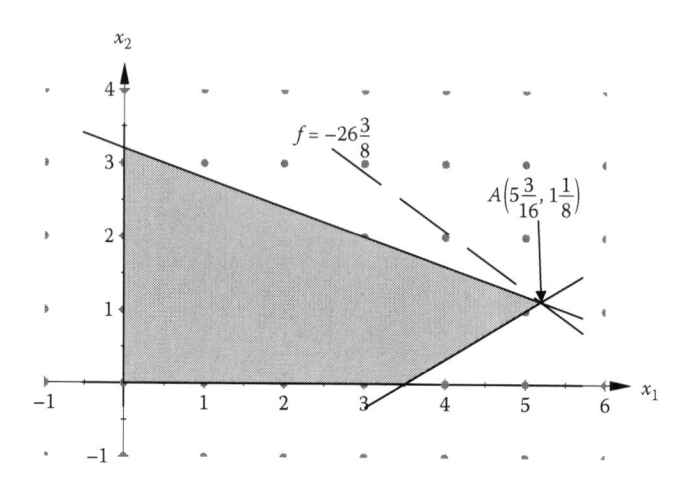

280 Optimization: Algorithms and Applications

FIGURE 10.4
Optimal noninteger solution.

The optimal value of the objective function is $-26\frac{3}{8}$ and occurs at $x_1 = 5\frac{3}{16}$ and $x_2 = 1\frac{1}{8}$ (Figure 10.4).

Because both variables are not integers, we branch and create two subproblems:

NODE 1
    Subproblem 1
        Minimize $f(x) = -4x_1 - 5x_2$
        subject to $2x_1 + 5x_2 \leq 16$
                      $2x_1 - 3x_2 \leq 7$
                      $x_1 \leq 5$

    Subproblem 2
        Minimize $f(x) = -4x_1 - 5x_2$
        subject to $2x_1 + 5x_2 \leq 16$
                      $2x_1 - 3x_2 \leq 7$
                      $x_1 \geq 6$

The MATLAB code (*subproblem1.m*) is executed with the initializations

```
A = [2 5 1 0 0;
     2 -3 0 1 0;
     1 0 0 0 1];
b = [16;7;5];
c = [-4;-5;0;0;0];
```

Integer Programming

```
basic_set = [3 4 5];
nonbasic_set = [1 2];
```

and the following output is obtained.

```
basic_set =
      3     4     5
nonbasic_set =
      1     2
Initial_Table =
      1     0     0     2     5    16
      0     1     0     2    -3     7
      0     0     1     1     0     5
Cost =
      0     0     0    -4    -5     0
```

```
basic_set =
      2     4     5
nonbasic_set =
      1     3
Table =
      1     0     0    2/5   1/5   16/5
      0     1     0   16/5   3/5   83/5
      0     0     1    1     0     5
Cost =
      0     0     0    -2     1     16
```

```
basic_set =
      2     4     1
nonbasic_set =
      3     5
Table =
      1     0     0    1/5  -2/5   6/5
      0     1     0    3/5 -16/5   3/5
      0     0     1    0     1     5
Cost =
      0     0     0    1     2     26
- - -SOLUTION- - -
basic_set =
      2     4     1
xb =
      6/5
      3/5
      5
zz =
      -26
```

The optimal value of the objective function is −26 and occurs at $x_1 = 5$ and $x_2 = 1\frac{1}{5}$ (Figure 10.5). As x_2 has a noninteger value, we need to branch here

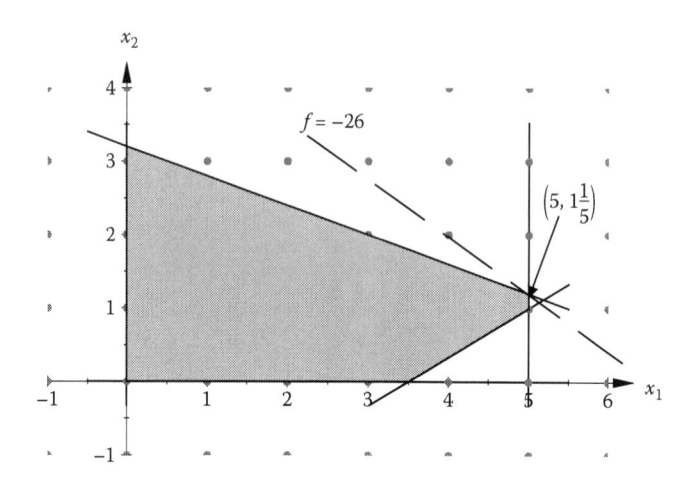

FIGURE 10.5
Noninteger solution.

(node 2). In a similar way, subproblem 2 can be solved (node 3). The solution to subproblem 2 results in an infeasible solution. No further branching is therefore required from subproblem 2.

Two further nodes (4 and 5) are created from node 2. Two new subproblems are

NODE 2
   Subproblem 1
      Minimize $f(x) = -4x_1 - 5x_2$
      subject to $2x_1 + 5x_2 \leq 16$
                   $2x_1 - 3x_2 \leq 7$
                   $x_2 \leq 1$
   Subproblem 2
      Minimize $f(x) = -4x_1 - 5x_2$
      subject to $2x_1 + 5x_2 \leq 16$
                   $2x_1 - 3x_2 \leq 7$
                   $x_2 \geq 2$

```
A = [2 5 1 0 0;
     2 -3 0 1 0;
     0 1 0 0 1];
b = [16;7;1];
c = [-4;-5;0;0;0];
basic_set = [3 4 5];
nonbasic_set = [1 2];
```

Integer Programming

and the following output is obtained.

```
basic
           3      4      5
nonbasic_set =
           1      2
Initial_Table =
           1      0      0      2      5     16
           0      1      0      2     -3      7
           0      0      1      0      1      1
Cost =
           0      0      0     -4     -5      0

basic_set =
           3      4      2
nonbasic_set =
           1      5
Table =
           1      0      0      2     -5     11
           0      1      0      2      3     10
           0      0      1      0      1      1
Cost =
           0      0      0     -4      5      5

basic_set =
           3      1      2
nonbasic_set =
           4      5
Table =
           1      0      0     -1     -8      1
           0      1      0    1/2    3/2      5
           0      0      1      0      1      1
Cost =
           0      0      0      2     11     25
- - -SOLUTION- - -
basic_set =
           3      1      2
xb =
           1
           5
           1
zz =
         -25
```

The optimal value of the objective function is −25 and occurs at $x_1 = 5$ and $x_2 = 1$ (Figure 10.6). Because this subproblem has an integer feasible solution, we fathom the node here and do not branch from here. Similarly, the solution of the second problem gives an objective function value of −22 and occurs

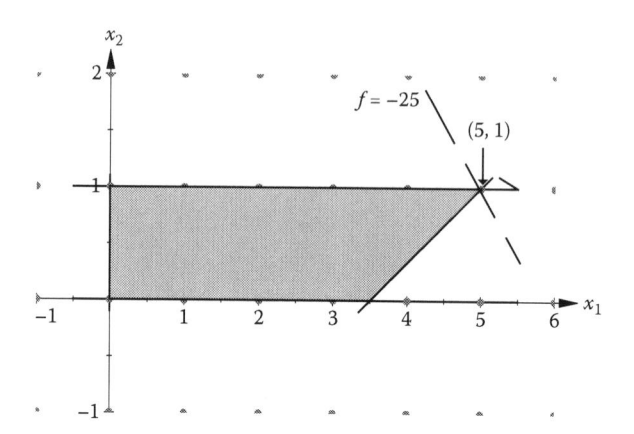

FIGURE 10.6
Feasible integer solution (optimal).

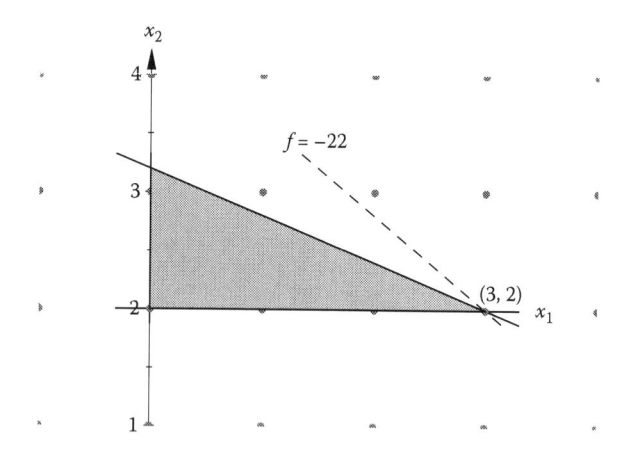

FIGURE 10.7
Feasible integer solution.

at $x_1 = 3$ and $x_2 = 2$ (Figure 10.7). Because the value of objective function in subproblem 2 is greater than the value of the objective function in subproblem 1, we also fathom the node 5. The optimal value of the original integer programming problem is therefore −25 and occurs at $x_1 = 5$ and $x_2 = 1$. The tree diagram for this problem is shown in Figure 10.8.

10.3.2 Evolutionary Method

The particle swarm optimization (PSO) method can be successfully used to solve integer programming problems. The method was elaborated in Chapter 5 and it successfully solved nonlinear constraint optimization problems (Chapter 6)

Integer Programming

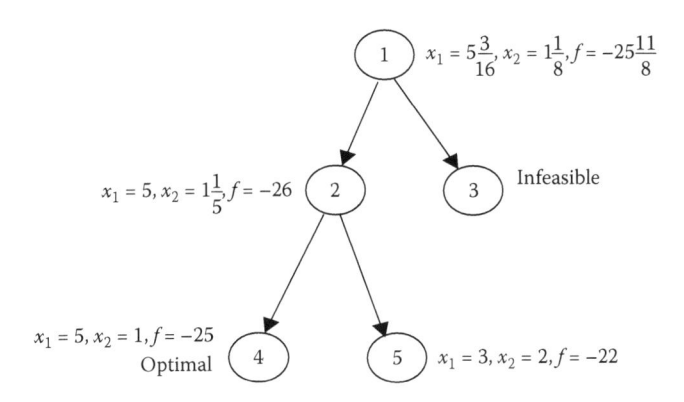

FIGURE 10.8
Tree diagram for the test problem.

as well as multiobjective problems (Chapter 7). In this method, the velocity of individual $v_{i,k}$ is updated using the equation

$$v_{i+1,k} = w_1 v_{i,k} + \phi_1 (p_{xik} - x_{i,k}) u_i + \phi_2 (g_{ix} - x_{i,k}) u_i \qquad (10.11)$$

where w_1, ϕ_1, and ϕ_2 are the tuning factors of the algorithm. The position of each individual is updated as

$$x_{i+1,k} = x_{i,k} + v_{i+1,k} \qquad (10.12)$$

For integer variables in the problem, we can round off the variable to the nearest integer (Laskari et al. 2002) as

$$x_{i+1,k} = \text{round}(x_{i+1,k}) \qquad (10.13)$$

The rest of the procedure remains same and is mentioned in Chapter 5. The constrained welded beam optimization problem (see Chapter 6) is again considered with a modification that some of the variables take integer values only. The optimal solution obtained using the PSO method for different versions of this problem is mentioned in Table 10.1.

TABLE 10.1
Optimal Solution to Different Welded Beam Problems

Welded Beam	Integer Variable Constraint	x_1^*	x_2^*	x_3^*	x_4^*	f^*
Problem 1	None	0.244	6.212	8.299	0.244	2.381
Problem 2	x_3, x_4	0.681	2.794	5	1	5.471
Problem 3	x_3	0.263	5.869	8	0.263	2.461
Problem 4	x_4	0.645	3.734	4.099	1	5.211
Problem 5	x_2	0.241	6	8.644	0.242	2.399
Problem 6	x_2, x_4	0.614	4	4.099	1	5.213

Chapter Highlights

- An all-integer programming problem contains design variables that can take integer values only.
- In mixed-integer programming problems, some decision variables are of the integer type and some can take fractional values or are of the continuous type.
- Optimization problems in which design variables can take only discrete values are referred to as discrete programming problems.
- There is also a special type of integer programming called a zero-one programming problem in which design variables can take a value of 0 or 1.
- In Gomory's cutting plane method, the linear integer programming problem is first solved using the simplex method by ignoring the integer requirement of the variables. If the variables at the optimal solution happen to be integers, the algorithm is terminated. Otherwise, some additional constraints are imposed on the problem.
- If there are n integer variables to be evaluated in a zero-one problem, an enumerated search would require 2^n evaluations of the objective function and constraints. Balas' method uses an implicit enumeration technique to find the optimal solution.
- In the branch-and-bound method, the optimization problem is solved with continuous variables, and the integer constraints are relaxed. If the solution obtained is integers, the algorithm is terminated as it represents the optimal solution of the integer problem. If one of the integer variables is continuous, then one has to solve two additional subproblems with additional constraints.
- The PSO method can be used to solve nonlinear mixed-integer programming problems with minor modifications.

Formulae Chart

Standard form of linear integer programming where Balas' method is used:

Minimize

$$z = c^T x$$

subject to
$$Ax = b$$
$$x \in \{0, 1\}$$
$$c \geq 0$$

Problems

1. Solve the following integer programming problem using the graphical method.

    Minimize $f(x) = -3x_1 - 2x_2$
    subject to
    $2x_1 + x_2 \leq 17$
    $2x_1 + 3x_2 \leq 40$
    $3x_1 + 3x_2 \leq 26$
    $x_1, x_2 \geq 0$

    where x_1 and x_2 are integers.

2. Solve the following integer programming problem using Gomory's cutting plane method.

    Minimize $f(x) = -x_1 + 2x_2$
    subject to
    $2x_1 + 2x_2 \leq 4$
    $6x_1 + 2x_2 \leq 9$
    $x_1, x_2 \geq 0$

    where x_1 and x_2 are integers.

3. A small wooden furniture manufacturer has specialized in two types of furniture: chairs and tables, both requiring two types of raw material. Chairs require 6 and 7 units of the first and second kind of raw material whereas tables require 14 and 7 units of the first and second kind of raw material. In a day, the manufacturer has a supply of 42 units and 35 units of two types of raw material. Profit analysis indicates that every unit of chair contributes Rs. 100 and every unit of table contributes Rs. 160. The manufacturer would like to know the optimum number of chairs and tables to be produced so as to maximize the profit (Shenoy et al. 1986). Formulate this as an integer-programming problem and solve it.

4. Solve the following zero-one programming problem (Shenoy et al. 1986) using Balas' method.

Minimize $16x_1 + 15x_2 + 17x_3 + 15x_4 + 40x_5 + 12x_6 + 13x_7 + 9x_8 + 12x_9$

subject to $13x_1 + 50x_2 + 7x_3 + 6x_4 + 36x_5 + 6x_6 + 46x_7 + 38x_8 + 18x_9 \le 50$

$3x_1 + 8x_2 + 6x_3 + 2x_4 + 34x_5 + 6x_6 + 4x_7 + 7x_8 + 3x_9 \le 20$

$x \in \{0, 1\}$

5. Solve the following integer programming problem using Gomory's cutting plane method.

Minimize $f(x) = -3x_1 - 5x_2$
subject to $2x_1 + 5x_2 \le 15$
$2x_1 - 2x_2 \le 5$
$x_1, x_2 \ge 0$

where x_1 and x_2 are integers.

6. Solve the following integer programming problem using Gomory's cutting plane method.

Minimize $f(x) = -4x_1 - 7x_2$
subject to $x_1 + x_2 \le 6$
$5x_1 + 9x_2 \le 50$
$x_1, x_2 \ge 0$

where x_1 and x_2 are integers.

7. Solve the following integer programming problem using Gomory's cutting plane method

Maximize $f(x) = 3x_1 + 2x_2$
subject to $2x_1 + x_2 \le 5$
$2x_1 - 7x_2 \le 4$
$x_1, x_2 \ge 0$

where x_1 and x_2 are integers.

8. Solve the following integer programming problem using the branch-and-bound method.

Maximize $f(x) = x_1 + 2x_2$
subject to $2x_1 + x_2 \le 4$
$3x_1 + 4x_2 \le 5$
$x_1, x_2 \ge 0$

where x_1 and x_2 are integers.

11

Dynamic Programming

11.1 Introduction

Dynamic programming is an optimization technique in which a complex optimization problem is divided into a number of *stages* (or subproblems) in which a policy decision has to be taken at each stage. The stages are solved sequentially, one by one. The stages generally represent a time-varying phenomenon such as the amount of inventory in a store. Dynamic programming thus refers to planning of a time-varying system. The series of interrelated decisions taken at each stage is done using the *state* information associated with that stage and has to be suitably linked with the next stage. The dimensionality of the problem increases with an increase in the number of states. The series of best policy decisions taken at each stage is referred to as the optimal policy of the optimization problem. The *principle of optimality* in dynamic programming states that the optimal decision at a given stage is independent of the optimal decisions taken in the previous stages. Typically in dynamic programming, the optimal decision pertaining to the last stage is taken first and then moved backward to the next stage and the process is continued until the first stage is reached. The technique of dynamic programming was developed by Richard Bellman in the 1950s. The method is used to solve a number of problems in different areas (Edwin and Gruber 1971; George 1963; Leondes and Smith 1970). The method, though easy to implement, has a serious drawback: the complexity of the problem increases with an increase in the number of variables. This is frequently referred to as the *curse of dimensionality* in dynamic programming. This chapter discusses aspects of deterministic and probabilistic dynamic programming.

11.2 Deterministic Dynamic Programming

In dynamic programming, when the current policy decision and the state completely determine the state of the next stage, it is called deterministic dynamic programming. Let the state at stage n be denoted by s_n. The policy decision x_n transforms this state to s_{n+1} at the next stage $n + 1$. The function $f^*_{n+1}(s_{n+1})$ is the optimal value of the objective function to which the contribution made by x_n decision is to be added (Figure 11.1). This provides the contribution of n stages and is given by $f_n(s_n, x_n)$. This function is optimized with respect to x_n to give $f^*_n(s_n) = f_n(s_n, x^*_n)$. The procedure is repeated by moving back one stage.

Let us take an example to explain the concept of dynamic programming. A person in a remote place A has to reach city I to withdraw money from an ATM. Though he has the option to select different paths to reach his goal, he is interested in finding the path that has a minimum distance to be covered. The intermediate villages where he can change his path are given by B, C, D, E, F, G, and H. The distance between the villages is given in Figure 11.2.

Before using dynamic programming, let us select the path that results in the minimum distance from one city to another. From village A, the minimum distance is 3 to village C. From village C, the minimum distance is 6 to village F. In this way the total distance traveled is 16 and the path is

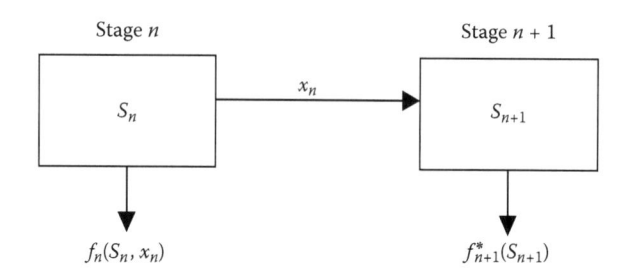

$$A \to C \to F \to G \to I$$

FIGURE 11.1
Structure of deterministic dynamic programming.

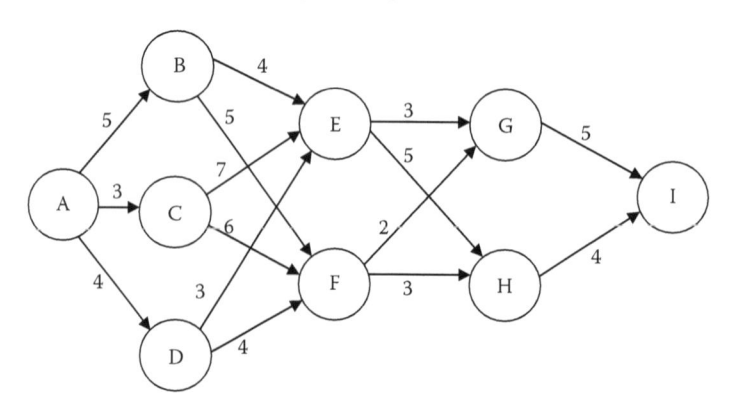

FIGURE 11.2
Distance (not to scale) between the villages.

Dynamic Programming

For an n stage problem in dynamic programming, the current stage is designated as n and the current state is s_n. The policy decision variable is given by x_n and the optimal policy is given by the recursive relationship

$$f_n^*(s_n) = \min\left\{c_{sx_n} + f_{n+1}^*(x_n)\right\} \tag{11.1}$$

where c_{sx_n} is the cost for stage n and $f_{n+1}^*(x_n)$ is the cost for stages $n + 1$ and higher. Equation 11.1 is minimized with respect to x_n. This is a four-stage problem and we start from the last stage ($n = 4$), as shown in Figure 11.3.

At this stage, a person can be either at G or H. If he is at G, the shortest distance (in fact, it is the only path) to reach the destination (I) is 5. Similarly, if he is at H, the shortest distance to reach the destination is 4. The results of stage 4 are summarized in Table 11.1.

Let's go back one stage ($n = 3$). At stage 3, a person can be either at E or F. If he is at E, he has two paths, to go either to G or H, and the distance to be covered is 3 and 5 respectively. The additional distance from G (or H) to I, which is computed in the last stage, is to be added at this stage. The distance covered for the route E–G–I is 8 and for E–H–I it is 9 (Figure 11.4). Similarly, one can compute distance for the path F–G–I and F–H–I. The results of stage 3 are given in Table 11.2.

Let's go back one more stage ($n = 2$). At stage 2, a person can be at B, C, or D. From here, his immediate destination can be E or F. The minimum distance from E and F to the destination was already computed in Table 11.1. The results for stage 2 are mentioned in Table 11.3.

In a similar manner results for stage 1 are summarized in Table 11.4.

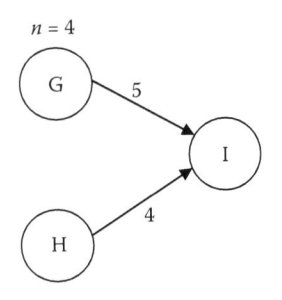

FIGURE 11.3
Stage 4.

TABLE 11.1
Stage 4

s	$f_4(s)$	x_4^*
G	5	I
H	4	I

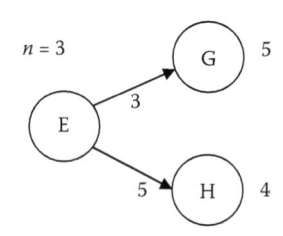

FIGURE 11.4
Stage 3.

TABLE 11.2

Stage 3

$$f_3(s,x_3) = c_{sx_3} + f_4^*(x_4)$$

s	G	H	$f_3^*(s)$	x_3^*
E	8	9	8	G
F	7	7	7	G, H

TABLE 11.3

Stage 2

$$f_2(s,x_2) = c_{sx_2} + f_3^*(x_3)$$

s	E	F	$f_2^*(s)$	x_2^*
B	12	12	12	E, F
C	15	13	13	F
D	11	11	11	E, F

TABLE 11.4

Stage 1

$$f_1(s,x_1) = c_{sx_1} + f_2^*(x_2)$$

s	B	C	D	$f_1^*(s)$	x_1^*
A	17	16	15	15	D

Thus the minimum distance from A to destination I is 15. There are three optimal paths for this problem:

$$A \rightarrow D \rightarrow E \rightarrow G \rightarrow I$$
$$A \rightarrow D \rightarrow F \rightarrow G \rightarrow I$$
$$A \rightarrow D \rightarrow F \rightarrow H \rightarrow I$$

Dynamic Programming

Example 11.1

Solve the following linear programming problem (LPP) using dynamic programming.

Maximize $\quad z = 2x_1 + 3x_2$
subject to
$$x_1 \leq 3$$
$$2x_2 \leq 11$$
$$2x_1 + 3x_2 \leq 12$$
$$x_1 \geq 0, x_2 \geq 0$$

This is a two-stage problem because it contains two interacting variables. The states in the problem are the right-hand side of the inequality constraints. For the first stage, the resources available for the first activity (x_1) are

$$s_1 = \{3, 11, 12\}$$

When x_1 is allocated, the remaining resources for the next state (Figure 11.5) will be

$$s_2 = \{3 - x_1, 11, 12 - 2x_1\}$$

The stage 2 problem can be written

Maximize $\quad z = 3x_2$
subject to
$$2x_2 \leq 11$$
$$3x_2 \leq 12$$

Thus maximum allocation of x_2 is limited by

$$\min\left\{\frac{11}{2}, \frac{12 - 2x_1}{3}\right\}$$

Stage 1

$s_1 = \{3, 11, 12\}$

$f_1(s_1, x_1)$

Stage 2

$s_2 = \{3 - x_1, 11, 12 - 2x_1\}$

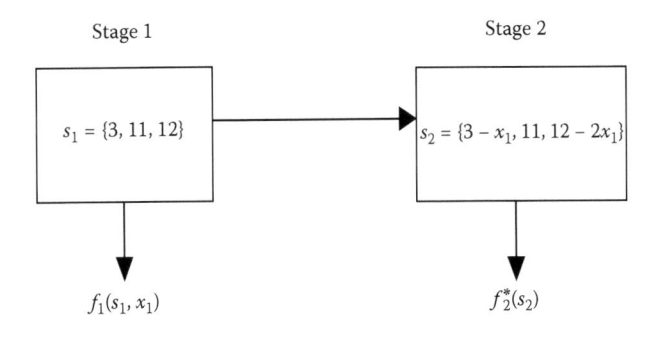

$f_2^*(s_2)$

FIGURE 11.5
Two-stage problem.

Clearly, the minimum of the two terms is

$$\frac{12-2x_1}{3}, x_1 \geq 0$$

Thus,

$$f_2^*(s_2) = 3 \min\left\{\frac{11}{2}, \frac{12-2x_1}{3}\right\} = 12 - 2x_1$$

Therefore,

$$f_1^*(s_1) = \left\{2x_1 + 3 \min\left\{\frac{11}{2}, \frac{12-2x_1}{3}\right\}\right\} = 2x_1 + 12 - 2x_1 = 12$$

Thus, the maximum value of the function is 12 and occurs at $x_1^* = 0$. Substituting the value of x_1 in one of the constraint equations gives $x_2^* = 4$.

11.3 Probabilistic Dynamic Programming

In deterministic dynamic programming, the state and decisions of the present stage completely determine the state of the next stage. In probabilistic dynamic programming, the state of the next stage is determined with some probability distribution. Let us take the following example, which is solved using the concept of probabilistic dynamic programming.

A milk vendor purchases six cases of milk from a dairy farm for Rs. 900 per case. He has three booths where he can sell the milk at Rs. 2000 per case. Any unsold milk of the day can be returned back to the dairy farm at a rate of Rs. 500 per case. The demand at the three booths has certain probabilities and is given in Table 11.5.

Find the optimal policy in allocating six cases of milk to different booths so as to maximize the profit.

To maximize profits, we need to maximize the revenue as cost is fixed. Like previous problems, the first step in the dynamic programming is to identify the stages, states, and decision policies. In this problem, number of stages refers to the number of booths. Thus, it is a three-stage problem. The state at each stage is the number of milk cases available for allocation and let it be denoted by s_i for the ith stage. Let the decision policy of allocating number of cases of milk to a particular booth be denoted by x_i. Let $r_i(x_i)$ represent the revenue earned by allocating x_i cases to ith store and $f_i(s_i)$ represent the maximum expected revenue earned by assigning x_i cases to the ith store. As with the earlier problems, we will start with the last stage. Before that, let us compute the elements of the revenue table $r_i(x_i)$ for $0 \leq x_i \leq 3$ as maximum

Dynamic Programming

TABLE 11.5
Demands from Different Booths

	Demand (in Number of Cases)	Probability
Booth 1	1	0.5
	2	0.4
	3	0.1
Booth 2	1	0.5
	2	0.3
	3	0.2
Booth 3	1	0.6
	2	0.2
	3	0.2

demand at any store is 3. Let us illustrate this by taking a case for booth 1 where two cases of milk are to be allocated. This is denoted by $r_1(2)$.

$$r_1(2) = 0.5 \times 2500 + 0.4 \times 4000 + 0.1 \times 4000 = 3250$$

In a similar manner, other elements of $r_i(x_i)$ can be constructed and are given in Table 11.6.

The state and decision policies for different stages are summarized in Tables 11.7 through 11.9.

TABLE 11.6
Revenue Earned by Allocating Resources

x_i	$r_1(x_1)$	$r_2(x_2)$	$r_3(x_3)$
0	0	0	0
1	2000	2000	2000
2	3250	3250	3500
3	3900	4050	3900

TABLE 11.7
Stage 3

		$r_3(x_3)$				
s_3	0	1	2	3	$f_3^*(s_3)$	x_3^*
0	0	–	–	–	0	0
1	0	2000	–	–	2000	1
2	0	2000	3500	–	3500	2
3	0	2000	3500	3900	3900	3
4	0	2000	3500	3900	3900	3
5	0	2000	3500	3900	3900	3
6	0	2000	3500	3900	3900	3

TABLE 11.8

Stage 2

| | \multicolumn{4}{c}{$r_2(x_2) + f_3(s_2 - x_2)$} | | | | |
s_2	0	1	2	3	$f_2^*(s_2)$	x_2^*
0	0	–	–	–	0	0
1	2000	2000	–	–	2000	0, 1
2	3500	4000	3250	–	4000	1
3	3900	5500	5250	4050	5500	1
4	3900	5900	6750	6050	6750	2
5	3900	5900	7150	7550	7550	3
6	3900	5900	7150	7950	7950	3

TABLE 11.9

Stage 1

| | \multicolumn{4}{c}{$r_1(x_1) + f_2(s_1 - x_1)$} | | |
s_1	0	1	2	3	$f_1^*(s_1)$	x_1^*
6	7950	9550	10,000	9400	10,000	3

The optimal policy is to allocate three cases of milk to booth 1, one case of milk to booth 2, and two cases of milk to booth 3.

Chapter Highlights

- Dynamic programming refers to planning of time-varying systems.
- In dynamic programming, a complex optimization problem is divided into a number of *stages* (or subproblems) in which a policy decision has to be taken at each stage.
- The series of interrelated decisions taken at each stage is done using the *state* information associated with that stage and has to be suitably linked with the next stage.
- The *principle of optimality* in dynamic programming states that the optimal decision at a given stage is independent of the optimal decisions taken in the previous stages.
- In dynamic programming, when the current policy decision and the state completely determine the state of the next stage, it is called deterministic dynamic programming.

Dynamic Programming

Formula Chart

Recursive relationship:

$$f_n^*(s_n) = \min\left\{c_{sx_n} + f_{n+1}^*(x_n)\right\}$$

Problems

1. Solve the following LPP using dynamic programming.

   i. Minimize $\quad z = 3x_1 - 2x_2$
       subject to $\quad x_1 + 2x_2 \leq 10$
       $\quad\quad\quad\quad\quad 2x_1 - x_2 \leq 5$
       $\quad\quad\quad\quad\quad -4x_1 + 3x_2 \geq 5$
       $\quad\quad\quad\quad\quad x_1, x_2 \geq 0$

   ii. Maximize $\quad z = 2x_1 + 5x_2$
       subject to $\quad 3x_1 + x_2 \leq 11$
       $\quad\quad\quad\quad\quad x_1 - x_2 \leq 6$
       $\quad\quad\quad\quad\quad -2x_1 + x_2 \leq 10$
       $\quad\quad\quad\quad\quad x_1, x_2 \geq 0$

   iii. Maximize $\quad z = 4x_1 + 5x_2$
        subject to $\quad 2x_1 + x_2 \leq 20$
        $\quad\quad\quad\quad\quad -3x_1 + 2x_2 \leq 25$
        $\quad\quad\quad\quad\quad -x_1 + x_2 \leq 30$
        $\quad\quad\quad\quad\quad x_1, x_2 \geq 0$

2. Solve the following integer programming problem using dynamic programming.

   Minimize $\quad f(x) = -3x_1 - 2x_2$
   subject to $\quad 2x_1 + x_2 \leq 17$
   $\quad\quad\quad\quad\quad 2x_1 + 3x_2 \leq 40$
   $\quad\quad\quad\quad\quad 3x_1 + 3x_2 \leq 26$
   $\quad\quad\quad\quad\quad x_1, x_2 \geq 0$

   where x_1 and x_2 are integers.

3. Find the optimal policy of the stagecoach problem (Figure 11.6) to minimize the distance from A to I.

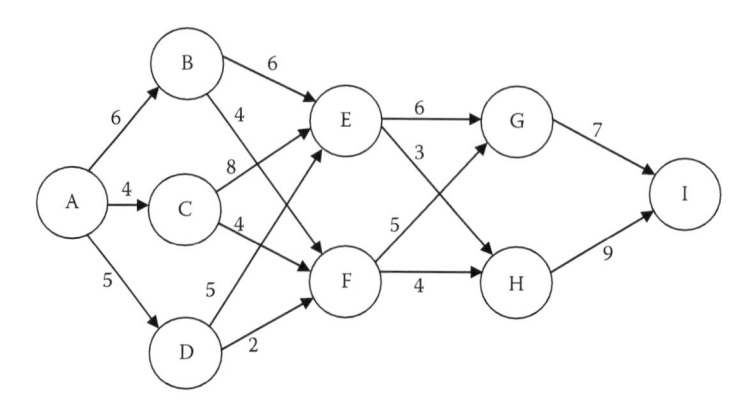

FIGURE 11.6
Stagecoach problem.

4. A system consists of three components (R_1, R_2, and R_3) arranged in series. The reliability of the system is given by

$$R = R_1 R_2 R_3$$

The reliability of each component can be increased by arranging (in parallel) itself to a similar component. If r_i is the reliability of each component, then reliability of the subsystem in parallel is given by

$$R_i = 1 - (1 - r_i)^{n_i}$$

where n_i is the number of components arranged in parallel. The costs of various components along with their reliabilities are given in Table 11.10.

Maximize the reliability of the system if an amount of $700 is available for investment.

TABLE 11.10

Cost and Reliability of Various Components

Component	Cost ($)	Reliability of Each Component
1	100	0.93
2	150	0.96
3	190	0.98

Bibliography

Chapter 1

Agnew, R.P. 1960. *Differential Equations*. New York: McGraw-Hill.

Arora, R.K., and Pradeep, K. 2003. Aerodynamic Shape Optimization of a Re-Entry Capsule. In *AIAA Atmospheric Flight Mechanics Conference and Exhibit*. AIAA-5394-2003, Texas.

Ashok, D.B., and Tirupathi, R.C. 2011. *Optimization Concepts and Applications in Engineering*. Cambridge, UK: Cambridge University Press.

Bellman, R. 1953. An Introduction to the Theory of Dynamic Programming. RAND Corp. Report.

Betts, J.T. 1998. Survey of Numerical Methods for Trajectory Optimization. *Guidance, Control and Dynamics* 21(2):193–207.

Burghes, D.N., and Wood, A.D. 1980. *Mathematical Models in the Social, Management and Life Sciences*. Hemstead, UK: Ellis Horwood.

Dantzig, G.B. 1949. Programming of Interdependent Activities: II Mathematical Model. *Econometrica* 17(3):200–211.

Dantzig, G.B. 1990. The Diet Problem. *Interfaces* 20(4):43–47.

Deb, K. 1995. *Optimization for Engineering Design: Algorithms and Examples*. Upper Saddle River, NJ: Prentice Hall.

Epperson, J.F. 2010. *An Introduction to Numerical Methods and Analysis*. Hoboken, NJ: John Wiley & Sons.

Gomory, R.E. 1958. Outline of an Algorithm for Integer Solutions to Linear Programs. *Bulletin of American Society* 64:275–278.

Griva, I., Nash, S.G., and Sofer, A. 2009. *Linear and Nonlinear Optimization*. Philadelphia: SIAM.

Hancock, H. 1917. *Theory of Maxima and Minima*. Boston: Ginn and Company.

Jaluria, Y. 2008. *Design and Optimization of Thermal Systems*. Boca Raton, FL: CRC Press.

Kantorovich, L.V. 1939. Mathematical Methods of Organizing and Planning Production. Leningrad State University.

Karush, W. 1939. Minima of Functions of Several Variables with Inequalities as Side Constraints. M.Sc. Dissertation. Illinois: University of Chicago.

King, J.R. 1975. *Production, Planning and Control: An Introduction to Quantitative Methods*. Oxford: Pergamon Press.

Kuhn, H.W., and Tucker, A.W. 1951. Nonlinear Programming. *Proceedings of the Second Berkeley Symposium on Mathematical Statistics and Probability*. University of California Press, 481–492.

Pravica, D.W., and Spurr, M.J. 2010. *Mathematical Modeling for the Scientific Method*. Burlington, MA: Jones and Bartlett Learning.

Rao, S.S. 2009. *Engineering Optimization: Theory and Practice.* Hoboken, NJ: John Wiley & Sons.
Schafer, M. 2006. *Computational Engineering: Introduction to Numerical Methods.* New York: Springer Science+Business Media.
Shepley, R.L. 1984. *Differential Equations.* New York: John Wiley & Sons.
Spivey, W.A., and Thrall, R.M. 1970. *Linear Optimization.* New York: Holt, Rinehart and Winston.
Venkataraman, P. 2009. *Applied Optimization with MATLAB Programming.* Hoboken, NJ: John Wiley & Sons.
Vinh, N.X. 1981. *Optimal Trajectories in Atmospheric Flight.* Amsterdam: Elsevier.
Watkins, D.S. 2010. *Fundamentals of Matrix Computations.* Hoboken, NJ: John Wiley & Sons.
Wismer, D.A., and Chattergy, R. 1978. *Introduction to Nonlinear Optimization: A Problem Solving Approach.* Amsterdam: Elsevier/North Holland.

Chapter 2

Deb, K. 1995. *Optimization for Engineering Design: Algorithms and Examples.* Upper Saddle River, NJ: Prentice Hall.
Dennis, J.E., and Schnabel, R.B. 1983. *Numerical Methods for Unconstrained Optimization and Nonlinear Equations.* Upper Saddle River, NJ: Prentice Hall.
Kincaid, D., and Cheney, W. 2009. *Numerical Analysis: Mathematics of Scientific Computing.* Providence, RI: American Mathematical Society.
King, R.M., and Mody, N.A. 2011. *Numerical and Statistical Methods for Biomedical Engineering: Applications in MATLAB.* Cambridge, UK: Cambridge University Press.
Philips, D.T., Ravindran, A., and Solberg, J.J. 1976. *Operations Research: Principles and Practice.* New York: John Wiley & Sons.
Reklaitis, G.V., Ravindran, A., and Ragsdell, K.M. 1983. *Engineering Optimization: Methods and Applications.* New York: John Wiley & Sons.

Chapter 3

Andreas, A., and Wu-Shey, L. 2007. *Practical Optimization: Algorithms and Engineering Applications.* New York: Springer Science+Business Media.
Ashok, D.B., and Tirupathi, R.C. 2011. *Optimization Concepts and Applications in Engineering.* Cambridge, UK: Cambridge University Press.
Cauchy, A.L. 1847. Methode generale pour la resolution des systemes d'equations simultanees. *Comptes Rendus de l'Académie des Sciences* 25:536–538.
Colville, A.R. 1968. *A Comparative Study on Nonlinear Programming Codes,* Report 320-2949, IBM. New York Scientific Centre.

Dennis, J.E., and Schnabel, R.B. 1983. *Numerical Methods for Unconstrained Optimization and Nonlinear Equations.* Upper Saddle River, NJ: Prentice Hall.

Fletcher, R., and Powell, M.J.D. 1963. A Rapidly Convergent Descent Method for Minimization. *The Computer Journal* 6:163–168.

Fletcher, R., and Reeves, C.M. 1964. Function Minimization by Conjugate Gradients. *The Computer Journal* 7(2):149–154.

Freudenstein, F., and Roth, B. 1963. Numerical Solutions of Systems of Nonlinear Equations. *Journal of ACM* 10(4):550–556.

Griva, I., Nash, S.G., and Sofer, A. 2009. *Linear and Nonlinear Optimization.* Philadelphia: SIAM.

Haftka, R.T., and Gurdal, Z.A. 1992. *Elements of Structural Optimization.* Dordrecht, the Netherlands: Kluwer.

Levenberg, K. 1944. A Method for the Solution of Certain Nonlinear Problems in Least Squares. *Quarterly Applied Mathematics* 2(2):164–168.

Marquardt, D. 1963. An Algorithm for Least Squares Estimation of Nonlinear Parameters. *SIAM Journal of Applied Mathematics* 11(2):431–441.

Nelder, J.A., and Mead, R. 1965. A Simplex Method for Function Minimization. *Computer Journal* 7(2):308–313.

Nowak, U., and Weimann, L. 1991. *A Family of Newton Codes for Systems of Highly Nonlinear Equation.* TR 91-10. Berlin: Konard Zuse Zentrumf. Informationstechn.

Powell, M.J.D. 1970. A Hybrid Method for Nonlinear Equations. Chapter 6. In: *Numerical Methods for Nonlinear Algebraic Equations.* P. Rabinowitz (ed.), Gordon and Breach, 87–114.

Rao, S.S. 2009. *Engineering Optimization: Theory and Practice.* Hoboken, NJ: John Wiley & Sons.

Reklaitis, G.V., Ravindran, A., and Ragsdell, K.M. 1983. *Engineering Optimization: Methods and Applications.* New York: John Wiley & Sons.

Rosenbrock, H.H. 1960. An Automatic Method for Finding the Greatest or Least Value of a Function. *The Computer Journal* 3(3):175–184.

Shanno, D.F. 1970. Conditioning of Quasi-Newton Methods for Function Minimization. *Mathematics of Computation* 24:647–656.

Venkataraman, P. 2009. *Applied Optimization with MATLAB Programming.* Hoboken, NJ: John Wiley & Sons.

Chapter 4

Barnes, E.R. 1986. A Variation on Karmarkar's Algorithm for Solving Linear Programming Problems. *Mathematical Programming* 36:174–182.

Dantzig, G.B. 1963. *Linear Programming and Extensions.* Princeton, NJ: Princeton University Press.

Fishback, P.E. 2010. *Linear and Nonlinear Programming with MAPLE: An Interactive Applications-Based Approach.* Boca Raton, FL: CRC Press.

Griva, I., Nash, S.G., and Sofer, A. 2009. *Linear and Nonlinear Optimization.* Philadelphia: SIAM.

Karmarkar, N. 1984. A New Polyline-Time Algorithm for Linear Programming, *Combinatorica* 4(4):373–395.

Rao, S.S. 2009. *Engineering Optimization: Theory and Practice*. Hoboken, NJ: John Wiley & Sons.
Spivey, W.A., and Thrall, R.M. 1970. *Linear Optimization*. New York: Holt, Rinehart and Winston.
Vanderbei, R.J., Meketon, M.S., and Freedman, B.A. 1986. A Modification of Karmarkar's Linear Programming Problem. *Algorithmica* 1:395–407.

Chapter 5

Davis, L. 1987. *Genetic Algorithm and Simulated Annealing*. London: Pitman.
Deb, K. 1995. *Optimization for Engineering Design: Algorithms and Examples*. Upper Saddle River, NJ: Prentice Hall.
Dorigo, M. 1992. *Optimization, Learning and Natural Algorithms*, PhD dissertation, Dipartimento di Elettronica, Politecnico di Milano. Italy.
Dorigo, M., Maniezzo, V., and Colorni, A. 1996. Ant System: Optimization by a Colony of Cooperating Agents. *IEEE Transactions on Systems, Man, and Cybernetics B* 26(1):29–41.
Gendreau, M., and Potvin, J.Y. 2010. *Handbook of Metaheuristics*, International Series in Operations Research and Management Science, Vol. 146. Heidelberg and New York: Springer Science+Business Media.
Glover, F. 1986. Future Paths for Integer Programming and Links to Artificial Intelligence. *Computers and Operations Research* 13:533–549.
Goldberg, D.E. 1989. *Genetic Algorithms in Search Optimization and Machine Learning*. Reading, MA: Addison-Wesley.
Goldberg, D.E., Deb, K., and Thierens, D. 1993. Towards a Better Understanding of Mixing in Genetic Algorithms. *Journal of the Society of Instrument and Control Engineers* 32(1):10–16.
Hartl, D.L. 1991. *Basic Genetics*. Burlington, MA: Jones & Bartlett.
Holland, J.H. 1992. *Adaptation in Natural and Artificial Systems*. Cambridge, MA: MIT Press.
Kennedy, J., and Eberhart, R. 2001. *Swarm Intelligence*. San Diego: Academic Press.
Laarhoven, P.J.M.V., and Aarts, E.H.L. 1987. *Simulated Annealing: Theory and Applications*. Dordrecht: Kluwer.
Laskari, E.C., Parsopoulos, K.E., and Vrahatis, M.N. 2002. Particle Swarm Optimization for Minmax Problems. *Evolutionary Computation Proceedings* 2:1576–1581.
Merkle, D., Middendorf, M., and Schmeck, H. 2002. Ant Colony Optimization for Resource-Constrained Project Scheduling. *IEEE Transactions on Evolutionary Computation* 6(4):333–346.
Michalewicz, Z. 1996. *Genetic Algorithm + Data Structures = Evolution Programs*. Heidelberg and New York: Springer-Verlag.
Mitchell, M. 1996. *An Introduction to Genetic Algorithms*. Cambridge, MA: MIT Press.
Pham, D.T., and Karaboga, D. 2000. *Intelligent Optimization Techniques: Genetic Algorithm, Tabu Search, Simulated Annealing and Neural Networks*. Heidelberg and New York: Springer Science+Business Media.

Reimann, M., Doerner, K., and Hartl, R.F. 2004. D-Ants: Savings Based Ants Divide and Conquer the Vehicle Routing Problems. *Computers & Operations Research* 31(4):563–591.

Shmygelska, A., and Hoos, H.H. 2005. An Ant Colony Optimization Algorithm for the 2D and 3D Hydrophobic Polar Protein Folding Problem. *BMC, Bioinformatics* 6:30.

Chapter 6

Byrd, R.H., Gilbert, J.C., and Nocedal, J. 2000. A Trust Region Method Based on Interior Point Techniques for Nonlinear Programming. *Mathematical Programming* 89(1):149–185.

Byrd, R.H., Schnabel, R.B., and Shultz, G.A. 1988. Approximate Solution of the Trust Region Problem by Minimization over Two-Dimensional Subspaces. *Mathematical Programming* 40(3):247–263.

Coleman, T.F., and Verma, A. 2001. A Preconditioned Conjugate Gradient Approach to Linear Equality Constrained Minimization. *Computational Optimization and Applications* 20(1):61–72.

Griva, I., Nash, S.G., and Sofer, A. 2009. *Linear and Nonlinear Optimization*. Philadelphia: SIAM.

Han, S.P. 1977. A Globally Convergent Method for Nonlinear Programming. *Journal of Optimization Theory and Applications* 22(3):297–309.

Haug, E.J., and Arora, J.S. 1979. *Applied Optimal Design: Mechanical and Structural Systems*. New York: John Wiley & Sons.

Horowitz, B., Guimaraes, L.J.N., and Afonso, S.M.B. 2008. A Concurrent Efficient Global Optimization Algorithm Applied to Engineering Problems. AIAA 2008-6010.

Moré, J.J., and Sorensen, D.C. 1983. Computing a Trust Region Step. *SIAM Journal on Scientific and Statistical Computing* 3:553–572.

Nocedal, J., and Wright, S.J. 2006. *Numerical Optimization*. Springer Series in Operations Research. New York: Springer Science+Business Media.

Powell, M.J.D. 1978. The Convergence of Variable Metric Methods for Nonlinearly Constrained Optimization Calculations. *Nonlinear Programming 3*. New York: Academic Press.

Ragsdell, K.M., and Philips, D.T. 1976. Optimal Design of a Class of Welded Structures Using Geometric Programming. *ASME Journal of Engineering for Industries* 98(3):1021–1025.

Rao, S.S. 2009. *Engineering Optimization: Theory and Practice*. Hoboken, NJ: John Wiley & Sons.

Reklaitis, G.V., Ravindran, A., and Ragsdell, K.M. 1983. *Engineering Optimization: Methods and Applications*. New York: John Wiley & Sons.

Sandgren, E. 1990. Nonlinear Integer and Discrete Programming in Mechanical Design Optimization. *ASME Journal of Mechanical Design* 112:223–229.

Thygeson, J.R., and Grossmann, E.D. 1970. Optimization of a Continuous Through-Circulation Dryer. *American Institute of Chemical Engineers Journal* 16:749–754.

Chapter 7

Adimurthy, V., Ramanan, R.V., and Pankaj, P. 2012. *Optimization in Aerospace Dynamics.* Bangalore, India: ISRO.

Arora, R.K., and Pradeep, K. 2003. Aerodynamic Shape Optimization of a Re-Entry Capsule. *AIAA Atmospheric Flight Mechanics Conference and Exhibit.* AIAA-5394-2003.

Binh, T., and Korn, U. 1997. MOBES: A Multi-Objective Evolution Strategy for Constrained Optimization Problems. *Proceedings of the Third International Conference on Genetic Algorithms* 176–182.

Chankong, V., and Haimes, Y.Y. 1983. *Multiobjective Decision Making: Theory and Methodology.* Amsterdam: North Holland.

Chernyi, G.G. 1961. *Introduction to Hypersonic Flow.* New York: Academic Press.

Deb, K. 2002. *Multi-Objective Optimization Using Evolutionary Algorithms.* Hoboken, NJ: John Wiley & Sons.

Knowles, J.D., and Corne, D.W. 2000. Approximating the Non-Dominated Front Using the Pareto Archived Evolution Strategy. *Evolutionary Computation* 8(2):149–172.

Miettinen, K. 1999. *Nonlinear Multiobjective Optimization.* Dordrecht: Kluwer.

Parsopoulos, K.E., and Vrahatis, M.N. 2002. Particle Swarm Optimization in Multiobjective Problems. *Proceedings of the ACM 2002 Symposium on Applied Computation*: 603–607.

Statnikov, R.B., and Matusov, J.B. 1995. *Multicriteria Optimization and Engineering,* Springer.

Veldhuizen, D.A.V. 1999. *Multiobjective Evolutionary Algorithms: Classification, Analyzes, and New Innovations.* PhD dissertation, Air Force Institute of Technology, Wright-Patterson AFB.

Zitzler, E., Deb, K., and Thiele, L. 2000. Comparison of Multiobjective Evolutionary Algorithms: Empirical Results. *Evolutionary Computation* 8(2):173–195.

Chapter 8

Abuo-El-Ata, M., Fergany, H., and El-Wakeel, M. 2003. Probabilistic Multi-Item Inventory Model with Varying Order Cost Under Two Restrictions: A Geometric Programming Approach. *International Journal of Production Economics* 83(3):223–231.

Beightler, C.S., Philips, D.T., and Wilde, D.J. 1967. *Foundations of Optimization.* Upper Saddle River, NJ: Prentice Hall.

Cao, B.Y. 2002. *Fuzzy Geometric Programming.* Dordrecht: Kluwer Academic.

Chen, T.Y. 1992. Structural Optimization Using Single-Term Posynomial Geometric Programming. *Computer and Structures* 45(5):911–918.

Chiang, M. 2005. *Geometric Programming for Communication Systems.* Hanover, MA: NOW Publishers.

Chu, C., and Wong, D.F. 2001. VLSI Circuit Performance Optimization by Geometric Programming. *Annals of Operations Research* 105:37–60.

Dawson, J., Boyd, S., Hershenson, M., and Lee, T.H. 2001. Optimal Allocation of Local Feedback in Multistage Amplifiers via Geometric Programming. *IEEE Transactions on Circuits and Systems* 48(1):1–11.

Dembo, R.S. 1976. A Set of Geometric Programming Test Problems and Their Solutions. *Mathematical Programming* 10:192–213.
Dembo, R.S. 1978. Dual to Primal Conversion in Geometric Programming. *Journal of Optimization Theory and Applications* 26:243–252.
Dey, S., and Roy, T.K. 2013. Optimization of Structural Design Using Geometric Programming. *International Journal of Engineering Research and Technology* 2(8).
Duffin, R.J., Peterson, E.L., and Zener, C. 1967. *Geometric Programming*. New York: John Wiley & Sons.
Fang, S.C., Rajasekara, J.R., and Tsao, H.S.J. 1997. *Entropy Optimization and Mathematical Programming*. Dordrecht: Kluwer Academic.
Hajela, P. 1986. Geometric Programming Strategies for Large Scale Structural Synthesis. *AIAA Journal* 24(7):1173–1178.
Hoburg, W., and Abbeel, P. 2012. Geometric Programming for Aircraft Design Optimization. 53rd AIAA/ASME/ASCE/AHS/ASC Structures, Structural Dynamics and Materials Conference, Hawaii. AIAA 2012-1680.
Jefferson, T.R., and Scott, C.H. 1979. Avenues of Geometric Programming—Applications. NZOR 7-1.
Ojha, A.K., and Biswal, K.K. 2010. Posynomial Geometric Programming Problems with Multiple Parameters. *Journal of Computing* 2(1).
Phillips, D.T., Ravindran, A., and Solberg, J.J. 1976. *Operations Research—Principles and Practice*. John Wiley & Sons.
Rao, S.S. 2009. *Engineering Optimization: Theory and Practice*. New Jersey: John Wiley & Sons.
Stoecker, W.F. 1971. *Design of Thermal System*. New York: McGraw-Hill.

Chapter 9

Allison, J.T., and Papalambros, P.Y. 2008. Consistency Constraint Allocation in Augmented Lagrangian Coordination. Distance Education and Training Council (DETC), 49823.
Anderson, V.L., and McLean, R.A. 1974. *Design of Experiments: A Realistic Approach*. New York: Marcel Dekker.
Balesdent, M., Berend, N., and Depince, P. 2010. Optimal Design of Expendable Launch Vehicles using Stage-Wise MDO Formulation. 13th AIAA/ISSMO Multidisciplinary Analysis Optimization Conference, Texas. AIAA 2010-9324.
Cornell, J.A., and Khuri, A.I. 1996. *Response Surfaces: Design and Analyses*. Marcel Dekker.
Dapeng, W., Gary, W.G., and Naterer, G.F. 2007. Extended Collaboration Pursuing Method for Solving Larger Multidisciplinary Design Optimization Problems. *AIAA Journal* 45(6):1208–1221.
Fletcher, R. 1981. *Practical Methods of Optimization: Constrained Optimization*. New York: John Wiley & Sons.
Geyer, P. 2009. Component-Oriented Decomposition for Multidisciplinary Design Optimization in Building Design. *Advanced Engineering Informatics* 23(1):12–31.
Goldberg, D.E. 1989. *Genetic Algorithm in Search, Optimization, and Machine Learning*. Reading, MA: Addison-Wesley.

He, Y., and McPhee, J. 2005. Multidisciplinary Optimization of Multi-Body Systems with Application to the Design of Rail Vehicles. *Multi-Body System Dynamics* 14(2):111–135.

Herskovits, J. 2004. A Mathematical Programming Algorithms for Multidisciplinary Design Optimization. 10th AIAA/ISSMO Multidisciplinary Analysis Optimization Conference, Texas. AIAA 2004-4502.

IIT Mumbai. 2004. Workshop on Optimization Based Design/Multidisciplinary Design Optimization (Course Notes), Trivandrum.

Kodiyalam, S., and Sobieszczanski-Sobieski, J. 2000. Bi-Level Integrated System Synthesis with Response Surfaces. *AIAA Journal* 38(8):1479–1485.

Kroo, I., and Manning, V. 2000. Collaborative Optimization: Status and Directions. 8th AIAA/ISSMO Multidisciplinary Analysis Optimization Conference, Texas. AIAA 2000-4721.

Manokaran, K., Vidya, G., and Sivaramakrishnan, A.E. 2009. Wing Planform Design Optimization for Reusable Launch Vehicle. *Journal of Aircraft* 46(2):726–730.

Martins, J.R.R.A., and Lambe, A.B. 2013. Multidisciplinary Design Optimization: A Survey of Architectures. *AIAA Journal* 51(9):2049–2075.

McAllister, C.D., and Simpson, T.W. 2003. Multidisciplinary Robust Design Optimization of an Internal Combustion Engine. *Journal of Mechanical Design* 125(1):124–130.

Sobieszczanski-Sobieski, J., Agte, J.S., and Sandusky, R.R., Jr. 2000. Bi-Level Integrated System Synthesis. *AIAA Journal* 38(1):164–172.

Sobieszczanski-Sobieski, J., and Haftka, R.T. 1996. Multidisciplinary Aerospace Design Optimization: Survey of Recent Developments. 34th AIAA Aerospace Sciences Meeting and Exhibit, Reno. AIAA 96-0711.

Tedford, N.P., and Martins, J.R.R.A. 2006. On the Common Structure of MDO Problems: A Comparison of Architectures. 11th AIAA/ISSMO Multidisciplinary Analysis Optimization Conference, Texas. AIAA 2006-7080.

Tedford, N.P., and Martins, J.R.R.A. 2010. Benchmarking Multidisciplinary Design Optimization Algorithms. *Optimization and Engineering* 11:159–183.

Xiaoqian Chen Li Yan, Wencai Luo, Ling Xu, Young Zhao, and Zhenguo Wang. 2006. Research on Theory and Application of Multidisciplinary Design Optimization of Flight Vehicles. 2nd AIAA MDO Specialist Conference, VA. AIAA 2006-1721.

Yoder, S., and Brockman, J. 1996. A Software Architecture for Collaborative Development and Solution of MDO Problems (Multidisciplinary Design Optimization). AIAA, NASA and ISSMO Symposium on Multidisciplinary Analysis and Optimization, VA. AIAA 96-4103.

Yushin Kim, Yong-Hee Jeon, and Dong-Ho Lee. 2006. Multi-Objective and Multi-Disciplinary Design Optimization of Supersonic Fighter Wing. *Journal of Aircraft* 43(3):817–824.

Chapter 10

Balas, E. 1965. An Additive Algorithm for Solving Linear Programs with Zero-One Variables. *Operations Research* 13(4):517–546.

Bricker, D.L. 1999. *Notes on Balas' Additive Algorithm.* Ames: University of Iowa Press.

Dakin, R.J. 1965. A Tree-Search Algorithm for Mixed-Integer Programming Problems. *Computer Journal* 8(3):250–255.
Garfinkel, R.S., and Nemhauser, G.L. 1972. *Integer Programming*. New York: John Wiley & Sons.
Gomory, R.E. 1960. *An Algorithm for the Mixed Integer Problem*. Rand Report, R.M. 25797.
Gomory, R.E., and Baumol, W.J. 1960. Integer Programming and Pricing. *Econometrica* 28(3):521–550.
Kennedy, J., and Eberhart, R.C. 2001. *Swarm Intelligence*. San Diego: Academic Press.
Land, A.H., and Doig, A. 1960. An Automatic Method for Solving Discrete Programming Problems. *Econometrica* 28:497–520.
Laskari, E.C., Parsopoulos, K.E., and Vrahatis, M.N. 2002. Particle Swarm Optimization for Integer Programming. *Evolutionary Computation Proceedings* 2:1576–1581.
Rao, S.S. 2009. *Engineering Optimization: Theory and Practice*. Hoboken, NJ: John Wiley & Sons.
Shenoy, G.V., Srivastava, U.K., and Sharma, S.C. 1986. *Operations Research for Management*. New Delhi: New Age International Publishers.
Wismer, D.A., and Chattergy, R. 1978. *Introduction to Nonlinear Optimization: A Problem Solving Approach*. Amsterdam: North Holland.

Chapter 11

Bellman, R.E. 1957. *Dynamic Programming*. Princeton, NJ: Princeton University Press.
Bertsekas, D.P. 1987. *Dynamic Programming: Deterministic and Stochastic Models*. Upper Saddle River, NJ: Prentice Hall.
Denardo, E.V. 1982. *Dynamic Programming Models and Applications*. Upper Saddle River, NJ: Prentice Hall.
Dreyfus, S. 2010. Modern Computational Applications of Dynamic Programming. *Journal of Industrial and Systems Engineering* 4(3):152–155.
Edwin, J.E., and Gruber, M.J. 1971. Dynamic Programming Application in Finance. *The Journal of Finance* 26(2):473–506.
George, R.E. 1963. Nuclear Rocket Thrust Optimization using Dynamic Programming. *AIAA Journal* 1(5):1159–1166.
Hadley, G. 1964. *Nonlinear and Dynamic Programming*. Reading, MA: Addison-Wesley.
Hillier, F.S., and Lieberman, G.J. 2010. *Introduction to Operations Research*. New York: McGraw-Hill.
Howard, R.A. 1960. *Dynamic Programming and Markov Processes*. John Wiley & Sons.
Leondes, C.T., and Smith, F.T. 1970. Optimization of Interplanetary Orbit Transfers by Dynamic Programming. *Journal of Spacecraft and Rockets* 7(5):558–564.
Powell, W.B. 2010. *Approximate Dynamic Programming: Solving the Curses of Dimensionality*. Hoboken, NJ: John Wiley & Sons.
Puterman, M.L. 2005. *Markov Decision Processes: Discrete Stochastic Dynamic Programming*. Hoboken, NJ: John Wiley & Sons.
Smith, D.K. 1991. *Dynamic Programming: A Practical Introduction*. Ellis Horwood.
Sniedovich, M. 1992. *Dynamic Programming*. New York: Marcel Dekker.

Appendix A

MATLAB, The MathWorks Inc., 3 Apple Hill Drive, Natick, MA, 01760-2098, USA, Tel: 508-647-7000, Fax: 508-647-7001, email: info@mathworks.com, web: mathworks.com.

Pratap, R. 2010. *Getting Started with MATLAB: A Quick Introduction for Scientists and Engineers.* Oxford and New York: Oxford University Press.

Appendix A: Introduction to MATLAB®

A.1 Introduction

MATLAB is a software package of The MathWorks Inc., for technical computing that does both computing and visualization with ease. It has a number of built-in functions that can be used by an individual's application. The acronym MATLAB stands for MATrix LABoratory. Matrices are the basic building blocks of MATLAB. Though MATLAB is primarily used for numerical computations, it also supports symbolic computations. The main advantage of MATLAB is the ease with which one can translate the idea into an application. MATLAB runs on almost all computer platforms, whether Microsoft Windows, Apple Macintosh or Unix. On Microsoft Windows, MATLAB can be started by double clicking the MATLAB shortcut icon. See Figure A.1 for a typical desktop of MATLAB.

Observe that the desktop has four windows: current folder, command window, workspace, and command history. The command prompt is shown by >>. All commands are to be typed here. The command history windows keep a record of the previously typed commands across multiple sessions. The previously typed command in this window can be double-clicked so that it can be executed again. All files listed in the left window correspond to the current folder directory. The file can be opened for editing by simply double-clicking on it. The type and size of the variables are shown in the workspace window (empty in this figure). There is a provision to select the variables and plot them.

A.2 Matrices and Arrays

Type the matrix A in the command prompt

```
>> A = [1 2 3; 4 -1 -2; 5 6 7]
```

Then press enter. The following output is displayed in the command window.

```
A =
     1     2     3
     4    -1    -2
     5     6     7
```

Observe that a declaration of dimensions of A is not required. Let us learn few more commands.

```
>> sum(A)
ans =
    10     7     8
```

The sum function adds the elements of each column. To get sum of each row

```
>> sum(A')
ans =
     6     1    18
```

where A' is the transpose of the matrix A. The diagonal elements can be obtained using the diag function.

```
>> diag(A)
ans =
     1
    -1
     7
```

Simultaneous use of functions in a single command is also permissible. For example,

```
>> sum(diag(A))
ans =
     7
```

Suppose we want to assign the element −2 in matrix A to a variable x. The element −2 is in second row and third column of A. Then

```
>> x = A(2,3)
x =
    -2
```

Consider the colon operator

```
>> x = 1:2:10
```

The output is a row vector containing numbers from 1 to 10 in steps of 2:

```
ans =
     1     3     5     7     9
```

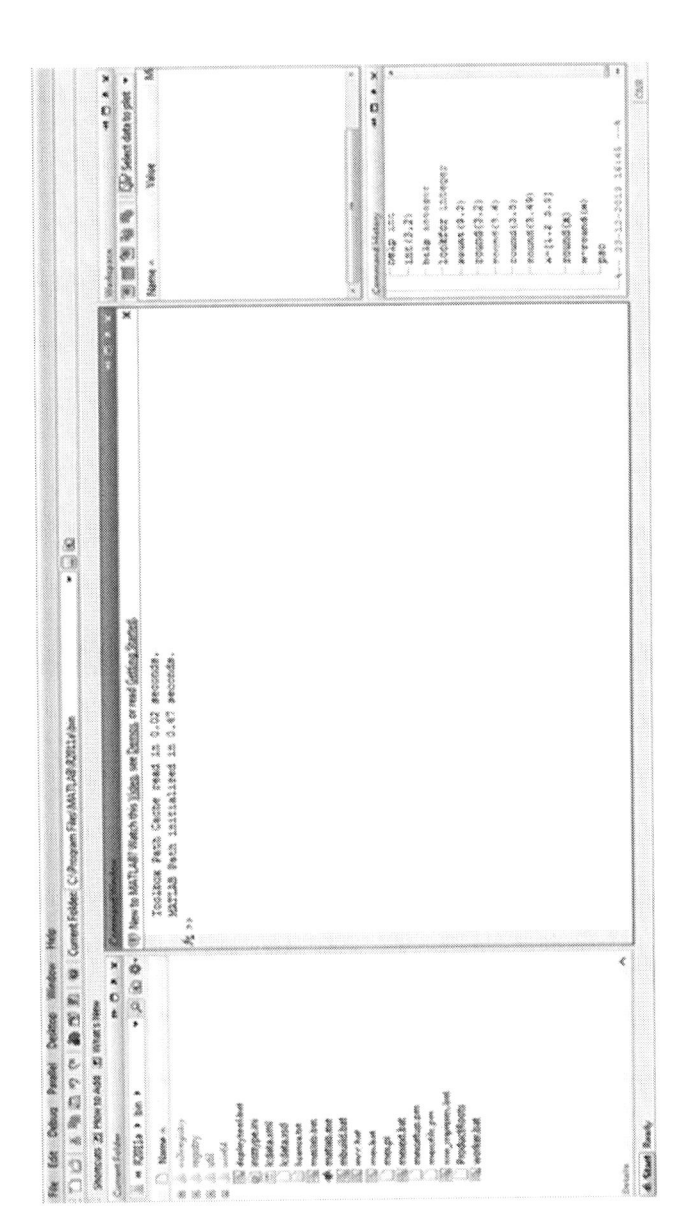

FIGURE A.1
Command window.

To get all rows (or columns) of a matrix, a colon operator can be used. For example, to get second column of A,

```
>> A(:,2)
ans =
     2
    -1
     6
```

To get third row of A

```
>> A(3,:)
ans =
     5     6     7
```

A.3 Expressions

MATLAB does not require variable type declarations. For example,

```
drag_coefficient = 0.6
```

The variables are case sensitive; that is, the variable LIFT is different from lift.

MATLAB uses conventional decimal notation. Scientific notation uses the letter e to specify a power-of-ten scale factor. Imaginary numbers use either i or j as a suffix. Some examples are

```
6      -999     0.0005    109.1237    1.60210e-20    9.123e23    7i
-6.28j    4e6i
```

MATLAB uses the following operators and the precedence follows standard mathematical rules.

- \+ Addition
- − Subtraction
- * Multiplication
- / Division
- \ Left division
- ^ Power
- () Specify evaluation order

Appendix A

The relational operators >, <, >=, <= consider only the real part for the purpose of comparison while the operator = = considers both real and imaginary parts.

Some elementary functions in MATLAB are

Trigonometric

sin	Sine
sind	Sine of argument in degrees
sinh	Hyperbolic sine
asin	Inverse sine
asind	Inverse sine, result in degrees
asinh	Inverse hyperbolic sine
cos	Cosine
cosd	Cosine of argument in degrees
cosh	Hyperbolic cosine
acos	Inverse cosine
acosd	Inverse cosine, result in degrees
acosh	Inverse hyperbolic cosine
tan	Tangent
tand	Tangent of argument in degrees
tanh	Hyperbolic tangent
atan	Inverse tangent
atan2	Four-quadrant inverse tangent
atanh	Inverse hyperbolic tangent
sec	Secant
secd	Secant of argument in degrees
sech	Hyperbolic secant
asec	Inverse secant
asecd	Inverse secant, result in degrees
asech	Inverse hyperbolic secant
csc	Cosecant
cscd	Cosecant of argument in degrees
csch	Hyperbolic cosecant
acsc	Inverse cosecant
acscd	Inverse cosecant, result in degrees
acsch	Inverse hyperbolic cosecant
cot	Cotangent
cotd	Cotangent of argument in degrees
coth	Hyperbolic cotangent
acot	Inverse cotangent

acotd	Inverse cotangent, result in degrees
acoth	Inverse hyperbolic cotangent
hypot	Square root of sum of squares

Exponential

exp	Exponential
expm1	Compute $\exp(x) - 1$ accurately
log	Natural logarithm
log1p	Compute $\log(1 + x)$ accurately
log10	Common (base 10) logarithm
log2	Base 2 logarithm and dissect floating point number
pow2	Base 2 power and scale floating point number
sqrt	Square root
nthroot	Real nth root of real numbers

Complex

abs	Absolute value
angle	Phase angle
complex	Construct complex data from real and imaginary parts
conj	Complex conjugate
imag	Complex imaginary part
real	Complex real part
isreal	True for real array

Rounding and Remainder

fix	Round toward zero
floor	Round toward minus infinity
ceil	Round toward plus infinity
round	Round toward nearest integer
mod	Modulus (signed remainder after division)
rem	Remainder after division
sign	Signum

MATLAB also provides values of useful constants.

pi	3.14159265...
i	Imaginary unit
j	Same as i
eps	Floating-point relative precision
realmin	Smallest floating-point number
realmax	Largest floating-point number
Inf	Infinity
NaN	Not-a-number

A.4 Matrix Operations

Examples of zeros, ones, and rand functions are given below.

```
>> Y = zeros(3,2)
Y =
        0        0
        0        0
        0        0
>> X = ones(2,3)
X =
        1        1        1
        1        1        1
>> Z = rand(2)
Z =
        0.1656   0.2630
        0.6020   0.6541
```

The rand function generates a uniformly generated random number between 0 and 1.

The matrix A can be saved in the same directory, for a later use, by the command:

```
>> save -ascii aa X
```

Sometimes it is necessary to clear all variables and functions from the command window. This is done with the command

```
>> clear all
```

Now if A is punched in the command prompt it results in an error.

```
>> A
```

??? Undefined function or variable 'A'.

To get back the saved value of matrix A, use the load command

```
>> A=load('aa')
A =
        1        2        3
        4       -1       -2
        5        6        7
```

To know about a function name, use help from the menu or simply type help functionnane in the command. For example,

```
>> help clc
```

CLC Clear command window.
CLC clears the command window and homes the cursor.

If one is not able to recollect the function name, use the lookfor command. For example, to get the name of absolute function:

```
>> lookfor absolute
```

abs	Absolute value
genelowvalfilter	Filters genes with low absolute expression levels
imabsdiff	Absolute difference of two images
meanabs	Mean of absolute elements of a matrix or matrices
sumabs	Sum of absolute elements of a matrix or matrices
mae	Mean absolute error performance function
sae	Sum absolute error performance function
dmae	Mean absolute error performance derivative function
circlepick	Pick bad triangles using an absolute tolerance
mad	Mean/median absolute deviation

The concatenation of the matrices is shown with the following example.

```
>> A = [1 2 3; 4 -1 -2; 5 6 7]
A =
         1         2         3
         4        -1        -2
         5         6         7
>> B = [8;9;10]
B =
         8
         9
        10
>> Z = [A B]
Z =
         1         2         3         8
         4        -1        -2         9
         5         6         7        10
```

Suppose we want to delete the second column of the Z matrix. This can be done by

```
>> Z(:,2)=[]
Z =
         1         3         8
         4        -2         9
         5         7        10
```

Appendix A 317

The inverse of the square matrix can be computed by

```
>> inv(Z)
ans =
        -0.3517     0.1102      0.1822
         0.0212    -0.1271      0.0975
         0.1610     0.0339     -0.0593
```

The eigenvalues of the square matrix are computed by

```
>> eig(Z)
ans =
        17.1878
        -2.3534
        -5.8344
```

Some of the array operators are

+ Addition
− Subtraction
.* Element-by-element multiplication
./ Element-by-element division
.\ Element-by-element left division
.^ Element-by-element power
.' Unconjugated array transpose

For example,

```
>> U = [1 2 3]
U =
         1      2      3
>> V = [-1 -2 -3]
V =
        -1     -2     -3
>> U.*V
ans =
        -1     -4     -9
```

The display of numbers is controlled by the format command. Typical commands are

```
format short
format long
```

The previous command can be brought back into the command prompt using the key ↑.

A.5 Plotting

If x and y are two vectors then plot(x, y) makes a graph. For example, consider the following example.

```
>> x = 0:0.01:2*pi;
>> y = cos(x);
>> y1 = sin(x);
>> plot(x,y,'--',x,y1,'r:')
>> xlabel('0 \leq x \leq 2\pi')
>> ylabel('Sine and Cosine functions')
>> legend('cos(x)', 'sin(x)')
>> title('Multiple Plots')
```

Figure A.2 is displayed on the desktop and can be edited using the figure menu.

The following example demonstrates how to make a contour plot (Figure A.3).

```
>> [X,Y] = meshgrid(-2:.01:2,-2:.01:3);
>> Z = X.^2+Y.^2;
>> v=[1;2;3;4;5;6;7;8;9;10;11;12];
>> [c,h] = contour(X,Y,Z,v); clabel(c,h);
```

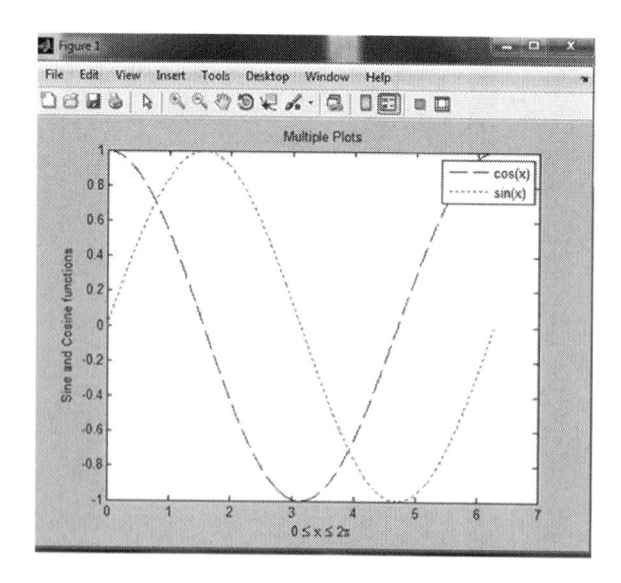

FIGURE A.2
Multiple plots.

Appendix A

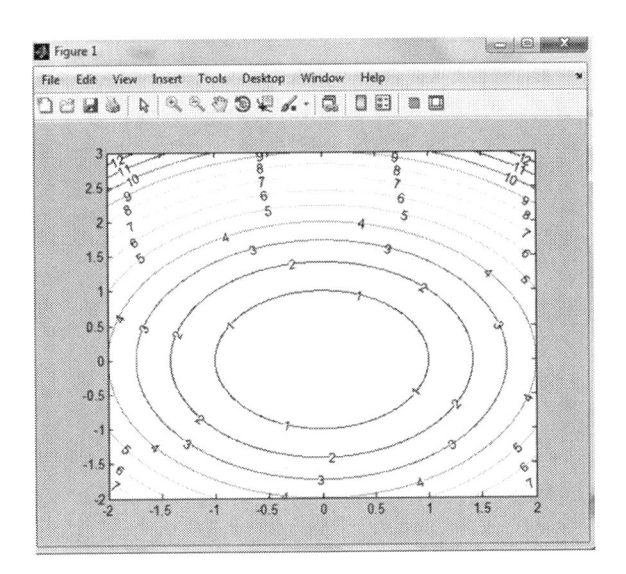

FIGURE A.3
Contour plot.

A.6 Programming

The if-else statement can be demonstrated through the following example.

```
>> for i = 1:6
x = rand(1);
if x<0.5
disp('x is less than 0.5')
else
disp('x is greater than 0.5')
end
end
```

Note that the first end is the end of the if statement and the second end is the end of the for loop. Note that when a semicolon is put at the end of a statement, it suppresses printing of the variable. The following output is displayed by running the above code.

```
x is greater than 0.5
x is greater than 0.5
x is less than 0.5
x is greater than 0.5
x is greater than 0.5
x is less than 0.5
```

A break statement is used for an early exit from a for or while loop.

Instead of running all the commands in the command window, one can create a script file with extension.m. For example, type edit test.m on the command prompt, resulting in opening of an empty file. Type the following contents into that file.

```
for i = 1:10
x = rand(1);
if x<0.5
disp('x is less than 0.5')
else
disp('x is greater than 0.5')
end
end
```

and save it. Then execute the script file by typing test in the command prompt. The script can also be executed by clicking on ▶ in the editor window.

Functions are also script files with extension.m, but they accept input arguments and return output arguments. The function name and file name should be the same. For example, a function springsystem.m takes the input x and y and returns an output z.

```
function z = springsystem(x,y)
```

It is important to note that a function without the arguments cannot be executed. For example, simply typing springsystem at the command prompt will result in an error. The input/output arguments can have different names while calling the function. For example,

```
k = springsystem(a,b)
```

is perfectly fine.

Appendix B: MATLAB® Code

Chapter 1

Code Name	Details
graph_examp12.m	Solves Example 1.2 using the graphical method
graph_examp14.m	Solves Example 1.4 using the graphical method
convexity.m	Plots some convex functions
derivative.m	Computes and plots first and second derivatives of a function
grad.m	Plots the gradient vector
positive_definite.m	Checks whether the square matrix is positive definite
quadr.m	Linear and quadratic approximations of a function
quadr_examp12.m	Linear and quadratic approximations of a function given in Example 1.2

```
%%%%%%%%%%%%%%%%%%%%%%%%%%%%%%%%%%%%%%%%%
% MATLAB code graph_examp12.m
%%%%%%%%%%%%%%%%%%%%%%%%%%%%%%%%%%%%%%%%%
%
% x1 -> radius of can
% x2 -> height of can
% area -> area of can
% pi -> MATLAB variable
% x,y,z -> array of design points
% vv -> user identified contour values
% cons_x2-> value of x2 when constraint is active
% contour -> MATLAB function to generate contours
% xlabel, ylabel, legend, hold on -> MATLAB functions
%
for x1 = 1:100
    for x2 = 1:200
        area = 2*pi*x1*x2 + 2*pi*x1*x1;
        x(x1,x2) = x1;
        y(x1,x2) = x2;
        z(x1,x2) = area;
    end
end
vv = [15000;26436;50000;70000;200000];
[c, h] = contour(x,y,z,vv); clabel(c, h);
hold on
```

```
for x1 = 10:100
    cons_x2 = 330000/(pi*x1*x1);
    plot(x1,cons_x2,'*')
    hold on
end
xlabel('x_1, mm')
ylabel('x_2, mm')
legend('Objective Function','Constraint')
%%%%%%%%%%%%%%%%%%%%%%%%%%%%%%%%%%%%%%%%


%%%%%%%%%%%%%%%%%%%%%%%%%%%%%%%%%%%%%%%%%
% MATLAB code graph_examp14.m
%%%%%%%%%%%%%%%%%%%%%%%%%%%%%%%%%%%%%%%%%
%
% L -> length of rod
% rho -> density of rod material
% d -> diameter of rod
% m -> mass of rod
% I -> moment of inertia
% k -> mass per unit length
% f1 -> frequency
% plot, xlabel, ylabel, legend -> MATLAB function
%
L = 1;
rho = 7800;
E = 2e11;

for d = 0.0:0.001:0.05
mass = (pi/4)*d*d*L*rho;
k = mass/L;
I = (pi/64)*d^4;
f1 = (1/(2*pi))*(3.5156/(L*L))*sqrt((E*I)/k);
plot(d,mass,'+')
hold on
plot(d,f1,'*')
hold on
end
xlabel('d, m')
ylabel('objective function (kg), constraint(Hz)')
legend('Objective Function','Constraint')
%%%%%%%%%%%%%%%%%%%%%%%%%%%%%%%%%%%%%%%%%


%%%%%%%%%%%%%%%%%%%%%%%%%%%%%%%%%
% MATLAB code convexity.m
%%%%%%%%%%%%%%%%%%%%%%%%%%%%%%%%%
%
% x -> independent variable
% y -> dependent variable
```

Appendix B

```
% plot, xlabel, ylabel, meshgrid -> MATLAB functions
%

    x = meshgrid(-2:0.01:2);
    y = x.^2;
    subplot(2,2,1), plot(x,y)
    xlabel('x')
    ylabel('y')
    hold on
    y = exp(x);
    subplot(2,2,2), plot(x,y)
    xlabel('x')
    ylabel('y')
    hold on
    y = exp(y);
    subplot(2,2,3), plot(x,y)
    xlabel('x')
    ylabel('y')
    hold on
    y = exp(x.^2);
    subplot(2,2,4), plot(x,y)
    xlabel('x')
    ylabel('y')
    hold on
%%%%%%%%%%%%%%%%%%%%%%%%%%%%%%%%%


%%%%%%%%%%%%%%%%%%%%%%%%%%%%%%%
% MATLAB code derivative.m
%%%%%%%%%%%%%%%%%%%%%%%%%%%%%%%
%
% delx -> delta-x
% fx -> f(x)
% deriv -> derivative of the function at xd
% sderiv -> second derivative of the function at xdd
% signchange -> change of derivative sign
% locatepoints -> point at which derivative changes sign
% plot, subplot, xlabel, ylabel, hold -> MATLAB functions
%
delx = 0.01;
x=0.1:delx:1.0;
fx = @(x)2*sin(5*x)+3*x.^3-2*x.^2+3*x-5;
subplot(311), plot(x,fx(x),'LineWidth',2)
hold on
ylabel('f(x)')
grid on
 for i = 2:length(x)-1
     xd(i-1) = x(i);
     deriv(i-1) = (fx(x(i+1))-fx(x(i-1)))/(2*delx);
 end

 subplot(312), plot(xd,deriv,'LineWidth',2)
 grid on
```

```
hold on
ylabel('f''(x)')
signchange = deriv(1:length(deriv)-1) .* deriv(2:length(deriv));
locatepoints = xd(find(signchange<0))
subplot(311), plot(locatepoints,fx(locatepoints),'r*')
subplot(312), plot(xd(find(signchange<0)),deriv(find(signchange<0)),'r*')
for ii = 2:length(xd)-1
      xdd(ii-1) = xd(ii);
      sderiv(ii-1) = (fx(xd(ii+1))+fx(xd(ii-1))-2*fx(xd(ii)))/
                     (delx*delx);
end
 subplot(313), plot(xdd,sderiv,'LineWidth',2)
grid on
hold on
subplot(313),  plot(xdd(find(signchange<0)),sderiv(find(signchange<0)),'r*')
xlabel('x')
ylabel('f''''(x)')
%%%%%%%%%%%%%%%%%%%%%%%%%%%%%%%%%%%




%%%%%%%%%%%%%%%%%%%%%%%%%%%%%%%%%%%%%%%%
% MATLAB code grad.m
%%%%%%%%%%%%%%%%%%%%%%%%%%%%%%%%%%%%%%%%
%
% x1 -> radius of can
% x2 -> height of can
% area -> area of can
% pi -> MATLAB variable
% x,y,z -> array of design points
% vv -> user identified contour values
% x1t, x2t -> identified point at which gradient required
% contour -> MATLAB function to generate contours
% xlabel, ylabel, legend, plot, hold on -> MATLAB functions
%
clear all
clc
for x1 = 1:100
    for x2 = 1:100
        area = 2*pi*x1*x2 + 2*pi*x1*x1;
        x(x1,x2) = x1;
        y(x1,x2) = x2;
        z(x1,x2) = area;
    end
end
vv = [5000,15000,30000,50000,70000,90000];
[c, h] = contour(x,y,z,vv); clabel(c, h);
hold on
x1t = 25;
x2t = 70.493;
slope = (x2t+2*x1t)/x1t;
i = 1;
```

Appendix B

```
for delx1 = -10:10
    delx2 = -slope*delx1;
    x11(i) = x1t+delx1;
    x22(i) = x2t+delx2;
    i = i+1;
end
plot(x11,x22,'r--')
hold on
i = 1;
for delx1 = -10:10
    delx2 = (1/slope)*delx1;
    x11(i) = x1t+delx1;
    x22(i) = x2t+delx2;
    i = i+1;
end
plot(x11,x22,'b+')
xlabel('x_1, mm')
ylabel('x_2, mm')
legend('Objective Function','Tangent','Gradient')
%%%%%%%%%%%%%%%%%%%%%%%%%%%%%%%%%%%%%%


%%%%%%%%%%%%%%%%%%%%%%%%%%%%%%%%%%%%%
% MATLAB code positive_definite.m
%%%%%%%%%%%%%%%%%%%%%%%%%%%%%%%%%%%%%
% H -> hessian matrix
% eig, fprintf -> MATLAB function
% eigenvalues -> of the hessian matrix
%
H =  [2 1 1;
      1 2 1;
      0 2 3];
eigenvalues = eig(H);
eigenvalues
if eigenvalues >= 0
    fprintf('The matrix is positive definite\n')
else
    fprintf('The matrix is not positive definite\n')
end
%%%%%%%%%%%%%%%%%%%%%%%%%%%%%%%%%%%%%%
```

```
%%%%%%%%%%%%%%%%%%%%%%%%%%%%%%%%%%%%%%%%
% MATLAB code quadr.m
%%%%%%%%%%%%%%%%%%%%%%%%%%%%%%%%%%%%%%%%
%
% x -> independent variable and symbolic variable (later)
% y -> exp(-x)
% syms -> symbolic object (MATLAB function)
% taylor -> MATLAB function
% subs -> symbolic substitution (MATLAB function)
% xlabel, ylabel, legend, plot, hold on -> MATLAB functions
%
x = -2:0.01:2;
y = exp(-x);
plot(x,y)
hold on
% Linear approximation
syms x
f = taylor(exp(-x),2);
x = -2:0.01:2;
z = subs(f);
plot(x,z,'r--')
% Quadratic approximation
syms x
f = taylor(exp(-x),3);
x = -2:0.01:2;
z = subs(f);
plot(x,z,'g--')
legend('exp(-x)','linear','quadratic')
xlabel('x')
ylabel('f(x)')
%%%%%%%%%%%%%%%%%%%%%%%%%%%%%%%%%%%%%%%%

%%%%%%%%%%%%%%%%%%%%%%%%%%%%%%%%%%%%%%%%
% MATLAB code quadr_examp12.m
%%%%%%%%%%%%%%%%%%%%%%%%%%%%%%%%%%%%%%%%
%
% x1 -> radius of can
% x2 -> height of can
% area -> area of can
% pi -> MATLAB variable
% x,y,z -> array of design points
% vv -> user identified contour values
% xit, x2t -> identified point at which gradient required
% contour -> MATLAB function to generate contours
% syms -> symbolic object (MATLAB function)
% subs -> symbolic substitution (MATLAB function)
% gradient -> analytical value
% hessian -> analytical value
% xlabel, ylabel, legend, plot, hold on -> MATLAB functions
%
```

Appendix B

```
clear all
clc
for x1 = 1:200
    for x2 = 1:200
        area = 2*pi*x1*x2 + 2*pi*x1*x1;
        x(x1,x2) = x1;
        y(x1,x2) = x2;
        z(x1,x2) = area;
    end
end
vv = [15000;50000;60000;70000;80000;90000;150000;200000];
[c, h] = contour(x,y,z,vv); clabel(c, h);
hold on
syms x1p x2p
gradient = [2*pi*x2p+4*pi*x1p;2*pi*x1p];
hessian = [4*pi 2*pi; 2*pi 0];
% Linear approximation
x1p = 60;
x2p = 72.629;
gf = subs(gradient);
for delx1 = 1:60
    for delx2 = 1:60
        x1 = x1p + delx1;
        x2 = x2p + delx2;
        area = 2*pi*x1p*x2p + 2*pi*x1p*x1p + gf'*[delx1;delx2]
               + 0.5*[delx1 delx2]*(hessian*[delx1;delx2]);
        [x1 x2 area]
        xx(delx1,delx2) = x1;
        yy(delx1,delx2) = x2;
        zz(delx1,delx2) = area;
    end
end
vv1 = [50000;60000;70000;80000;90000];
[c, h] = contour(xx,yy,zz,vv1,'rd','LineWidth',3);
xlabel('x_1, mm')
ylabel('x_2, mm')
legend('Objective Function','Quadratic approx.')
%%%%%%%%%%%%%%%%%%%%%%%%%%%%%%%%%%%%%%%%%%%
```

Chapter 2

Code Name	Details
exhaustive.m	Exhaustive search to locate the minimum of the test problem
bisection.m	Bisection method
func.m	Objective function to be coded here
newtonraphson.m	Newton–Raphson method
secant.m	Secant method
cubic.m	Cubic polynomial fit
golden.m	Golden section method

```
%%%%%%%%%%%%%%%%%%%%%%%%%%%%%%%%%%%%%%%%
% MATLAB code exhaustive.m
%%%%%%%%%%%%%%%%%%%%%%%%%%%%%%%%%%%%%%%%
%
% delta -> step size for search
% T -> independent variable, temperature
% U -> cost function
% uvec -> vector of cost function evaluated at
%          different temperatures
% minu -> minimum of cost function
% min -> MATLAB function
%
clear all
clc
uvec=[];
delta = 0.01;
for T = 40:delta:90
    U = 204165.5/(330-2*T) + 10400/(T-20);
    uvec = [uvec U];
    plot(T,U)
    hold on
end
xlabel('T');ylabel('U');
[minu,i]= min(uvec);
fprintf('Minimum Cost =   %6.2f\n ',minu)
fprintf('occurs at T =   %6.2f\n ',40+(i-1)*delta)
%%%%%%%%%%%%%%%%%%%%%%%%%%%%%%%%%%%%%%%%


%%%%%%%%%%%%%%%%%%%%%%%%%%%%%%%%%%%%%%%%
% MATLAB code bisection.m
%%%%%%%%%%%%%%%%%%%%%%%%%%%%%%%%%%%%%%%%
%
% a -> lower bound of the design variable
% b -> upper bound of the design variable
```

Appendix B 329

```
% alpha -> midpoint of a and b
% delx -> ?x for central difference method
% derivative -> derivative using central difference method
% derivative_alpha -> derivative at x = alpha
% abs -> absolute of a number, MATLAB function
%
clear all
clc
a = 40;
b = 90;
epsilon = 0.01;
delx = 0.01;
fprintf('     a            b      \n')
fprintf('-------------------------\n')
for i= 1:100
 fprintf(' %7.3f     %8.3f \n',a,b)
 alpha = (a+b)/2;
 derivative = (func(a+delx) - func(a-delx) )/(2*delx);
 derivative_alpha = (func(alpha+delx)- func(alpha-delx))/
     (2*delx);
if (derivative*derivative_alpha) < 0
 b = alpha;
else
 a = alpha;
end
if abs(a-b) < epsilon
 break;
end
end
fprintf('-------------------------\n')
fprintf('x* = %7.3f      Minimum =    %8.3f\n',a,func(a))
fprintf('Number of function calls =    %3d\n',4*i)
%
%%%%%%%%%%%%%%%%%%%%%%%%%%%%%%%%%%%%%%%


%%%%%%%%%%%%%%%%%%%%%%%%%%%%%%%%%%%%%%%
% MATLAB code func.m
%%%%%%%%%%%%%%%%%%%%%%%%%%%%%%%%%%%%%%%
%
% x -> input variable to the function
% fx -> output from the function
%
function fx = func(x)
    fx = 204165.5/(330-2*x) + 10400/(x-20);
%%%%%%%%%%%%%%%%%%%%%%%%%%%%%%%%%%%%%%%
```

```
%%%%%%%%%%%%%%%%%%%%%%%%%%%%%%%%%%%%%%%%
% MATLAB code newtonraphson.m
%%%%%%%%%%%%%%%%%%%%%%%%%%%%%%%%%%%%%%%%
%
% x -> initial guess of design variable
% delx -> ?x for central difference method
% derivative -> derivative using central difference method
% sec_derivative -> second derivative
% epsilon -> constant used to terminate the program
% xprev -> value of x stored from previous iteration
%
clear all
clc
x = 45;
delx = 0.01;
epsilon = 0.01;
fprintf('     x      f(x)       Deriv. Second deriv.\n')
fprintf('----------------------------------------\n')
for i = 1:100
 derivative = (func(x+delx) - func(x-delx))/(2*delx);
 sec_derivative =(func(x+delx)+func(x-delx)-2*func(x))/
   (delx*delx);
fprintf('%8.3f %8.3f %8.3f %8.3f\n',x,func(x),derivative,
 sec_derivative)
 xprev = x;
 x = x- derivative/sec_derivative;
 if abs(x-xprev) < epsilon
    break;
 end
end
fprintf('----------------------------------------\n')
fprintf('Number of function calls =     %3d\n',5*i)
%%%%%%%%%%%%%%%%%%%%%%%%%%%%%%%%%%%%%%%%


%%%%%%%%%%%%%%%%%%%%%%%%%%%%%%%%%%%%%%%%
% MATLAB code secant.m
%%%%%%%%%%%%%%%%%%%%%%%%%%%%%%%%%%%%%%%%
%
% a -> lower bound of the design variable
% b -> upper bound of the design variable
% alpha -> midpoint of a and b
% delx -> ?x for central difference method
% derivative -> derivative using central difference method
% derivative_alpha -> derivative at x = alpha
% abs -> absolute of a number, MATLAB function
% flag -> set the flag when minimum is bracketed
%
clear all
clc
```

Appendix B

```
a = 40;
b = 90;
epsilon = 0.001;
delx = 0.01;
flag = 0;
fprintf('   Alpha         Deriv. \n')
fprintf('------------------------\n')
for i = 1:100
 alpha = (a+b)/2;
 derivative = (func(a+delx) - func(a-delx))/(2*delx);
derivative_alpha = (func(alpha+delx)-func(alpha-delx))/
 (2*delx);
 if (derivative*derivative_alpha) < 0
    b = alpha;
    flag = 1;
 else
    a = alpha;
 end
 if flag == 1
    break;
 end
end
 for j = 1:100
  fprintf('  %7.3f      %8.3f \n',alpha,derivative_alpha)
  derivative_a = (func(a+delx) - func(a-delx))/(2*delx);
  derivative_b = (func(b+delx) - func(b-delx))/(2*delx);
  alpha = b - derivative_b*(b-a)/(derivative_b-derivative_a);
  derivative_alpha = (func(alpha+delx) - func(alpha-delx))/
    (2*delx);
  if derivative_alpha > 0
     b = alpha;
  else
     a = alpha;
  end
  if abs(derivative_alpha) < epsilon
     break;
  end
 end
end
fprintf('-------------------------\n')
fprintf('x* = %7.3f      Minimum = %8.3f\n',alpha,func(alpha))
fprintf('Number of function calls =   %3d\n',4*i+6*j)
%
%%%%%%%%%%%%%%%%%%%%%%%%%%%%%%%%%%%%%%%
```

```
%%%%%%%%%%%%%%%%%%%%%%%%%%%%%%%%%%%%%%%%
% MATLAB code cubic.m
%%%%%%%%%%%%%%%%%%%%%%%%%%%%%%%%%%%%%%%%
%
% a -> lower bound of the design variable
% b -> upper bound of the design variable
% alpha -> midpoint of a and b
% delx -> ?x for central difference method
% derivative -> derivative using central difference method
% derivative_alpha -> derivative at x = alpha
% abs -> absolute of a number, MATLAB function
% flag -> set the flag when minimum is bracketed
% derivative_a -> derivative at point a
% derivative_b -> derivative at b
%
a = 40;
b = 90;
delx = 0.01;
flag = 0;
epsilon= 0.001;
fprintf('      a            b     \n')
fprintf('-------------------------\n')
for i = 1:100
 alpha = (a+b)/2;
 derivative = (func(a+delx) - func(a-delx))/(2*delx);
 derivative_alpha = (func(alpha+delx)-func(alpha-delx))/
    (2*delx);
if (derivative*derivative_alpha) < 0
    b = alpha;
    flag = 1;
 else
    a = alpha;
 end
 if flag == 1
    break;
 end
end
for j = 1:100
  fprintf('  %7.3f      %8.3f \n',a,b)
 derivative_a = (func(a+delx) - func(a-delx))/(2*delx);
 derivative_b = (func(b+delx) - func(b-delx))/(2*delx);
 z = 3*(func(a)-func(b))/(b-a) + derivative_a + derivative_b;
 w = ((b-a)/abs(b-a))*sqrt(z*z-derivative_a*derivative_b);
 mew = (derivative_b+w-z)/(derivative_b-derivative_a+2*w);
if mew <= 1
    x_opt = b - mew*(b-a);
 else
    x_opt = a;
 end
```

Appendix B

```matlab
    alpha1 = (func(x_opt+delx) - func(x_opt-delx) )/(2*delx);
    if abs(alpha1) < epsilon
      break;
    else
      if (derivative_a*alpha1) < 0
      b = x_opt;
    else
      a = x_opt;
     end
    end
   end
   fprintf('-------------------------\n')
   fprintf('x* =   %7.3f       Minimum =   %8.3f\n',x_opt,func ...
      (x_opt))
   fprintf('Number of function calls =    %3d\n',4*i+8*j)
   %
   %%%%%%%%%%%%%%%%%%%%%%%%%%%%%%%%%%%%%%%%


   %%%%%%%%%%%%%%%%%%%%%%%%%%%%%%%%%%%%%
   % MATLAB code golden.m
   %%%%%%%%%%%%%%%%%%%%%%%%%%%%%%%%%%%%%
   %
   % a -> lower bound of the design variable
   % b -> upper bound of the design variable
   % alpha -> midpoint of a and b
   % falpha1 -> function value at x = alpha1
   % falpha2 -> function value at x = alpha2
   % epsilon -> constant used to terminate the algorithm
   % abs -> absolute of a number, MATLAB function
   % tau -> 2-golden number
   %
   clear all
   clc
   a = 40;
   b = 90;
   epsilon = 0.00001;
   tau = 0.381967;
   alpha1 = a*(1-tau) + b*tau;
   alpha2 = a*tau + b*(1-tau);
   falpha1 = func(alpha1);
   falpha2 = func(alpha2);
   fprintf('     a             b     \n')
   fprintf('-----------------------\n')
   for i = 1:100
       fprintf(' %7.3f     %8.3f \n',a,b)
       if falpha1 > falpha2
           a = alpha1;
           alpha1 = alpha2;
           falpha1 = falpha2;
           alpha2 = tau*a + (1-tau)*b;
           falpha2 = func(alpha2);
       else
           b = alpha2;
```

```
            alpha2 = alpha1;
            falpha2 = falpha1;
            alpha1 = tau*b + (1-tau)*a;
            falpha1 = func(alpha1);
        end
    if abs(func(alpha1)-func(alpha2)) < epsilon
        break;
    end
end
fprintf('-------------------------\n')
fprintf('x* =  %7.3f        Minimum =   %8.3f\n',alpha1,func(alpha1))
fprintf('Number of function calls =     %3d\n',2+i)
%
%%%%%%%%%%%%%%%%%%%%%%%%%%%%%%%%%%%%%%%%%


%%%%%%%%%%%%%%%%%%%%%%%%%%%%%%%%%%%%%%%%%
% MATLAB code func.m
%%%%%%%%%%%%%%%%%%%%%%%%%%%%%%%%%%%%%%%%%
% objective function to be coded here
% different test functions
%
function fx = func(x)
%     fx = 204165.5/(330-2*x) + 10400/(x-20);
%     fx = 3*x^4+(x-1)^2;
%     fx = -4*x*sin(x);
%     fx = 2*(x-3)^2+exp(0.5*x*x);
      fx = 3*(x)^2+12/(x^3)-5;
%     fx = 2*x*x+16/x;
%
%%%%%%%%%%%%%%%%%%%%%%%%%%%%%%%%%%%%%%%%%
```

Appendix B

Chapter 3

Code Name	Details
golden_funct1.m	Golden section method for a multivariable function
func_multivar.m	Objective function to be coded here
rosenbrock.m	Plot of Rosenbrock function
springsystem.m	Finds minimum of the spring system problem
steep_des.m	Steepest descent method
grad_vec.m	Gradient vector computation
contour_testproblem.m	Plots contour of the test problem function
newton.m	Newton's method
hessian.m	Computes Hessian matrix
modified_newton.m	Modified Newton's method
levenbergmarquardt.m	Levenberg–Marquardt's method
conjugate.m	Conjugate gradient method
DFP.m	Davidon–Fletcher–Powell (DFP) method
BFGS.m	Broyden–Fletcher–Goldfarb–Shanno (BFGS) method
powell.m	Powell's conjugate direction method
neldermead.m	Nelder–Mead algorithm
\Robotics\	Directory containing codes for problems in robotics

```
%%%%%%%%%%%%%%%%%%%%%%%%%%%%%%%%%%%%%%%
% MATLAB code golden_funct1.m
%%%%%%%%%%%%%%%%%%%%%%%%%%%%%%%%%%%%%%%
%
% a -> lower bound of the design variable
% b -> upper bound of the design variable
% falpha1 -> function value at x = alpha1
% falpha2 -> function value at x = alpha2
% epsilon -> constant used to terminate the algorithm
% abs -> absolute of a number, MATLAB function
% tau -> 2-golden number
% func_multivar -> returns the value of a multivariable
%                  function
%
%
function [alpha1,falpha1] = golden_funct1(x,search)
a = -5;
b = 5;
tau = 0.381967;
epsilon = 1e-5;
alpha1 = a*(1-tau) + b*tau;
alpha2 = a*tau + b*(1-tau);
falpha1 = func_multivar(x+alpha1*search);
```

```
falpha2 = func_multivar(x+alpha2*search);
for i= 1:1000
  if falpha1 > falpha2
       a = alpha1;
       alpha1 = alpha2;
       falpha1 = falpha2;
       alpha2 = tau*a + (1-tau)*b;
       falpha2 = func_multivar(x+alpha2*search);
    else
       b = alpha2;
       alpha2 = alpha1;
       falpha2 = falpha1;
       alpha1 = tau*b + (1-tau)*a;
       falpha1 = func_multivar(x+alpha1*search);
    end
    if abs(func_multivar(x+alpha1*search)-
    func_multivar(x+alpha2*search)) < epsilon
    break;
    end;
end
%
%%%%%%%%%%%%%%%%%%%%%%%%%%%%%%%%%%%%%%%%


%%%%%%%%%%%%%%%%%%%%%%%%%%%%%%%%%%%%%%%%
% MATLAB code func_multivar.m
%%%%%%%%%%%%%%%%%%%%%%%%%%%%%%%%%%%%%%%%
%
function fx = func_multivar(x)
    fx = 100*(x(2)-x(1)^2)^2 + (1-x(1))^2;
%
%%%%%%%%%%%%%%%%%%%%%%%%%%%%%%%%%%%%%%%%


%%%%%%%%%%%%%%%%%%%%%%%%%%%%%%%%%%%%%%%%
% MATLAB code rosenbrock.m
%%%%%%%%%%%%%%%%%%%%%%%%%%%%%%%%%%%%%%%%
%
% plots the Rosenbrock's function
%
clear all
clc
[x1,x2] = meshgrid(-2:.03:2,-2:.03:2);
z = 100*(x2-x1.^2).^2+(1-x1).^2
surf(x1,x2,z)
shading interp
view (170,20)
xlabel('x1')
```

Appendix B

```matlab
ylabel('x2')
zlabel('f(x1,x2)')
%
%%%%%%%%%%%%%%%%%%%%%%%%%%%%%%%%%%%%%%%%

%%%%%%%%%%%%%%%%%%%%%%%%%%%%%%%%%%%%%
% MATLAB code springsystem.m
%%%%%%%%%%%%%%%%%%%%%%%%%%%%%%%%%%%%%
%
clear all
clc
zprev = inf;
i = 0;
j = 0;
for x = -1:0.01:1
    i = i+1;
    for y = -1:0.01:1
        j = j+1;
        z = 100*(sqrt(x^2+(y+1)^2)-1)^2 + 90*(sqrt(x^2+(y-
            1)^2)-1)^2 -(20*x+40*y);
    if z < zprev
        zprev = z;
        xbest = x;
        ybest = y;
    end
    end
end
fprintf('Minimum Potential =   %7.4f\n ',zprev)
fprintf('occurs at x1,x2 =    %10.4f %10.4f\n',xbest,ybest)
%
%%%%%%%%%%%%%%%%%%%%%%%%%%%%%%%%%%%%%%%%


%%%%%%%%%%%%%%%%%%%%%%%%%%%%%%%%%%%%%%
% MATLAB code steep_des.m
%%%%%%%%%%%%%%%%%%%%%%%%%%%%%%%%%%%%%%
%
% n_of_var -> number of design variables
% x = [-1.5 1.5] -> starting value of x
% epsilon1,epsilon2 -> constants used for terminating the
%                     algorithm
% delx -> required for gradient computation
% falpha_prev -> function value at first/previous iteration
% deriv -> gradient vector
% search -> search direction (set to negative of gradient)
%
clear all
clc
n_of_var = 2;
x = [-3 2];
```

```
epsilon1 = 1e-6;
epsilon2 = 1e-6;
delx = 1e-3;
falpha_prev = func_multivar(x);
fprintf('Initial function value =   %7.4f\n ',falpha_prev)
fprintf(' No.         x-vector         f(x)        Deriv \n')
fprintf('_____\n')
for i = 1:3000
deriv = grad_vec(x,delx,n_of_var);
search = -deriv;
[alpha,falpha] = golden_funct1(x,search);
if abs(falpha-falpha_prev)<epsilon1 || norm(deriv)<epsilon2
    break;
end
falpha_prev = falpha;
x = x + alpha*search;
fprintf('%3d %8.3f %8.3f  % 8.3f   %8.3f
      \n',i,x,falpha,norm(deriv))
end
fprintf('_____\n')
%
%%%%%%%%%%%%%%%%%%%%%%%%%%%%%%%%%%%%%%%


%%%%%%%%%%%%%%%%%%%%%%%%%%%%%%%%%%%%%%%
% MATLAB code grad_vec.m
%%%%%%%%%%%%%%%%%%%%%%%%%%%%%%%%%%%%%%%
%
% compute gradient vector using central difference method
% xvec, xvec1 -> vector of design variables
% deriv(i) -> derivative w.r.t. ith variable
%
function deriv = grad_vec(x,delx,n_of_var)
xvec = x;
xvec1 = x;
for i = 1:length(x)
    xvec = x;
    xvec1 = x;
    xvec(i) = x(i) + delx;
    xvec1(i) = x(i) - delx;
deriv(i) = (func_multivar(xvec) - func_multivar(xvec1))/
    (2*delx);
end
%
%%%%%%%%%%%%%%%%%%%%%%%%%%%%%%%%%%%%%%%
```

Appendix B

```
%%%%%%%%%%%%%%%%%%%%%%%%%%%%%%%%%%%%%%
% MATLAB code contour_testproblem.m
%%%%%%%%%%%%%%%%%%%%%%%%%%%%%%%%%%%%%%
%
% plots contour of the test problem
% surfc -> Matlab function
clear all
clc
i = 0;
j = 0;
for x = -5:.02:5
    i = i+1;
     for y = -5:.02:5
        j = j+1;
        z(i,j) = 100*(sqrt(x^2+(y+1)^2)-1)^2 + 90*(sqrt(x^2+
                 (y - 1)^2)-1 )^2 -(20*x+40*y);
          t1(i,j) = x;
          t2(i,j) = y;
      end
      j = 0;
end
surfc(t1,t2,z)
shading interp
xlabel('x1')
ylabel('x2')
zlabel('f(x1,x2)')
%
%%%%%%%%%%%%%%%%%%%%%%%%%%%%%%%%%%%%%%


%%%%%%%%%%%%%%%%%%%%%%%%%%%%%%%%%%%%%%
% MATLAB code newton.m
%%%%%%%%%%%%%%%%%%%%%%%%%%%%%%%%%%%%%%
%
% n_of_var -> number of design variables
% x = [-3 2] -> starting value of x
% epsilon1, epsilon2 -> constant used for terminating
%                      the algorithm
% delx -> required for gradient computation
% f_prev -> function value at first/previous iteration
% deriv -> gradient vector
% sec_deriv -> hessian matrix
%
clear all
clc
n_of_var = 2;
x = [-3 2];
epsilon1 = 1e-7;
epsilon2 = 1e-7;
delx = 1e-3;
```

```
f_prev = func_multivar(x);
fprintf('Initial function value =   %7.4f\n',f_prev)
fprintf('No.       x-vector        f(x)        Deriv \n')
fprintf('_____\n')
for i = 1:50
    f_prev = func_multivar(x);
    deriv = grad_vec(x,delx,n_of_var);
    sec_deriv = hessian(x,delx,n_of_var);
     x = (x' - inv(sec_deriv)*deriv')';
     f = func_multivar(x);
       if abs(f-f_prev)<epsilon1 || norm(deriv)<epsilon2
         break;
       end
        fprintf('%3d %8.3f %8.3f  % 8.3f  %8.3f
        \n',i,x,f,norm(deriv))
end
fprintf('%3d %8.3f %8.3f  % 8.3f  %8.3f  \n',i,x,f,norm(deriv))
fprintf('_____\n')
%
%%%%%%%%%%%%%%%%%%%%%%%%%%%%%%%%%%%%%%%%%%


%%%%%%%%%%%%%%%%%%%%%%%%%%%%%%%%%%%%%%%%%%
% MATLAB code hessian.m
%%%%%%%%%%%%%%%%%%%%%%%%%%%%%%%%%%%%%%%%%%
%
%compute hessian matrix
% sec_deriv -> second derivative matrix
% func_multivar() -> multivariable function
% temp -> temporary variable
% Note that n_of_var = length(x)
%
function sec_deriv = hessian(x,delx,n_of_var)
for i = 1:length(x)
  for j = 1:length(x)
        if i == j
     temp = x;
     temp(i) = x(i) + delx;
     term1 = func_multivar(temp);
     temp(i) = x(i) - delx;
     term2 = func_multivar(temp);
     term3 = func_multivar(x);
     sec_deriv(i,j) = (term1-2*term3+term2)/(delx^2);
        else
     temp = x;
     temp(i) = x(i) + delx;
     temp(j) = x(j) + delx;
     term1 = func_multivar(temp);
     temp = x;
```

Appendix B

```
        temp(i) = x(i) + delx;
        temp(j) = x(j) - delx;
        term2 = func_multivar(temp);
        temp = x;
        temp(i) = x(i) - delx;
        temp(j) = x(j) + delx;
        term3 = func_multivar(temp);
        temp = x;
        temp(i) = x(i) - delx;
        temp(j) = x(j) - delx;
        term4 = func_multivar(temp);
        sec_deriv(i,j) = (term1-term2-term3+term4)/(4*delx^2);
                end
    end
end
%
%%%%%%%%%%%%%%%%%%%%%%%%%%%%%%%%%%%%%%%%%%


%%%%%%%%%%%%%%%%%%%%%%%%%%%%%%%%%%%%%%%%
% MATLAB code modified_newton.m
%%%%%%%%%%%%%%%%%%%%%%%%%%%%%%%%%%%%%%%%
%
% n_of_var -> number of design variables
% x = [-3 2] -> starting value of x
% epsilon1, epsilon2 -> constant used for terminating
%                      the algorithm
% delx -> required for gradient computation
% falpha_prev -> function value at first/previous iteration
% deriv -> gradient vector
% sec_deriv -> hessian matrix
% search -> search direction (vector)
clear all
clc
n_of_var = 2;
x = [-3 2];
epsilon1 = 1e-7;
epsilon2 = 1e-7;
delx = 1e-3;
f_prev = func_multivar(x);
fprintf('Initial function value =  %7.4f\n ',f_prev)
fprintf('No.      x-vector       f(x)       Deriv \n')
fprintf('_____\n')
for i = 1:20
    falpha_prev = func_multivar(x);
    deriv = grad_vec(x,delx,n_of_var);
    sec_deriv = hessian(x,delx,n_of_var);
    search = -inv(sec_deriv)*deriv';
    [alpha,falpha] = golden_funct1(x,search');
```

```matlab
        if abs(falpha-falpha_prev)<epsilon1 || ...
            norm(deriv)<epsilon2
    break;
    end
    falpha_prev = falpha;
    x = x + alpha*search';
    f = func_multivar(x);
    fprintf('%3d %8.3f %8.3f  % 8.3f  %8.3f ...
        \n',i,x,falpha,norm(deriv))
end
fprintf('%3d %8.3f %8.3f  % 8.3f  %8.3f ...
    \n',i,x,falpha,norm(deriv))
fprintf('_____\n')
%
%%%%%%%%%%%%%%%%%%%%%%%%%%%%%%%%%%%%%%%%


%%%%%%%%%%%%%%%%%%%%%%%%%%%%%%%%%%%%%%%%
% MATLAB code levenbergmarquardt.m
%%%%%%%%%%%%%%%%%%%%%%%%%%%%%%%%%%%%%%%%
%
% n_of_var -> number of design variables
% x = [-3 2] -> starting value of x
% lambda -> initially set to a large value
% epsilon1, epsilon2 -> constant used for terminating
%                      the algorithm
% delx -> required for gradient computation
% f_prev -> function value at first/previous iteration
% deriv -> gradient vector
% sec_deriv -> hessian matrix
% search -> search direction (vector)
%
clear all
clc
n_of_var = 2;
x = [-3 2];
lambda = 1e3;
epsilon1 = 1e-7;
epsilon2 = 1e-7;
delx = 1e-3;
f_prev = func_multivar(x);
fprintf('Initial function value = %7.4f\n ',f_prev)
fprintf(' No.       x-vector       f(x)      Deriv \n')
fprintf('_____\n')
for i = 1:100
    f_prev = func_multivar(x);
    deriv = grad_vec(x,delx,n_of_var);
    sec_deriv = hessian(x,delx,n_of_var);
    search = -inv(sec_deriv+lambda*eye(length(x)))*deriv';
    x = x + search';
```

```
    f = func_multivar(x);
      if f < f_prev
        lambda = lambda/2;
      else
        lambda = 2*lambda;
      end
      if abs(f-f_prev)<epsilon1 || norm(deriv)<epsilon2
        break;
      end
      fprintf('%3d %8.3f %8.3f  % 8.3f  %8.3f
              \n',i,x,f,norm(deriv))
end
fprintf('_____\n')
%
%%%%%%%%%%%%%%%%%%%%%%%%%%%%%%%%%%%%%%%


%%%%%%%%%%%%%%%%%%%%%%%%%%%%%%%%%%%%
% MATLAB code conjugate.m
%%%%%%%%%%%%%%%%%%%%%%%%%%%%%%%%%%%%
%
% n_of_var -> number of design variables
% x = [-3 2] -> starting value of x
% epsilon1, epsilon2 -> constant used for terminating
%                       the algorithm
% delx -> required for gradient computation
% falpha_prev -> function value at first/previous iteration
% deriv -> gradient vector
% search -> search direction (vector)
%
clear all
clc
n_of_var = 2;
x = [-3 2];
epsilon1 = 1e-7;
epsilon2 = 1e-7;
delx = 1e-3;
falpha_prev = func_multivar(x);
fprintf('Initial function value = %7.4f\n ',falpha_prev)
fprintf('No.      x-vector       f(x)       Deriv \n')
fprintf('_____\n')
for i = 1:300
        if i==1
          deriv_prev = grad_vec(x,delx,n_of_var);
          search_prev = -deriv_prev;
          [alpha,falpha] = golden_funct1(x,search_prev);
          if norm(deriv_prev)<epsilon2
        break;
```

```
            end
                x = x + alpha*search_prev;
                falpha_prev = func_multivar(x);
            else
                deriv = grad_vec(x,delx,n_of_var);
              search = -deriv + ...
              ((norm(deriv)^2)/(norm(deriv_prev)^2))*search_prev;
                [alpha,falpha] = golden_funct1(x,search);
                   if abs(falpha-falpha_prev)<epsilon1 ||
                   norm(deriv)<epsilon2
              break;
              end
                 deriv_prev = deriv;
                 search_prev = search;
                 x = x + alpha*search;
                 falpha_prev = func_multivar(x);
            end
         fprintf('%3d %8.3f %8.3f  % 8.3f   %8.3f
         \n',i,x,falpha,norm(deriv_prev))
end
fprintf('%3d %8.3f %8.3f  % 8.3f   %8.3f
\n',i,x,falpha,norm(deriv))
fprintf('_____\n')
%%%%%%%%%%%%%%%%%%%%%%%%%%%%%%%%%%%%%%%%

%%%%%%%%%%%%%%%%%%%%%%%%%%%%%%%%%%%%%%%%
% MATLAB code DFP.m
%%%%%%%%%%%%%%%%%%%%%%%%%%%%%%%%%%%%%%%%
%
% n_of_var -> number of design variables
% x = [-1.5 1.5] -> starting value of x
% epsilon1, epsilon2 -> constant used for terminating
%                      the algorithm
% delx -> required for gradient computation
% falpha_prev -> function value at first/previous iteration
% deriv -> gradient vector
% deltag -> difference in gradient vector (over previous
  iteration)
% A -> approximation of inverse of the hessian matrix
% search -> search direction
%
clear all
clc
n_of_var = 2;
x = [-3 2];
A = eye(length(x));
epsilon1 = 1e-7;
epsilon2 = 1e-7;
delx = 1e-3;
falpha_prev = func_multivar(x);
```

Appendix B

```
fprintf('Initial function value =    %7.4f\n ',falpha_prev)
fprintf(' No.        x-vector         f(x)       Deriv \n')
fprintf('_____\n')
for i = 1:100
    if i==1
    deriv_prev = grad_vec(x,delx,n_of_var);
    search = -deriv_prev;
    [alpha,falpha] = golden_funct1(x,search);
    if abs(falpha-falpha_prev)<epsilon1
    break;
    end
    falpha_prev = falpha;
    x = x + alpha*search;
    fprintf('%3d %8.3f %8.3f  % 8.3f   %8.3f  \n',i,x,falpha_
        prev,norm(deriv_prev))
    else
        deltax = (alpha*search);
        if i>2
            deltax = deltax';
        end
        deriv = grad_vec(x,delx,n_of_var);
        deltag = deriv-deriv_prev;
        term1 = (deltax'*deltax)/(deltax*deltag');
        term2 = (A*deltag'*deltag*A)/(deltag*A*deltag');
        A = A + term1 - term2;
        search = -A*deriv';
        [alpha,falpha] = golden_funct1(x,search');
        fprintf('%3d %8.3f %8.3f  % 8.3f   %8.3f  \n',i,x+alpha
            *search',falpha,norm(deriv))
        if abs(falpha-falpha_prev)<epsilon1 ||
            norm(deriv)<epsilon2
         break;
        end
        falpha_prev = falpha;
        deriv_prev = deriv;
        x = x+alpha*search';
    end
    end
fprintf('_____\n')
%
%%%%%%%%%%%%%%%%%%%%%%%%%%%%%%%%%%%%%%%%


%%%%%%%%%%%%%%%%%%%%%%%%%%%%%%%%%%%%%%%%
% MATLAB code BFGS.m
%%%%%%%%%%%%%%%%%%%%%%%%%%%%%%%%%%%%%%%%
% n_of_var -> number of design variables
% x = [-1.5 1.5] -> starting value of x
% epsilon1, epsilon2 -> constant used for terminating the
%                      algorithm
```

```matlab
% delx -> required for gradient computation
% falpha_prev -> function value at first/previous iteration
% deriv -> gradient vector
% deltag -> difference in gradient vector (over previous
%           iteration)
% A -> approximation of the hessian matrix
% search -> search direction
%
clear all
clc
n_of_var = 2;
x = [-3 2];
delx = 0.001;
A = eye(length(x));
epsilon1 = 1e-6;
epsilon2 = 1e-6;
delx = 1e-3;
falpha_prev = func_multivar(x);
fprintf('Initial function value =    %7.4f\n ',falpha_prev)
fprintf(' No.       x-vector        f(x)        Deriv \n')
fprintf('_____\n')
for i = 1:50
    if i==1
    deriv_prev = grad_vec(x,delx,n_of_var);
    search = -deriv_prev;
    [alpha,falpha] = golden_funct1(x,search);
    if abs(falpha-falpha_prev)<0.001
    break;
    end
    falpha_prev = falpha;
    x = x + alpha*search;
    fprintf('%3d %8.3f %8.3f  % 8.3f  %8.3f  \n',i,x,falpha_
        prev,norm(deriv_prev))
    else
        deltax = (alpha*search);
        if i>2
            deltax = deltax';
            search = search';
        end
        deriv = grad_vec(x,delx,n_of_var);
        deltag = deriv-deriv_prev;
        term1 = (deltag'*deltag)/(deltag*deltax');
        term2 = (deriv_prev'*deriv_prev)/(deriv_prev*search');
        A = A + term1 + term2;
        search = -inv(A)*deriv';
        [alpha,falpha] = golden_funct1(x,search');
fprintf('%3d %8.3f %8.3f  % 8.3f  %8.3f  \n',i,x+alpha*search',
    falpha,norm(deriv))
        if abs(falpha-falpha_prev)<epsilon1 ||
            norm(deriv)<epsilon2
```

Appendix B

```
            break;
        end
        falpha_prev = falpha;
        deriv_prev = deriv;
        x = x+alpha*search';
    end
end
fprintf('_____\n')
%
%%%%%%%%%%%%%%%%%%%%%%%%%%%%%%%%%%%%%%%%%


%%%%%%%%%%%%%%%%%%%%%%%%%%%%%%%%%%%%%%%%%
% MATLAB code powell.m
%%%%%%%%%%%%%%%%%%%%%%%%%%%%%%%%%%%%%%%%%
% n_of_var -> number of design variables
% x = [-3 2] -> starting value of x
% epsilon -> constant used for terminating the algorithm
% term -> linearly independent search directions
% falpha_prev -> function value at first/previous iteration
% search -> search direction
%
clear all
clc
n_of_var = 2;
x = [-3 2];
epsilon = 1e-6;
falpha_prev = func_multivar(x);
fprintf('Initial function value =   %7.4f\n ',falpha_prev)
fprintf(' No.       x-vector      f(x)    \n')
fprintf('_____\n')
for i = 1:n_of_var
    for j = 1:n_of_var+1
        if (i==j)
            term(i,j)=1;
        else
            term(i,j) = 0;
        end
    end
end
for i = 1: n_of_var
    search{i} = (term(:,i))';
    [alpha,falpha] = golden_funct1(x,search{i});
    x = x + alpha*search{i};
end
search{i+1} = (term(:,i+1))';
for k = 1:200
    xini = x;
    i = 1;
    while i<n_of_var+1
```

```matlab
        [alpha,falpha] = golden_funct1(x,search{i});
        x = x + alpha*search{i};
        i = i+1;
    end
        if abs(falpha-falpha_prev) < epsilon
            break;
        end
    search{i} = (x-xini);
    [alpha,falpha] = golden_funct1(x,search{i});
    x = x + alpha*search{i};
    temp = search;
    for i = 1:n_of_var
    search{i} = temp{i+1};
    end
    falpha_prev = falpha;
    fprintf('%3d %8.3f %8.3f  % 8.3f  \n',k,x,falpha)
end
fprintf('_____\n')
%
%%%%%%%%%%%%%%%%%%%%%%%%%%%%%%%%%%%%%%%%%%%


%%%%%%%%%%%%%%%%%%%%%%%%%%%%%%%%%%%%%%%%%%
% MATLAB code neldermead.m
%%%%%%%%%%%%%%%%%%%%%%%%%%%%%%%%%%%%%%%%%%
% n_of_var -> number of design variables
% lb, ub -> lower/upper bound of variable
% (optional for generating initial feasible points randomly)
% ybest -> best value of the objective function in the iteration
% ysecondbest -> second best value of the objective function
% yworst -> worst value of the objective function in the
%           iteration
% xworst -> corresponding value of the variable for yworst
% xc -> centroid of the polygon
% fcentroid -> function value at xc
% deviation -> sum square deviation of function values from
%              centroid
% xr => reflected point
% freflec => function value at reflected point
% xe => expanded point
% fexp => function value at expanded point
% xcont => contracted point
% fcont => function value at contracted point
%
clear all
clc
n_of_var = 2;
epsilon = 1e-4;
alpha = 1;
gamma = 2;
```

Appendix B

```
rho = -0.5;
lb = [-1 -1];
ub = [1 1];
fprintf('  Iteration     Deviation         f(x)     \n')
fprintf('_____\n')
for JJ = 1:50
for i = 1:length(lb)
    for j = 1:n_of_var+1
        a(i,j) = lb(i) + (ub(i)-lb(i))*rand;
    end
end
if JJ~=1
    a = x';
end
for i = 1:n_of_var+1
    for j = 1:n_of_var
        x(i,j) = a(j,i);
    end
    fval(i) = func_multivar(x(i,:));
end
[yworst,I] = max(fval);
[ybest,II] = min(fval);
% compute centroid
for i = 1:length(lb)
    sum(i) = 0;
    for j = 1:n_of_var+1
        if (j ~= I)
        sum(i) = sum(i) + a(i,j);
        else
            xworst(:,i) = a(i,j);
        end
    end
end
xc = sum./n_of_var;
fcentroid = func_multivar(xc);
sum1 = 0;
for i = 1:n_of_var+1
    sum1 = sum1 + (fcentroid-fval(i))^2;
end
deviation = sqrt(sum1/(n_of_var+1));
    if deviation < epsilon
        break;
    end
fval(I) = [];
[ysecondworst,Isw] = max(fval);
xr = xc + alpha*(xc-xworst);
freflec = func_multivar(xr);
if freflec < ybest
    %expansion
```

```matlab
        xe = xc + gamma*(xc-xworst);
        fexp = func_multivar(xe);
        if fexp < freflec
            x(I,:) = xe;
        else
            x(I,:) = xr;
        end
 else
    if freflec < ysecondworst
        x(I,:) = xr;
    else
      xcont = xc + rho*(xc-xworst);
      fcont = func_multivar(xcont);
      if fcont < yworst
          x(I,:) = xcont;
      end
    end
 end
fprintf('%3d %15.4f %15.3f    \n',JJ,deviation,ybest)
end
fprintf('_____\n')
xc
%
%%%%%%%%%%%%%%%%%%%%%%%%%%%%%%%%%%%%%%%%%


%%%%%%%%%%%%%%%%%%%%%%%%%%%%%%%%%%%%%%
% MATLAB code robotics_nominal_traj.m
%%%%%%%%%%%%%%%%%%%%%%%%%%%%%%%%%%%%%%
%
% Generates nominal trajectory for the robotics arm problem
clear all
clc
% generate 100 points in t
t = -pi:.063:pi
px = 30*cos(t);
py = 100*sin(t);
pz = 10*t + 66.04;
plot3(px,py,pz,'b-','LineWidth',3)
xlabel('px')
ylabel('py')
zlabel('pz')
%
%%%%%%%%%%%%%%%%%%%%%%%%%%%%%%%%%%%%%%
```

Appendix B 351

```
%%%%%%%%%%%%%%%%%%%%%%%%%%%%%%%%%%%%%%
% MATLAB code robotics_optimized_traj.m
%%%%%%%%%%%%%%%%%%%%%%%%%%%%%%%%%%%%%%
%
% Generates optimized trajectory for the robotics arm problem
function [] = generate_optimized_traj(x)
d1 = 66.04;
d3 = 14.91;
d4 = 43.31;
a2 = 43.18;
a3 = 2.03;
for i = 1:100
t = -pi + (i-1)*0.063;
theta1 = x(i);
theta2 = x(i+100);
theta3 = x(i+200);
c1 = cos(theta1);
c2 = cos(theta2);
s1 = sin(theta1);
s2 = sin(theta2);
c23 = cos(theta2+theta3);
s23 = sin(theta2+theta3);
f1(i) = c1*(a2*c2 + a3*c23 - d4*s23) - d3*s1;
f2(i) = s1*(a2*c2 + a3*c23 - d4*s23) + d3*c1;
f3(i) = d1 - a2*s2 - a3*s23 -d4*c23;
end
 plot3(f1,f2,f3,'r*')
end
%
%%%%%%%%%%%%%%%%%%%%%%%%%%%%%%%%%%%%%
```

Chapter 4

Code Name	Details
simplex.m	Simplex method
initial_cost.m	Computes cost coefficients for the nonbasic variables
phase1.m	Phase I of the simplex method
remove_variable.m	Removes user specified column from the nonbasic set
phase1_without_initialization.m	Phase I of the simplex method without initializing A matrix and b vector.
dual.m	Dual simplex method
interior.m	Affine scaling method

```matlab
%%%%%%%%%%%%%%%%%%%%%%%%%%%%%%%%%%%%%%%%%%%%%%%%%%%%%
% MATLAB code simplex.m
%%%%%%%%%%%%%%%%%%%%%%%%%%%%%%%%%%%%%%%%%%%%%%%%%%%%%
%
% The matrix A and b corresponds to equation Ax=b
% c -> vector of cost coefficients
% basic_set -> set of basic variables
% nonbasic_set -> setof nonbasic variables
% B -> matrix containing basic variable columns of A
% N -> matrix containing nonbasic variable columns of A
% xb -> basic variables
% y -> simplex multipliers
% cb -> cost coefficients of basic variables
% cn -> cost coefficients of nonbasic variables
%
clear all
clc
format rational
format compact
A = [3 1 1 0 0;
     1 2 0 1 0;
     1 0 0 0 1];
b = [10;8;3];
c = [-6;-7;0;0;0];
 basic_set = [3 4 5];
 nonbasic_set = [1 2];
for i = 1:length(basic_set)
    B(:,i) = A(:,basic_set(i));
    cb(i) = c(basic_set(i));
end
for i = 1:length(nonbasic_set)
    N(:,i) = A(:,nonbasic_set(i));
    cn(i) = c(nonbasic_set(i));
end
cn_cap = cn;
cb_ini = cb;
b_cap = b;
zz1 = 0;
fprintf('\n _____ \n')
basic_set
nonbasic_set
Initial_Table = [B N b_cap]
Cost =[cb cn_cap -zz1]
for i = 1:3
      [minvalue entering_basic_variable] = min(cn_cap);
entering_column = inv(B)*A(:,nonbasic_set(entering_basic_
    variable));
ratios = b_cap'./entering_column';
[min_ratio leaving_basic_variable] = min(ratios);
while min_ratio<0
```

Appendix B 353

```
    ratios(leaving_basic_variable) = inf;
    [min_ratio leaving_basic_variable] = min(ratios);
end
temp_basic_set = basic_set;
temp_nonbasic_set = nonbasic_set;
temp_cb = cb;
temp_cn = cn;
basic_set(leaving_basic_variable) = temp_nonbasic_
    set(entering_basic_variable);
nonbasic_set(entering_basic_variable) = temp_basic_
    set(leaving_basic_variable);
cb(leaving_basic_variable) = temp_cn(entering_basic_variable);
cn(entering_basic_variable) = temp_cb(leaving_basic_variable);
aa(nonbasic_set) = cn;
cn = aa(sort(nonbasic_set));
nonbasic_set = sort(nonbasic_set);
for ii = 1:length(basic_set)
    B(:,ii) = A(:,basic_set(ii));
end
for ii = 1:length(nonbasic_set)
    N(:,ii) = A(:,nonbasic_set(ii));
end
xb = inv(B)*b;
y = cb*inv(B);
cn_cap = cn-y*N;
b_cap = xb;
zz = zz1+cb*xb;
fprintf('\n _____ \n')
basic_set
nonbasic_set
Table = [eye(length(B)) inv(B)*N  b_cap]
Cost = [cb_ini cn_cap -zz]
if cn_cap >= 0
     break;
 end
end
fprintf('\n ------SOLUTION------\n')
basic_set
xb
zz
%
%%%%%%%%%%%%%%%%%%%%%%%%%%%%%%%%%%%%%%%%%%%%%

%%%%%%%%%%%%%%%%%%%%%%%%%%%%%%%%%%%%%%%%%%
% MATLAB code initial_cost.m
%%%%%%%%%%%%%%%%%%%%%%%%%%%%%%%%%%%%%%%%%%
%
cb = [0 1 1];
cn = [0 0 0];
N = [3 2 0;
```

```matlab
     2 -4 -1;
     3  4  0];
B = [0 1 0;0 0 1;1 0 0];
y = cb*inv(B);
cn_cap = cn-y*N
%%%%%%%%%%%%%%%%%%%%%%%%%%%%%%%%%%%%%%%%%%%%%%%%%%%

%%%%%%%%%%%%%%%%%%%%%%%%%%%%%%%%%%%%%%%%%%%%%%%%%%%
% MATLAB code phase1.m
%%%%%%%%%%%%%%%%%%%%%%%%%%%%%%%%%%%%%%%%%%%%%%%%%%%
%
% The matrix A and b corresponds to Ax=b
% c -> vector of cost coefficients
% basic_set -> set of basic variables
% nonbasic_set -> setof nonbasic variables
% B -> matrix containing basic variable columns of A
% N -> matrix containing nonbasic variable columns of A
% xb -> basic variables
% y -> simplex multipliers
% cb -> cost coefficients of basic variables
% cn -> cost coefficients of nonbasic variables
%
clear all
clc
format rational
format compact
A = [3  2  0 0 1 0;
     2 -4 -1 0 0 1;
     3  4  0 1 0 0];
b = [10;3;16];
c = [-5;2;1;0;0;0];
basic_set = [5 6 4];
nonbasic_set = [1 2 3];
for i = 1:length(basic_set)
    B(:,i) = A(:,basic_set(i));
    cb(i) = c(basic_set(i));
end
for i = 1:length(nonbasic_set)
    N(:,i) = A(:,nonbasic_set(i));
    cn(i) = c(nonbasic_set(i));
end
cn_cap = cn;
cb_ini = cb;
b_cap = b;
zz1 = 91/8;
fprintf('\n _____ \n')
basic_set
nonbasic_set
Initial_Table = [B N b_cap]
Cost = [cb cn_cap -zz1]
```

Appendix B

```matlab
for i = 1:1
    [minvalue entering_basic_variable] = min(cn_cap);
entering_column = inv(B)*A(:,nonbasic_set(entering_basic_
    variable));
ratios = b_cap'./entering_column';
[min_ratio leaving_basic_variable] = min(ratios);
while min_ratio<0
    ratios(leaving_basic_variable) = inf;
    [min_ratio leaving_basic_variable] = min(ratios);
end
temp_basic_set = basic_set;
temp_nonbasic_set = nonbasic_set;
temp_cb = cb;
temp_cn = cn;
basic_set(leaving_basic_variable) = temp_nonbasic_
    set(entering_basic_variable);
nonbasic_set(entering_basic_variable) = temp_basic_
    set(leaving_basic_variable);
cb(leaving_basic_variable) = temp_cn(entering_basic_variable);
cn(entering_basic_variable) = temp_cb(leaving_basic_variable);
aa(nonbasic_set) = cn;
cn = aa(sort(nonbasic_set));
nonbasic_set = sort(nonbasic_set);
for ii = 1:length(basic_set)
    B(:,ii) = A(:,basic_set(ii));
end
for ii = 1:length(nonbasic_set)
    N(:,ii) = A(:,nonbasic_set(ii));
end
xb = inv(B)*b;
y = cb*inv(B);
cn_cap = cn-y*N;
b_cap = xb;
zz = zz1+cb*xb;
fprintf('\n _____ \n')
basic_set
nonbasic_set
Table = [eye(length(B)) inv(B)*N  b_cap]
Cost = [cb_ini cn_cap -zz]
end
%
%%%%%%%%%%%%%%%%%%%%%%%%%%%%%%%%%%%%%%%%%%%%%%%%%%

%%%%%%%%%%%%%%%%%%%%%%%%%%%%%%%%%%%%%%%%%%%%%%%%%%
% MATLAB code remove_variable.m
%%%%%%%%%%%%%%%%%%%%%%%%%%%%%%%%%%%%%%%%%%%%%%%%%%
%
% This program removes user specified column from
% the nonbasic set
%
```

```matlab
remove_column = 3;
nonbasic_set(remove_column) = [];
N(:,remove_column) = [];
cn(remove_column) = [];
cn_cap = cn-y*N;
fprintf('\n ----Table after removing artificial
    variable------\n')
basic_set
nonbasic_set
Initial_Table = [eye(length(B)) inv(B)*N  b_cap]
Cost = [cb_ini cn_cap -zz]
%%%%%%%%%%%%%%%%%%%%%%%%%%%%%%%%%%%%%%%%%%%%%%


%%%%%%%%%%%%%%%%%%%%%%%%%%%%%%%%%%%%%%%%%%%%%%
% MATLAB code dual.m
%%%%%%%%%%%%%%%%%%%%%%%%%%%%%%%%%%%%%%%%%%%%%%
% The matrix A and b corresponds to equation Ax=b
% c -> vector of cost coefficients
% basic_set -> set of basic variables
% nonbasic_set -> set of nonbasic variables
% B -> matrix containing basic variable columns of A
% N -> matrix containing nonbasic variable columns of A
% xb -> basic variables
% y -> simplex multipliers
% cb -> cost coefficients of basic variables
% cn -> cost coefficients of nonbasic variables
%
clear all
clc
format rational
format compact
A = [-1  0  1  0  0  0;
      0 -1  0  1  0  0;
     -2 -1  0  0  1  0;
     -1 -3  0  0  0  1];
b = [-3;-4;-25;-26];
c = [9; 8; 0; 0; 0;0];
basic_set = [3 4 5 6];
nonbasic_set = [1 2];
for i = 1:length(basic_set)
    B(:,i) = A(:,basic_set(i));
    cb(i) = c(basic_set(i));
end
for i = 1:length(nonbasic_set)
    N(:,i) = A(:,nonbasic_set(i));
    cn(i) = c(nonbasic_set(i));
end

cn_cap = cn;
cb_ini = cb;
```

Appendix B

```
b_cap = b;
zz = 0;
fprintf('\n                                        \n')
basic_set
nonbasic_set
Initial_Table = [B N b_cap]
Cost = [cb cn_cap zz]
for i = 1:4
      [minvalue leaving_basic_variable] = min(b_cap);
      mat1 = inv(B)*N;
entering_row = mat1(leaving_basic_variable,:);
ratios = -1*(cn_cap'./entering_row');
[min_ratio entering_basic_variable] = min(ratios);
while min_ratio<0
    ratios(entering_basic_variable) = inf;
    [min_ratio entering_basic_variable] = min(ratios);
end
temp_basic_set = basic_set;
temp_nonbasic_set = nonbasic_set;
temp_cb = cb;
temp_cn = cn;
basic_set(leaving_basic_variable) = temp_nonbasic_
    set(entering_basic_variable);
nonbasic_set(entering_basic_variable) = temp_basic_
    set(leaving_basic_variable);
  cb(leaving_basic_variable) = temp_cn(entering_basic_
      variable);
  cn(entering_basic_variable) = temp_cb(leaving_basic_
      variable);
 aa(nonbasic_set) = cn;
 cn = aa(sort(nonbasic_set));
nonbasic_set = sort(nonbasic_set);
for ii = 1:length(basic_set)
    B(:,ii) = A(:,basic_set(ii));
end
for ii = 1:length(nonbasic_set)
    N(:,ii) = A(:,nonbasic_set(ii));
end
xb = inv(B)*b;
y = cb*inv(B);
cn_cap = cn-y*N;
b_cap = xb;
zz = cb*xb;
fprintf('\n                                        \n')
basic_set
nonbasic_set
Table = [eye(length(B)) inv(B)*N  b_cap]
Cost = [cb_ini cn_cap -zz]
if b_cap >= 0
    break;
```

```
end
end
fprintf('\n ------FINAL SOLUTION------\n')
basic_set
xb
zz
%%%%%%%%%%%%%%%%%%%%%%%%%%%%%%%%%%%%%%%%%%%%%%%%%%%


%%%%%%%%%%%%%%%%%%%%%%%%%%%%%%%%%%%%%%%%%%%%%%%%%%%
% MATLAB code interior.m
%%%%%%%%%%%%%%%%%%%%%%%%%%%%%%%%%%%%%%%%%%%%%%%%%%%
%
% Affine scaling method
%
clear all
clc
A = [3 1;
     1 2;
     1 0];
b = [10;8;3];
c = [6;7];
x = [0;0];
obj_prev = c'*x;
gamma = 0.9;
tolerance = 1e-5;
for i = 1:10
    vk = b-A*x;
    dv = diag(vk);
    hx = inv(A'*dv^-2*A)*c;
    hv = -A*hx;
    for j = 1:length(hv)
        if hv(j)<0
            var(j) = -vk(j)/hv(j);
        else
            var(j) = inf;
        end
    end
    alpha = gamma*min(var);
    x = x + alpha*hx;
    objective = c'*x;
    if abs(objective-obj_prev)<tolerance
        break;
    end
    obj_prev = objective;
end
objective
x
%
%%%%%%%%%%%%%%%%%%%%%%%%%%%%%%%%%%%%%%%%%%%%%%%%%%%
```

Appendix B

Chapter 5

Code Name	Details
prob.m	Genetic algorithm (GA; main program)
in.m	Inputs to GA
roulett.m	Roulette wheel selection
tournament.m	Tournament selection
func.m	Test function to be included here (for GA)
simann.m	Simulated annealing
func1.m	Objective function to be included here (for PSO and simulated annealing)
pso.m	Particle swarm optimization (PSO)

```
%%%%%%%%%%%%%%%%%%%%%%%%%%%%%%%%%%%%%%%%%%%%%
%    File name prob.m
%    Genetic algorithm - main program
%%%%%%%%%%%%%%%%%%%%%%%%%%%%%%%%%%%%%%%%%%%%%
%
clear all
clc
format long g;
% Read the input file
in;
% INITIALIZATION OF STRINGS
string = 0;
for i = 1:n_of_v
string = string+n_of_bits(i);
end
for j = 1:n_of_p
    for i = 1:string
    r(j,i) = rand;
        if r(j,i)< 0.5
        r(j,i) = 0;
        else
        r(j,i) = 1;
        end
    end
end
% MAIN LOOP
for g = 1:n_of_g
% Decoded value of r (with left bit as MSB)
deci = cell(n_of_v,1);
decoded = cell(n_of_v,1);
dum1 = 1;
dummy = n_of_bits(1);
for i = 1:n_of_v
deci{i} = bi2de(r(:,dum1:dummy),'left-msb');
dum1 = dum1+n_of_bits(i);
while dummy<string
dummy = dummy+n_of_bits(i+1);
end
```

```
% NORMALIZE TO THE VARIABLE RANGE
x1(:,i) = deci{i};
decoded{i} = range(i,1)+((range(i,2)-range(i,1))/(2^n_of_bits(i)-
            1))*x1(:,i);
xxx(:,i) = decoded{i};
end
% FUNCTION EVALUATION
for i = 1:n_of_p
[fitness1(i),constraint(i,:)] = func(xxx(i,:));
end
fitness = fitness1';
for hh = 1:length(fitness)
    if fitness(hh) < 0
        flag1 = 1;
    end
end
if flag1 == 1
    [factor,indices] = min(fitness);
    fitness1 = -factor+fitness;
end
% CALLING ROULETTE WHEEL
if tourni_flag ~= 1
    if problem == 'min'
    fitness2 = 1./(1+fitness1);
    end
[best_fit(g),indi(g)] = max(fitness2);
best_var(g,:) = xxx(indi(g),:);
if problem == 'min'
    best_fit(g) = fitness(indi(g));
end
% CUMULATIVE PROBABILITY
s = sum(fitness2);
cum_prob = fitness2/s;
roulett;
else
[best_fit(g),indi(g)] = min(fitness);
average_fitness = mean(fitness);
best_var(g,:) = xxx(indi(g),:);
% CALLING TOURNAMENT SELECTION
tournament;

% IF THIS IS A CONSTRAINT PROBLEM THEN WE HAVE TO USE THIS
    if n_of_c>=0
    best_fit(g) = min_fit;
    best_var(g,:) = xxx(indi(g),:);
    end
end
% CROSSOVER
for k = 1:2:n_of_p
parent1 = r_new(round(random('unif',0.5,n_of_p+0.5)),:);
parent2 = r_new(round(random('unif',0.5,n_of_p+0.5)),:);
 if multi_crossover == 0
cross_o_pos = round(random('unif',1.5,string+0.5-1));
child1(1:cross_o_pos) = parent2(1:cross_o_pos);
child1(cross_o_pos+1:string) = parent1(cross_o_pos+1:string);
```

Appendix B

```
child2(1:cross_o_pos) = parent1(1:cross_o_pos);
child2(cross_o_pos+1:string) = parent2(cross_o_pos+1:string);
else
pois_ra = rand(1);
if(pois_ra<0.1353)no_of_cross_over = 0;
end
if(pois_ra>=0.1353 & pois_ra<0.4059)no_of_cross_over = 1;
end
if(pois_ra>=0.4059 & pois_ra<0.6865)no_of_cross_over = 2;
end
if(pois_ra>=0.6865 & pois_ra<0.8769)no_of_cross_over = 3;
end
if(pois_ra>=0.8769)no_of_cross_over = 3;
end
switch no_of_cross_over
case 0
child1(1:string) = parent1(1:string);
child2(1:string) = parent2(1:string);
case 1
cross_o_pos = round(random('unif',1.5,string+0.5-1));
child1(1:cross_o_pos) = parent2(1:cross_o_pos);
child1(cross_o_pos+1:string) = parent1(cross_o_pos+1:string);
child2(1:cross_o_pos) = parent1(1:cross_o_pos);
child2(cross_o_pos+1:string) = parent2(cross_o_pos+1:string);
case 2
cross_o_pos1 = round(random('unif',1.5,string+0.5-1));
cross_o_pos2 = round(random('unif',1.5,string+0.5-1));
while (cross_o_pos2 == cross_o_pos1)
cross_o_pos2 = round(random('unif',1.5,string+0.5-1));
end
cross_sor = [cross_o_pos1 cross_o_pos2];
cross_sort = sort(cross_sor);
child1(1:cross_sort(1)) = parent1(1:cross_sort(1));
child1(cross_sort(1)+1:cross_sort(2)) = parent2(cross_sort(1)+1:cross_
    sort(2));
child1(cross_sort(2)+1:string) = parent1(cross_sort(2)+1:string);
child2(1:cross_sort(1)) = parent2(1:cross_sort(1));
child2(cross_sort(1)+1:cross_sort(2)) = parent1(cross_sort(1)+1:cross_
    sort(2));
child2(cross_sort(2)+1:string) = parent2(cross_sort(2)+1:string);
case 3
cross_o_pos1 = round(random('unif',1.5,string+0.5-1));
cross_o_pos2 = round(random('unif',1.5,string+0.5-1));
while (cross_o_pos2 == cross_o_pos1)
cross_o_pos2 = round(random('unif',1.5,string+0.5-1));
end
cross_o_pos3 = round(random('unif',1.5,string+0.5-1));
while (cross_o_pos3 == cross_o_pos1 & cross_o_pos3 == cross_o_pos2)
cross_o_pos3 = round(random('unif',1.5,string+0.5-1));
end
cross_sor = [cross_o_pos1 cross_o_pos2 cross_o_pos3];
cross_sort = sort(cross_sor);
child1(1:cross_sort(1)) = parent1(1:cross_sort(1));
child1(cross_sort(1)+1:cross_sort(2)) = parent2(cross_sort(1)+1:cross_
    sort(2));
```

```
child1(cross_sort(2)+1:cross_sort(3)) = parent1(cross_sort(2)+1:cross_
    sort(3));
child1(cross_sort(3)+1:string) = parent2(cross_sort(3)+1:string);
child2(1:cross_sort(1)) = parent2(1:cross_sort(1));
child2(cross_sort(1)+1:cross_sort(2)) = parent1(cross_sort(1)+1:cross_
    sort(2));
child2(cross_sort(2)+1:cross_sort(3)) = parent2(cross_sort(2)+1:cross_
    sort(3));
child2(cross_sort(3)+1:string) = parent1(cross_sort(3)+1:string);
end
end
r(k,:) = child1;
r(k+1,:) = child2;
end
% MUTATION
for i = 1:n_of_p
pr_m = random('unif',0,1);
for j = 1:string
    if pr_m<mut_prob
    if r(i,j) == 0
    r(i,j) = 1;
    else
    r(i,j) = 0;
    end
    end
end
end
[min_best_fit,ind] = min(best_fit);
[g best_var(ind,:) min(best_fit)];
if g >= 2
    if abs(last_gen_best-min(best_fit))<epsilon
        flag = flag+1;
    else
        flag = 0;
    end
end
if flag>stall_gen
    break;
end
last_gen_best = min(best_fit);
fprintf('%4i %9.4f %9.4f \n',g, best_var(ind,:), min(best_fit))
end % END OF MAIN LOOP
[min_best_fit,ind] = min(best_fit);
best_var(ind,:)
min_best_fit
%
%%%%%%%%%%%%%%%%%%%%%%%%%%%%%%%%%%%%%%%%%%%%%%%%%%%%


%%%%%%%%%%%%%%%%%%%%%%%%%%%%%%%%%%%%%%%%%%%%%%%
%    File name in.m
%    Input parameters for Genetic algorithm
%%%%%%%%%%%%%%%%%%%%%%%%%%%%%%%%%%%%%%%%%%%%%%%
%
problem = 'min'; % If roulette wheel is used to minimize
```

Appendix B

```
n_of_v = 1;              % number of variables
n_of_g = 10000;          % maximum number of generations
n_of_p = 10;             % population size
range(1,:) = [40 90];    % variable bound
n_of_bits(1) = 15;       % number of bits
cross_prob = 0.9;        % crossover probability
multi_crossover = 1;     % use multi-crossover
mut_prob = 0.02;         % mutation probability
tourni_flag = 0;         % use roulette wheel
epsilon = 1e-7;          % function tolerance
flag = 0;                % stall generations flag
flag1 = 0;               % scaling flag
stall_gen = 500;         % stall generations for termination
n_of_c = 0;              % for constraint handling
%
%%%%%%%%%%%%%%%%%%%%%%%%%%%%%%%%%%%%%%%%%%%%%%%%%

%%%%%%%%%%%%%%%%%%%%%%%%%%%%%%%%%%%%%%%%%%%%%%%%%
%    File name roulett.m
%    Roulette wheel selection
%%%%%%%%%%%%%%%%%%%%%%%%%%%%%%%%%%%%%%%%%%%%%%%%%
%
slot(1) = 0;
for ii = 2:n_of_p+1
    slot(ii) = cum_prob(ii-1)+slot(ii-1);
end
% COPY GENERATION
for kk = 1:n_of_p
    pr = rand;
    for iii = 1:n_of_p+1
        if (pr>slot(iii)) & (pr<slot(iii+1))
        st_t_c(kk) = iii;
        end
    end
end
for kkk = 1:n_of_p
    r_new(kkk,:) = r(st_t_c(kkk),:);
end
%
%%%%%%%%%%%%%%%%%%%%%%%%%%%%%%%%%%%%%%%%%%%%%%%%%

%%%%%%%%%%%%%%%%%%%%%%%%%%%%%%%%%%%%%%%%%%%%%%%%%
%    File name tournament.m
%    Tournament selection
%    Also modifies the function
%    in the presence of constraints
%%%%%%%%%%%%%%%%%%%%%%%%%%%%%%%%%%%%%%%%%%%%%%%%%
%
```

```
infeasible_flag = 1;
for i = 1:n_of_p
        if constraint(i)>= 0
            feas_tag(i) = 1;
            fit(i) = fitness(i);
        else
            feas_tag(i) = 0;
            fit(i) = -100000000000;
        end
end
for i = 1:n_of_p
if(feas_tag(i) == 1)
infeasible_flag = 0;
end
end
if(infeasible_flag == 1)
fit(1) = 1000000000000;
end
max_fit = max(fit);
for i = 1:n_of_p
    if feas_tag(i)== 0
        for j = 1:n_of_c
        if (constraint(i,j)<0)
    fitness(i) = max_fit+abs(constraint(i,j));
        end
        end
    end
end
[min_fit,indices] = min(fitness);
for i = 1:n_of_p
    pr = round(random('unif',0.5,n_of_p+0.5));
    while pr == i
    pr = round(random('unif',0.5,n_of_p+0.5));
    end
    if feas_tag(i) == feas_tag(pr)
        if fitness(i) <= fitness(pr)
        r_new(i,:) = r(i,:);
        else
        r_new(i,:) = r(pr,:);
        end
     else
        if feas_tag(i) == 1
        r_new(i,:) = r(i,:);
        else
        r_new(i,:) = r(pr,:);
        end
    end
end
%%%%%%%%%%%%%%%%%%%%%%%%%%%%%%%%%%%%%%%%%%%%%%%%%%%%%
```

Appendix B

```matlab
%%%%%%%%%%%%%%%%%%%%%%%%%%%%%%%%%%%%%%%%%%%%%%%%%
%   File name func.m
%   Enter the function to be optimized
%%%%%%%%%%%%%%%%%%%%%%%%%%%%%%%%%%%%%%%%%%%%%%%%%
%
function [y,constr] = func(x)
y = 204165.5/(330-2*x) + 10400/(x-20);
constr(1) = 10;
%
%%%%%%%%%%%%%%%%%%%%%%%%%%%%%%%%%%%%%%%%%%%%%%%%%


%%%%%%%%%%%%%%%%%%%%%%%%%%%%%%%%%%%%%%%%%%%%%%%%%
%   File name simann.m
%   Simulated annealing algorithm
%%%%%%%%%%%%%%%%%%%%%%%%%%%%%%%%%%%%%%%%%%%%%%%%%
%
% lb -> lower bound of variables
% ub -> upper bound of variables
% x(i) -> design variables
% rand -> random number from 0 to 1
% perturb_x(i) -> perturbation on design variables
%
clear all
clc
format long
epsilon = 0.002;
flag = 0;
lb = [-5.12 -5.12];
ub = [5.12 5.12];
for i = 1:length(lb)
    x(i) = lb(i) + (ub(i)-lb(i))*rand;
end
[E_old,constr] = func1(x);
bestx = x;
best_obj = E_old;
for j = 1:10000
    flag = flag+1;
    for i = 1:length(x)
        perturb_x(i) = epsilon*x(i)*rand;
        if rand < 0.5
    perturb_x(i) = -perturb_x(i);
        end
    end
x = x + perturb_x;
    for i = 1:length(x)
        if x(i)<lb(i) | x(i)>ub(i)
        x(i) = lb(i) + (ub(i)-lb(i))*rand;
        end
end
    end
```

```
[E_new,constr] = func1(x);
    if E_new < E_old
    E_old = E_new;
    else
    if exp(-(E_new-E_old)/E_old)> rand
      E_old = E_new;
    end
    x = x - perturb_x;
end
    px(j) = j;
    py(j) = E_new;

    if E_new < best_obj
        best_obj = E_new;
        bestx = x;
        flag = 0;
    end
[j bestx best_obj]
    if flag > 1000
    break;
    end
end
%%%%%%%%%%%%%%%%%%%%%%%%%%%%%%%%%%%%%%%%%%%%%%%%

%%%%%%%%%%%%%%%%%%%%%%%%%%%%%%%%%%%%%%%%%%%%%%%
%   File name func1.m
%   Enter the function to be optimized
%%%%%%%%%%%%%%%%%%%%%%%%%%%%%%%%%%%%%%%%%%%%%%%
%
function [y,constr] = func1(x)
y = 20 + x(1)*x(1)-10*cos(2*pi*x(1)) + x(2)*x(2)-
    10*cos(2*pi*x(2));
constr(1) = 10;
%
%%%%%%%%%%%%%%%%%%%%%%%%%%%%%%%%%%%%%%%%%%%%%%%

%%%%%%%%%%%%%%%%%%%%%%%%%%%%%%%%%%%%%%%%%%%%%%%
%   File name pso.m
%   Particle Swarm Optimization algorithm
%%%%%%%%%%%%%%%%%%%%%%%%%%%%%%%%%%%%%%%%%%%%%%%
%
% lb -> lower bound of variables
% ub -> upper bound of variables
% x -> position of individual
% v -> velocity of individual
% rand -> random number from 0 to 1
% fitness -> fitness of individual
% pbest -> best fitness achieved by individual
% gbest -> best fitness of group
```

Appendix B

```
% pop -> population size
% phi_1, phi_2 -> tuning parameters
% nmax -> maximum number of iterations
%
clear all
clc
format long
pop = 20;
phi_1 = 1.05;
phi_2 = 1.1;
nmax = 100;
weight = linspace(1,0.3,nmax);
lb = [-500 -500];
ub = [500 500];
for i = 1:length(lb)
    for j = 1:pop
        x(i,j) = lb(i) + (ub(i)-lb(i))*rand;
        v(i,j) = 0;
    end
end
for i = 1:pop
    fitness(i) = func1(x(:,i));
    pbest(i) = fitness(i);
    px(i,:) = x(:,i);
end
[gbest, location] = min(fitness);
gx = x(:,location);
plot3(px(:,1),px(:,2),pbest,'r*')
    grid on
    xlabel('x1')
    ylabel('x2')
    zlabel('f(x)')
for i = 1:nmax
    for j = 1:pop
v(:,j) = weight(i)*v(:,j) + phi_1*rand*(px(j,:)'-x(:,j)) +
        phi_2*rand*(gx-x(:,j));
        x(:,j) = x(:,j) + v(:,j);
        for k = 1:length(x(:,j))
            if x(k,j) < lb(k) || x(k,j) > ub(k)
                x(k,j) = lb(k) + (ub(k)-lb(k))*rand;
            end
        end
        fitness(j) = func1(x(:,j));
        if fitness(j) < pbest(j)
            pbest(j) = fitness(j);
            px(j,:) = x(:,j);
        end
    end
    [gbest, location] = min(pbest);
    gx = x(:,location);
```

```
    [gx' gbest]
    plot3(px(:,1),px(:,2),pbest,'r*')
    grid on
    xlabel('x1')
    ylabel('x2')
    zlabel('f(x)')
    axis([-500 500 -500 500 -1000 0])
    pause(0.2)
end
%%%%%%%%%%%%%%%%%%%%%%%%%%%%%%%%%%%%%%%%%%%%%

%%%%%%%%%%%%%%%%%%%%%%%%%%%%%%%%%%%%%%%%%%%%%
%   File name func1.m
%   Enter the function to be optimized
%%%%%%%%%%%%%%%%%%%%%%%%%%%%%%%%%%%%%%%%%%%%%
%
function y = func1(x)
y = -x(1)*sin(sqrt(abs(x(1)))) -x(2)*sin(sqrt(abs(x(2))));
%
%%%%%%%%%%%%%%%%%%%%%%%%%%%%%%%%%%%%%%%%%%%%%
```

Chapter 6

Code Name	Details
DFP.m (main program)	Davidon–Fletcher–Powell (DFP) method (see Chapter 3)
grad_vec.m (function)	Gradient vector computation (see Chapter 3)
golden_funct1.m (function)	Golden section method (see Chapter 3)
func1.m (function)	Computes value of objective function
constr.m (function)	Computes value of constraint function
pso.m (main program)	Particle swarm optimization (PSO) method to solve welded beam problem
func1.m (function)	Computes value of objective function
constr.m (function)	Computes value of constraint function
ALM.m (main program)	Augmented Lagrangian method
func1.m (function)	Computes value of augmented objective function
sqp.m (main program)	Sequential quadratic programming method
func_val.m (function)	Computes augmented Lagrangian function
func_val1.m (function)	Computes function value
eqconstr_val.m (function)	Computes equality constraints value
ineqconstr_val.m (function)	Computes inequality constraints value
grad_vec_f.m	Computes gradient vector of the objective function
grad_vec_eqcon.m (function)	Computes gradient vector for equality constraints
grad_vec_ineqcon.m (function)	Computes gradient vector for inequality constraints
hessian.m (function)	Computes Hessian matrix (see Chapter 3)

Appendix B

```
%%%%%%%%%%%%%%%%%%%%%%%%%%%%%%%%%%%%%%%
% MATLAB code func1.m
%%%%%%%%%%%%%%%%%%%%%%%%%%%%%%%%%%%%%%%
%
% y -> objective function
% penalty -> penalty term
%
function y = func1(x,scale_factor)
y = (x(1)-1)^2 + (x(2)-5)^2;
penalty = 0.0;
[h,g] = constr(x);
for i = 1:length(h)
    if h(i)~=0
        penalty = penalty + h(i)^2;
    end
end
for i = 1:length(g)
    if g(i)>0
        penalty = penalty + g(i)^2;
    end
end
y = y+penalty*scale_factor;
%%%%%%%%%%%%%%%%%%%%%%%%%%%%%%%%%%%%%%%


%%%%%%%%%%%%%%%%%%%%%%%%%%%%%%%%%%%%%%%
% MATLAB code constr.m
%%%%%%%%%%%%%%%%%%%%%%%%%%%%%%%%%%%%%%%
%
% define your constraints here
% g(1), g(2)... -> inequality constraints
% h(1), h(2), ...-> equality constraints
%
function [h,g] = constr(x)
h(1) = 0;
g(1) = -x(1)^2 + x(2) -4;
g(2) = -(x(1)-2)^2 + x(2) -3;
%%%%%%%%%%%%%%%%%%%%%%%%%%%%%%%%%%%%%%%


%%%%%%%%%%%%%%%%%%%%%%%%%%%%%%%%%%%%%%%%%
%   File name pso.m
%   Particle Swarm Optimization algorithm
%   Welded beam problem
%%%%%%%%%%%%%%%%%%%%%%%%%%%%%%%%%%%%%%%%%
%
% lb -> lower bound of variables
% ub -> upper bound of variables
% x -> position of individual
% v -> velocity of individual
% rand -> random number from 0 to 1
```

```
% fitness -> fitness of individual
% pbest -> best fitness achieved by individual
% gbest -> best fitness of group
% nmax -> maximum number of iterations
%
clear all
clc
format long
pop = 200;
phi_1 = 1.05;
phi_2 = 1.1;
nmax = 3000;
scale_factor = 10000000;
weight = linspace(1,0.3,nmax);
fprintf('_____\n')
lb = [0.1 0.1 0.1 0.1];
ub = [2 10 10 2];
for i = 1:length(lb)
    for j = 1:pop
        x(i,j) = lb(i) + (ub(i)-lb(i))*rand;
        v(i,j) = 0;
    end
end
for i = 1:pop
   fitness(i) = func1(x(:,i),scale_factor);
   pbest(i) = fitness(i);
   px(i,:) = x(:,i);
end
[gbest, location] = min(fitness);
gx = x(:,location);
for i = 1:nmax
    for j = 1:pop
      v(:,j) = weight(i)*v(:,j) + phi_1*rand*(px(j,:)'-x(:,j))
               + phi_2*rand*(gx-x(:,j));
      x(:,j) = x(:,j) + v(:,j);
      for k = 1:length(x(:,j))
          if x(k,j) < lb(k) || x(k,j) > ub(k)
              x(k,j) = lb(k) + (ub(k)-lb(k))*rand;
          end
      end
      fitness(j) = func1(x(:,j),scale_factor);
      if fitness(j) < pbest(j)
          pbest(j) = fitness(j);
          px(j,:) = x(:,j);
      end
    end
   [gbest, location] = min(pbest);
   gx = x(:,location);
   [gx' gbest];
```

Appendix B

```matlab
        fprintf('%3d %8.3f %8.3f %8.3f %8.3f % 8.3f
            \n',i,gx,gbest)
end
fprintf('_____ \n')
%%%%%%%%%%%%%%%%%%%%%%%%%%%%%%%%%%%%%%


%%%%%%%%%%%%%%%%%%%%%%%%%%%%%%%%%%%%%%%
% MATLAB code func1.m
%%%%%%%%%%%%%%%%%%%%%%%%%%%%%%%%%%%%%%%
%
% y -> objective function
% penalty -> penalty term
%
function y = func1(x,scale_factor)
y = 1.10471*x(1)*x(1)*x(2) + 0.04811*x(3)*x(4)*(14+x(2));
penalty = 0.0;
[h,g] = constr(x);
for i = 1:length(h)
    if h(i)~=0
        penalty = penalty + h(i)^2;
    end
end
for i = 1:length(g)
    if g(i)>0
        penalty = penalty + g(i)^2;
    end
end
y = y+penalty*scale_factor;
%
%%%%%%%%%%%%%%%%%%%%%%%%%%%%%%%%%%%%%%%


%%%%%%%%%%%%%%%%%%%%%%%%%%%%%%%%%%%%%
% MATLAB code constr.m
%%%%%%%%%%%%%%%%%%%%%%%%%%%%%%%%%%%%%
%
% define your constraints here
% g(1), g(2)... -> inequality constraints
% h(1), h(2), ...-> equality constraints
%
function [h,g] = constr(x)
h(1) = 0;
load = 6000;
length = 14;
modulusE = 30e6;
modulusG = 12e6;
tmax = 13600;
sigmamax = 30000;
delmax = 0.25;
tdash = load/(sqrt(2)*x(1)*x(2));
R = sqrt(x(2)*x(2)/4 + ((x(1)+x(3))/2)^2);
M = load*(length + x(2)/2);
J = 2* ((x(1)*x(2)/sqrt(2)) * (x(2)^2/12 +((x(1)+x(3))/2)^2));
tdashdash = M*R/J;
tx = sqrt(tdash^2 + 2*tdash*tdashdash*x(2)/(2*R) + tdashdash^2);
```

```
sigmax = 6*load*length/(x(4)*x(3)^2);
delx = 4*load*length^3/(modulusE*x(4)*x(3)^3);
pcx = (4.013*sqrt(modulusE*modulusG*x(3)^2*x(4)^6/36)/(length^2)) *
      (1- (x(3)/(2*length))*sqrt(modulusE/(4*modulusG)));
g(1) = tx/tmax -1;
g(2) = sigmax/sigmamax -1;
g(3) = x(1) - x(4);
g(4) = (.10471*x(1)*x(1) + 0.04811*x(3)*x(4)*(14+x(2)))/5 -1;
g(5) = 0.125 - x(1);
g(6) = delx/delmax -1;
g(7) = load/pcx -1;
g(8) = x(1)/2 -1;
g(9) = x(4)/2 -1;
g(10) = -x(1) + 0.1;
g(11) = -x(4) + 0.1;
g(12) = x(2)/10 -1;
g(13) = x(3)/10 -1;
g(14) = -x(2) + 0.1;
g(15) = -x(3) + 0.1;
%%%%%%%%%%%%%%%%%%%%%%%%%%%%%%%%%%%%%%%%


%%%%%%%%%%%%%%%%%%%%%%%%%%%%%%%%%%%%%%%%
% MATLAB code ALM.m
%%%%%%%%%%%%%%%%%%%%%%%%%%%%%%%%%%%%%%%%
%
% n_of_var -> number of design variables
% x = [0 1 1] -> starting value of x
% epsilon1, epsilon2 -> constant used for terminating
%                      the algorithm
% delx -> required for gradient computation
% falpha_prev -> function value at first/previous iteration
% deriv -> gradient vector
% deltag -> difference in gradient vector (over previous iteration)
% A -> approximation of inverse of the hessian matrix
% search -> search direction
% LAMBDA, BETA -> Lagrange Multipliers
% RK -> penalty parameter
%
clear all
clc
n_of_var = 2;
n_of_eqcons = 1;
n_of_iqcons = 2;
scale_factor = 1;
global LAMBDA RK BETA EQCONSTR ICONSTR FVALUE
LAMBDA = zeros(1,n_of_eqcons);
BETA = zeros(1,n_of_iqcons);
x = [0 1 1];
RK = x(3);
A = eye(length(x));
epsilon1 = 1e-6;
epsilon2 = 1e-6;
delx = 1e-3;
checkconstr = zeros(1,n_of_iqcons);
falpha_prev = func1(x,scale_factor);
```

Appendix B

```
fprintf('Initial function value =   %7.4f\n ',FVALUE)
fprintf('  No.       x-vector       rk       f(x)      |Cons.| \n')
fprintf('_____\n')
for i = 1:30
    if i==1
    deriv_prev = grad_vec(x,delx,n_of_var,scale_factor);
    search = -deriv_prev;
    [alpha,falpha] = golden_funct1(x,search,scale_factor);
    if abs(falpha-falpha_prev)<epsilon1
    break;
    end
    falpha_prev = falpha;
    x = x + alpha*search;
    yyy = func1(x,scale_factor);
    LAMBDA = LAMBDA + 2*RK*EQCONSTR;
    BETA = BETA + 2*RK*(max([ICONSTR; -BETA./(2*RK)]));
    checkconstr1 = max([ICONSTR;checkconstr]);
fprintf('%3d %8.3f %8.3f  % 8.3f  % 8.3f % 8.3f  \n',i,x,FVALUE,
    norm([EQCONSTR checkconstr1]))
    else
        deltax = (alpha*search);
        if i>2
            deltax = deltax';
        end
        deriv = grad_vec(x,delx,n_of_var,scale_factor);
        deltag = deriv-deriv_prev;
        term1 = (deltax'*deltax)/(deltax*deltag');
        term2 = (A*deltag'*deltag*A)/(deltag*A*deltag');
        A = A + term1 - term2;
        search = -A*deriv';
        [alpha,falpha] = golden_funct1(x,search',scale_factor);
        checkconstr1 = max([ICONSTR;checkconstr]);
        fprintf('%3d %8.3f %8.3f  % 8.3f  % 8.3f % 8.3f  \n',i,x,FVALUE,
            norm([EQCONSTR checkconstr1]))
    if abs(falpha-falpha_prev)<epsilon1 ||  norm(deriv)<epsilon2
        break;
    end
    falpha_prev = falpha;
    deriv_prev = deriv;
    x = x+alpha*search';
    yyy = func1(x,scale_factor);
    LAMBDA = LAMBDA + 2*RK*EQCONSTR;
    BETA = BETA + 2*RK*(max([ICONSTR; -BETA./(2*RK)]));
        end
    end
fprintf('_____\n\n')
if LAMBDA>=0 & BETA>=0
    fprintf('KKT Conditions are satisfied \n\n')
end
fprintf('Lagrange Multipliers: \n\n')
disp([LAMBDA BETA])
%
%%%%%%%%%%%%%%%%%%%%%%%%%%%%%%%%%%%%%%%%
```

```
%%%%%%%%%%%%%%%%%%%%%%%%%%%%%%%%%%%%%%%%
% MATLAB code func1.m
%%%%%%%%%%%%%%%%%%%%%%%%%%%%%%%%%%%%%%%%
%
function y = func1(x,scale_factor)
global LAMBDA RK BETA EQCONSTR ICONSTR FVALUE
y = (x(1)-1)^2 + (x(2)-5)^2;
h(1) = 0.0;
g(1) = -x(1)^2 + x(2) -4;
g(2) = -(x(1)-2)^2 + x(2) -3;
EQCONSTR = h;
ICONSTR = g;
FVALUE = y;
y = y + LAMBDA.*EQCONSTR + RK.*EQCONSTR^2 + sum(BETA.*
    max([ICONSTR; -BETA./(2*RK)])) + sum(RK*(max([ICONSTR;
    -BETA./(2*RK)])).^2);
%
%%%%%%%%%%%%%%%%%%%%%%%%%%%%%%%%%%%%%%%%


%%%%%%%%%%%%%%%%%%%%%%%%%%%%%%%%%%%%%%%%
% MATLAB code sqp.m
%%%%%%%%%%%%%%%%%%%%%%%%%%%%%%%%%%%%%%%%
%
% n_of_var -> number of design variables
% x = [-1.5 1.5] -> starting value of x
% epsilon1 -> constant used for terminating the algorithm
% delx -> required for gradient computation
% falpha_prev -> function value at first/previous iteration
% deriv -> gradient vector
% quadprog -> MATLAB function to solve quadratic programming
% LAMBDA -> Lagrange multipliers
%
clear all
clc
warning off
n_of_var = 2;
n_of_eqcons = 1;
n_of_iqcons = 1;
scale_factor = 10;
global LAMBDA RK BETA EQCONSTR ICONSTR FVALUE
LAMBDA = zeros(1,n_of_eqcons);
BETA = zeros(1,n_of_iqcons);
X = [10 -5];
RK = 1;
A = eye(length(X));
epsilon1 = 1e-6;
delx = 1e-3;
```

Appendix B

```
checkconstr = zeros(1,n_of_iqcons);
fprintf('      No.         x-vector        f(x)     |Cons.| \n')
fprintf('_____\n')
checkconstr = zeros(1,n_of_iqcons);
for i = 1:3
deriv_f = grad_vec_f(X,delx,n_of_var,scale_factor);
sec_deriv_f = hessian(X,delx);
deriv_eqcon = grad_vec_eqcon(X,delx,n_of_eqcons);
deriv_ineqcon = grad_vec_ineqcon(X,delx,n_of_iqcons);
options = optimset('Display','off');
x = quadprog(sec_deriv_f,deriv_f,deriv_ineqcon,-ineqconstr_
    val(X),deriv_eqcon,-eqconstr_val(X),[],[],X,options);
fprev = func_val(X);
X = X+x';
yyy = func_val(X);
LAMBDA = LAMBDA + 2*RK*EQCONSTR;
BETA = BETA + 2*RK*(max([ICONSTR; -BETA./(2*RK)]));
fnew = func_val(X);
checkconstr1 = max([ineqconstr_val(X);checkconstr]);
disp([i X FVALUE norm([checkconstr1 eqconstr_val(X)])]);
if abs(fnew-fprev) < epsilon1
    break
end
end
fprintf('_____\n')
%
%%%%%%%%%%%%%%%%%%%%%%%%%%%%%%%%%%%%%%%%%


%%%%%%%%%%%%%%%%%%%%%%%%%%%%%%%%%%%%%%%%%
% MATLAB code func_val.m
%%%%%%%%%%%%%%%%%%%%%%%%%%%%%%%%%%%%%%%%%
% computes augmented Lagrangian function value
%
function y = func_val(x)
global LAMBDA RK BETA EQCONSTR ICONSTR FVALUE
y = (x(1)-1)^2 + (x(2)-2)^2;
g = ineqconstr_val(x);
h = eqconstr_val(x);
EQCONSTR = h;
ICONSTR = g;
FVALUE = y;
y = y + LAMBDA*EQCONSTR' + RK*EQCONSTR*EQCONSTR' + sum(BETA.*
    max([ICONSTR; -BETA./(2*RK)])) + sum(RK*(max([ICONSTR;
    -BETA./(2*RK)])).^2);
%
%%%%%%%%%%%%%%%%%%%%%%%%%%%%%%%%%%%%%%%%%
```

```
%%%%%%%%%%%%%%%%%%%%%%%%%%%%%%%%%%%%%%%%
% MATLAB code func_val1.m
%%%%%%%%%%%%%%%%%%%%%%%%%%%%%%%%%%%%%%%%
% computes function value
%
function y = func_val1(x)
y = (x(1)-1)^2 + (x(2)-2)^2;
%
%%%%%%%%%%%%%%%%%%%%%%%%%%%%%%%%%%%%%%%%


%%%%%%%%%%%%%%%%%%%%%%%%%%%%%%%%%%%%%%%%
% MATLAB code eqconstr_val.m
%%%%%%%%%%%%%%%%%%%%%%%%%%%%%%%%%%%%%%%%
% computes value of equality constraint
%
function h = eqconstr_val(x)
h(1) = 2*x(1)-x(2);
%
%%%%%%%%%%%%%%%%%%%%%%%%%%%%%%%%%%%%%%%%


%%%%%%%%%%%%%%%%%%%%%%%%%%%%%%%%%%%%%%%%
% MATLAB code ineqconstr_val.m
%%%%%%%%%%%%%%%%%%%%%%%%%%%%%%%%%%%%%%%%
% computes value of inequality constraint
%
function g = ineqconstr_val(x)
g(1) = x(1)-5;
%
%%%%%%%%%%%%%%%%%%%%%%%%%%%%%%%%%%%%%%%%


%%%%%%%%%%%%%%%%%%%%%%%%%%%%%%%%%%%%%%%%
% MATLAB code grad_vec_f.m
%%%%%%%%%%%%%%%%%%%%%%%%%%%%%%%%%%%%%%%%
% computes gradient vector (obj. function)
%
function deriv = grad_vec_f(x,delx,n_of_var,scale_factor)
xvec = x;
xvec1 = x;
for i = 1:length(x)
    xvec = x;
    xvec1 = x;
    xvec(i) = x(i) + delx;
    xvec1(i) = x(i) - delx;
    deriv(i) = (func_val1(xvec) - func_val1(xvec1))/(2*delx);
end
%
%%%%%%%%%%%%%%%%%%%%%%%%%%%%%%%%%%%%%%%%
```

Appendix B

```matlab
%%%%%%%%%%%%%%%%%%%%%%%%%%%%%%%%%%%%%%%%%%
% MATLAB code grad_vec_eqcon.m
%%%%%%%%%%%%%%%%%%%%%%%%%%%%%%%%%%%%%%%%%%
% computes gradient vector (eq. constraint)
%
function deriv = grad_vec_eqcon(x,delx,n_of_eqcons)
xvec = x;
xvec1 = x;
for j = 1:n_of_eqcons
for i = 1:length(x)
    xvec = x;
    xvec1 = x;
    xvec(i) = x(i) + delx;
    xvec1(i) = x(i) - delx;
    h = eqconstr_val(xvec);
    h1 = eqconstr_val(xvec1);
    deriv(j,i) = (h(j) - h1(j))/(2*delx);
end
end
%
%%%%%%%%%%%%%%%%%%%%%%%%%%%%%%%%%%%%%%%%%%


%%%%%%%%%%%%%%%%%%%%%%%%%%%%%%%%%%%%%%%%%%
% MATLAB code grad_vec_ineqcon.m
%%%%%%%%%%%%%%%%%%%%%%%%%%%%%%%%%%%%%%%%%%
% computes gradient vector (ineq. constraint)
%
function deriv = grad_vec_ineqcon(x,delx,n_of_iqcons)
xvec = x;
xvec1 = x;
for j = 1:n_of_iqcons
for i = 1:length(x)
    xvec = x;
    xvec1 = x;
    xvec(i) = x(i) + delx;
    xvec1(i) = x(i) - delx;
    g = ineqconstr_val(xvec);
    g1 = ineqconstr_val(xvec1);
    deriv(j,i) = (g(j) - g1(j))/(2*delx);
end
end
%
%%%%%%%%%%%%%%%%%%%%%%%%%%%%%%%%%%%%%%%%%%
```

Chapter 7

Code Name	Details
sqp.m (main program)	Sequential quadratic programming (SQP) method modified for weighted sum approach
func_val.m (function)	Computes augmented Lagrangian function value
func_val1.m (function)	Computes function value
sqp.m (main program)	SQP method modified for solving multiobjective problems using ε-constraint technique
func_val.m (function)	Computes augmented Lagrangian function value
func_val1.m (function)	Computes function value
ineqconstr_val.m	Computes inequality constraint value
pso.m (main program)	Particle swarm optimization (PSO) method with dynamic weights
func1.m (function)	Computes value of objective function
sqp.m (main program)	Main program for solving reentry problem
dynamics.m (function)	Computes area, volume, and X_{cp}

```
%%%%%%%%%%%%%%%%%%%%%%%%%%%%%%%%%%%%%
% MATLAB code sqp.m
%%%%%%%%%%%%%%%%%%%%%%%%%%%%%%%%%%%%%
%
% n_of_var -> number of design variables
% x = [0.1 0.1] -> starting value of x
% epsilon1 -> constants used for terminating the algorithm
% delx -> required for gradient computation
% falpha_prev -> function value at first/previous iteration
% deriv -> gradient vector
% quadprog -> MATLAB function to solve quadratic programming
% LAMBDA -> Lagrange multipliers
clear all
clc
warning off
n_of_var = 2;
n_of_eqcons = 1;
n_of_iqcons = 1;
scale_factor = 1;
global LAMBDA RK BETA EQCONSTR ICONSTR FVALUE W1 W2
LAMBDA = zeros(1,n_of_eqcons);
BETA = zeros(1,n_of_iqcons);
X = [0.1 0.1];
RK = 1;
A = eye(length(X));
epsilon1 = 1e-6;
delx = 1e-3;
for kk = 1:101
    X = [0.1 0.1];
    W1 = (kk-1)/100;
    W2 = 1 - W1;
```

Appendix B

```
checkconstr = zeros(1,n_of_iqcons);
fprintf('    No.         x-vector         f(x)         |Cons.| \n')
fprintf('_____\n')
checkconstr = zeros(1,n_of_iqcons);
for i = 1:10
deriv_f = grad_vec_f(X,delx,n_of_var,scale_factor);
sec_deriv_f = hessian(X,delx);
deriv_eqcon = grad_vec_eqcon(X,delx,n_of_eqcons);
deriv_ineqcon = grad_vec_ineqcon(X,delx,n_of_iqcons);
options = optimset('Display','off');
x = quadprog(sec_deriv_f,deriv_f,deriv_ineqcon,-ineqconstr_val(X),
deriv_eqcon,-eqconstr_val(X),[],[],X,options);
fprev = func_val(X);
X = X+x';
yyy = func_val(X);
LAMBDA = LAMBDA + 2*RK*EQCONSTR;
BETA = BETA + 2*RK*(max([ICONSTR; -BETA./(2*RK)]));
fnew = func_val(X);
checkconstr1 = max([ineqconstr_val(X);checkconstr]);
disp([i X FVALUE norm([checkconstr1 eqconstr_val(X)])]);
if abs(fnew-fprev) < epsilon1
    break
end
end
fprintf('_____\n')
plot(0.5*(X(1)^2+X(2)^2) , 0.5*((X(1)-1)^2 + (X(2)-3)^2),'r*')
hold on
end
xlabel('f1');
ylabel('f2');
%
%%%%%%%%%%%%%%%%%%%%%%%%%%%%%%%%%%%%%%%%%%


%%%%%%%%%%%%%%%%%%%%%%%%%%%%%%%%%%%%%
% MATLAB code func_val.m
%%%%%%%%%%%%%%%%%%%%%%%%%%%%%%%%%%%%%
% computes augmented Lagrangian function value
%
function y = func_val(x)
global LAMBDA RK BETA EQCONSTR ICONSTR FVALUE W1 W2
y = W1*0.5*(x(1)^2+x(2)^2) + W2*0.5*((x(1)-1)^2 + (x(2)-3)^2);
g = ineqconstr_val(x);
h = eqconstr_val(x);
EQCONSTR = h;
ICONSTR = g;
FVALUE = y;
y = y + LAMBDA*EQCONSTR' + RK*EQCONSTR*EQCONSTR' + sum(BETA.*
    max([ICONSTR; -BETA./(2*RK)])) + sum(RK*(max([ICONSTR;
    -BETA./(2*RK)])).^2);
%
%%%%%%%%%%%%%%%%%%%%%%%%%%%%%%%%%%%%%
```

```
%%%%%%%%%%%%%%%%%%%%%%%%%%%%%%%%%%%%%%%
% MATLAB code func_val1.m
%%%%%%%%%%%%%%%%%%%%%%%%%%%%%%%%%%%%%%%
% computes function value
%
function y = func_val1(x)
global W1 W2
y = W1*0.5*(x(1)^2+x(2)^2) + W2*0.5*((x(1)-1)^2 + (x(2)-3)^2);
%
%%%%%%%%%%%%%%%%%%%%%%%%%%%%%%%%%%%%%%%


%%%%%%%%%%%%%%%%%%%%%%%%%%%%%%%%%%%%%%%
% MATLAB code sqp.m modified for eps-constraints method
%%%%%%%%%%%%%%%%%%%%%%%%%%%%%%%%%%%%%%%
%
% n_of_var -> number of design variables
% x = [-1.5 1.5] -> starting value of x
% epsilon1 -> constants used for terminating the algorithm
% delx -> required for gradient computation
% falpha_prev -> function value at first/previous iteration
% deriv -> gradient vector
% quadprog -> MATLAB function to solve quadratic programming
% LAMBDA -> Lagrange multipliers
clear all
clc
warning off
n_of_var = 1;
n_of_eqcons = 1;
n_of_iqcons = 5;
scale_factor = 1;
global LAMBDA RK BETA EQCONSTR ICONSTR FVALUE W1 W2
LAMBDA = zeros(1,n_of_eqcons);
BETA = zeros(1,n_of_iqcons);
X = [1 1];
RK = 1;
A = eye(length(X));
epsilon1 = 1e-6;
delx = 1e-3;
for kk = 1:100
    W1 = (kk-1)/100;
    W2 = 1 - W1;
checkconstr = zeros(1,n_of_iqcons);
fprintf('       No.        x-vector        f(x)     |Cons.| \n')
fprintf('_____\n')
checkconstr = zeros(1,n_of_iqcons);
for i = 1:10
deriv_f = grad_vec_f(X,delx,n_of_var,scale_factor);
sec_deriv_f = hessian(X,delx);
deriv_eqcon = grad_vec_eqcon(X,delx,n_of_eqcons);
deriv_ineqcon = grad_vec_ineqcon(X,delx,n_of_iqcons);
options = optimset('Display','off');
x = quadprog(sec_deriv_f,deriv_f,deriv_ineqcon,-ineqconstr_val(X),...
        deriv_eqcon,-eqconstr_val(X),[],[],X,options);
fprev = func_val(X);
X = X+x';
```

Appendix B

```
yyy = func_val(X);
LAMBDA = LAMBDA + 2*RK*EQCONSTR;
BETA = BETA + 2*RK*(max([ICONSTR; -BETA./(2*RK)]));
fnew = func_val(X);
checkconstr1 = max([ineqconstr_val(X);checkconstr]);
disp([i X FVALUE norm([checkconstr1 eqconstr_val(X)])]);
if abs(fnew-fprev) < epsilon1
    break
end
end
fprintf('_____\n')
plot(X(1), (1+X(2)^2-X(1)-0.1*sin(3*pi*X(1))),'r*')
hold on
end
xlabel('f1')
ylabel('f2')
%
%%%%%%%%%%%%%%%%%%%%%%%%%%%%%%%%%%%%%%


%%%%%%%%%%%%%%%%%%%%%%%%%%%%%%%%%%%%%%
% MATLAB code func_val.m
%%%%%%%%%%%%%%%%%%%%%%%%%%%%%%%%%%%%%%
% computes augmented Lagrangian function value
%
function y = func_val(x)
global LAMBDA RK BETA EQCONSTR ICONSTR FVALUE W1 W2
y = (1+x(2)^2-x(1)-0.1*sin(3*pi*x(1)));
g = ineqconstr_val(x);
h = eqconstr_val(x);
EQCONSTR = h;
ICONSTR = g;
FVALUE = y;
y = y + LAMBDA*EQCONSTR' + RK*EQCONSTR*EQCONSTR' + sum(BETA.*
    max([ICONSTR; -BETA./(2*RK)])) + sum(RK*(max([ICONSTR; -BETA./
    (2*RK)])).^2);
%
%%%%%%%%%%%%%%%%%%%%%%%%%%%%%%%%%%%%%%


%%%%%%%%%%%%%%%%%%%%%%%%%%%%%%%%%%%%%%
% MATLAB code ineqconstr_val.m
%%%%%%%%%%%%%%%%%%%%%%%%%%%%%%%%%%%%%%
% computes value of inequality constraint
function g = ineqconstr_val(x)
global LAMBDA RK BETA EQCONSTR ICONSTR FVALUE W1 W2
g(1) = x(1)-1;
g(2) = -x(1);
g(3) = x(2)-2;
g(4) = -x(2)-2;
g(5) = x(1)- W1;
%%%%%%%%%%%%%%%%%%%%%%%%%%%%%%%%%%%%%%
```

```
%%%%%%%%%%%%%%%%%%%%%%%%%%%%%%%%%%%%%%%%%%%%%%%%%
%    File name pso.m
%    Particle Swarm Optimization algorithm with dynamic weights
%%%%%%%%%%%%%%%%%%%%%%%%%%%%%%%%%%%%%%%%%%%%%%%%%
%
% lb -> lower bound of variables
% ub -> upper bound of variables
% x -> position of individual
% v -> velocity of individual
% rand -> random number from 0 to 1
% fitness -> fitness of individual
% pbest -> best fitness achieved by individual
% gbest -> best fitness of group
%
clear all
clc
format long
global W1 W2
pop = 200;
phi_1 = 0.5;
phi_2 = 0.5;
nmax = 120;
weight = linspace(1,0.4,nmax);
lb = [0 -2];
ub = [1 2];
W1 = 0;
W2 = 1;
for i = 1:length(lb)
    for j = 1:pop
        x(i,j) = lb(i) + (ub(i)-lb(i))*rand;
        v(i,j) = 0;
    end
end
for i = 1:pop
    fitness(i) = func1(x(:,i));
    pbest(i) = fitness(i);
    px(i,:) = x(:,i);
end
[gbest, location] = min(fitness);
gx = x(:,location);
 for i = 1:nmax
    W1 = abs(sin(2*pi*i/150));
    W2 = 1-W1;
    for j = 1:pop
      v(:,j) = weight(i)*v(:,j) + phi_1*rand*(px(j,:)'-x(:,j)) + ...
            phi_2*rand*(gx-x(:,j));
        x(:,j) = x(:,j) + v(:,j);
       for k = 1:length(x(:,j))
            if x(k,j) < lb(k) || x(k,j) > ub(k)
                x(k,j) = lb(k) + (ub(k)-lb(k))*rand;
```

Appendix B

```
                end
            end
            fitness(j) = func1(x(:,j));
            if fitness(j) < pbest(j)
        pbest(j) = fitness(j);
        px(j,:) = x(:,j);
        end
            F1(j) = x(1,j);
            F2(j) = (1+x(2,j)^2-x(1,j)-0.1*sin(3*pi*x(1,j)));
end
    [gbest, location] = min(pbest);
    gx = x(:,location);
    [gx' gbest];
        plot(F1,F2,'r*')
    pause(0.1)
end
%
%%%%%%%%%%%%%%%%%%%%%%%%%%%%%%%%%%%%%%%%%%%%%


%%%%%%%%%%%%%%%%%%%%%%%%%%%%%%%%%%%%%%%%%%%
%   File name func1.m
%   Enter the function to be optimized
%%%%%%%%%%%%%%%%%%%%%%%%%%%%%%%%%%%%%%%%%%%
%
function [y] = func1(x)
global W1 W2
y = W1*x(1) + W2*(1+x(2)^2-x(1)-0.1*sin(3*pi*x(1)));
%%%%%%%%%%%%%%%%%%%%%%%%%%%%%%%%%%%%%%%%%%%


%%%%%%%%%%%%%%%%%%%%%%%%%%%%%%%%%%%%%%
% MATLAB code sqp.m
%%%%%%%%%%%%%%%%%%%%%%%%%%%%%%%%%%%%%%
%
% n_of_var -> number of design variables
% x = [-1.5 1.5] -> starting value of x
% epsilon1 -> constant used for terminating the algorithm
% delx -> required for gradient computation
% falpha_prev -> function value at first/previous iteration
% deriv -> gradient vector
% quadprog -> MATLAB function to solve quadratic programming
% LAMBDA -> Lagrange multipliers
clear all
clc
warning off
n_of_var = 5;
n_of_eqcons = 1;
n_of_iqcons = 12;
scale_factor = 1;
```

```
global LAMBDA RK BETA EQCONSTR ICONSTR FVALUE W1 W2
LAMBDA = zeros(1,n_of_eqcons);
BETA = zeros(1,n_of_iqcons);
X = [0.5 25 31 0.5 0.5];
RK = 1;
A = eye(length(X));
epsilon1 = 1e-2;
delx = 1e-3;
for kk = 640:20:1630
    W1 = (kk-1)/100;
checkconstr = zeros(1,n_of_iqcons);
fprintf('       No.        x-vector           f(x)       |Cons.| \n')
fprintf('_____\n')
checkconstr = zeros(1,n_of_iqcons);
for i = 1:30
deriv_f = grad_vec_f(X,delx,n_of_var,scale_factor);
sec_deriv_f = hessian(X,delx);
deriv_eqcon = grad_vec_eqcon(X,delx,n_of_eqcons);
deriv_ineqcon = grad_vec_ineqcon(X,delx,n_of_iqcons);
options = optimset('Display','off');
x = quadprog(sec_deriv_f,deriv_f,deriv_ineqcon,-ineqconstr_
    val(X),deriv_eqcon,-eqconstr_val(X),[],[],X,options);
fprev = func_val(X);
X = X+x';
yyy = func_val(X);
LAMBDA = LAMBDA + 2*RK*EQCONSTR;
BETA = BETA + 2*RK*(max([ICONSTR; -BETA./(2*RK)]));
fnew = func_val(X);
checkconstr1 = max([ineqconstr_val(X);checkconstr]);
disp([i X FVALUE norm([checkconstr1 eqconstr_val(X)])]);
if abs(fnew-fprev) < epsilon1
    break
end
end
fprintf('_____\n')
[ xcp, area, volume] = dynamics(X);
 plot(xcp, area,'k*','LineWidth',1.5)
hold on
end
xlabel('X_{cp}')
ylabel('A')
%%%%%%%%%%%%%%%%%%%%%%%%%%%%%%%%%%%%%%%%%


%%%%%%%%%%%%%%%%%%%%%%%%%%%%%%%%%%%%%%%%%
% MATLAB code dynamics.m
%%%%%%%%%%%%%%%%%%%%%%%%%%%%%%%%%%%%%%%%%
function [xcp, area, volume] = dynamics(X)
```

Appendix B

```
beta1 = [0.075759311956522
    0.001381832173914
    0.005825624927536
    0.087880851017943
    0.079788071083505
    0.103099119565217
    0.000091411652174
   -0.000083773739130
    0.024378102714516
    0.052872440763746];
cm = -0.278527718840579+beta1'*[X(1);X(2);X(3);X(4);X(5);
    X(1)^2;X(2)^2;X(3)^2;X(4)^2;X(5)^2];
beta2 = [0.150130422826087
   -0.003965447971014
    0.014019523043478
    0.103639584627328
    0.108001867667355
    0.047906228260870
    0.000186851275362
   -0.000219940956522
   -0.000485274327121
    0.016298096388315];
cn = -0.314286112399353+beta2'*[X(1);X(2);X(3);X(4);X(5);
    X(1)^2;X(2)^2;X(3)^2;X(4)^2;X(5)^2];
xcp = cm/cn;
r1 = X(1)*cos(deg2rad(X(2)));
r2 = X(1)*cos(deg2rad(X(2))) + X(4)*tan(deg2rad(X(2)));
r3 = r2 + X(5)*tan(deg2rad(X(3)));
area = 2*pi*X(1)*X(1)*(1-sin(deg2rad(X(2)))) +
       pi*(r1+r2)*sqrt((r2-r1)^2 + X(4)^2) +
       pi*(r3+r2)*sqrt((r3-r2)^2 + X(5)^2) + pi*r3^2;
cap = (pi*r1*(1-sin(deg2rad(X(2))))/6) * (3*r1*r1 +
      (r1*(1-sin(deg2rad(X(2))))^2));
volume = cap + 0.3333*pi*X(4)*(r2^2 + r1^2 + r2*r1) +
         0.3333*pi*X(5)*(r3^2 + r2^2 + r2*r3);
end
%%%%%%%%%%%%%%%%%%%%%%%%%%%%%%%%%%%%%%%%%
```

Chapter 9

Code Name	Details
sqp.m (main program)	Sequential quadratic programming (SQP) method (for multidisciplinary design optimization [MDO] application)
discipline1.m (function)	Output from first discipline

discipline2.m (function)	Output from first discipline
eqconstr_val.m (function)	Computes equality constraints value
ineqconstr_val.m (function)	Computes inequality constraints value
func_val1.m (function)	Computes function value
func_val.m (function)	Computes augmented Lagrangian function
grad_vec_eqcon.m (function)	Computes gradient vector for equality constraints
grad_vec_ineqcon.m (function)	Computes gradient vector for inequality constraints
hessian.m (function)	Computes Hessian matrix (see Chapter 3)

```
%%%%%%%%%%%%%%%%%%%%%%%%%%%%%%%%%%%%%
% MATLAB code sqp.m
%%%%%%%%%%%%%%%%%%%%%%%%%%%%%%%%%%%%%
%
% n_of_var -> number of design variables
% x = [-1.5 1.5] -> starting value of x
% epsilon1 -> constant used for terminating the algorithm
% delx -> required for gradient computation
% falpha_prev -> function value at first/previous iteration
% deriv -> gradient vector
% quadprog -> MATLAB function to solve quadratic programming
% LAMBDA -> Lagrange multipliers
clear all
clc
warning off
n_of_var = 5;
n_of_eqcons = 2;
n_of_iqcons = 4;
scale_factor = 10;
global LAMBDA RK BETA EQCONSTR ICONSTR FVALUE
LAMBDA = zeros(1,n_of_eqcons);
BETA = zeros(1,n_of_iqcons);
X = [1 2 5 1 0];
RK = 1;
A = eye(length(X));
epsilon1 = 1e-6;
delx = 1e-3;
checkconstr = zeros(1,n_of_iqcons);
fprintf('     No.          x-vector            f(x)         |Cons.| \n')
fprintf('_____\n')
checkconstr = zeros(1,n_of_iqcons);
for i = 1:30
deriv_f = grad_vec_f(X,delx,n_of_var,scale_factor);
sec_deriv_f = hessian(X,delx);
deriv_eqcon = grad_vec_eqcon(X,delx,n_of_eqcons);
deriv_ineqcon = grad_vec_ineqcon(X,delx,n_of_iqcons);
options = optimset('Display','off');
x = quadprog(sec_deriv_f,deriv_f,deriv_ineqcon,-ineqconstr_val(X),...
    deriv_eqcon,-eqconstr_val(X),[],[],X,options);
fprev = func_val(X);
X = X+x';
yyy = func_val(X);
LAMBDA = LAMBDA + 2*RK*EQCONSTR;
BETA = BETA + 2*RK*(max([ICONSTR; -BETA./(2*RK)]));
```

Appendix B 387

```
fnew = func_val(X);
checkconstr1 = max([ineqconstr_val(X);checkconstr]);
disp([i X FVALUE norm([checkconstr1 eqconstr_val(X)])]);
if abs(fnew-fprev) < epsilon1
    break
end
end
fprintf('_____\n')
%
%%%%%%%%%%%%%%%%%%%%%%%%%%%%%%%%%%%%%%%%%%


%%%%%%%%%%%%%%%%%%%%%%%%%%%%%%%%%%%%%%%%%%
% MATLAB code discipline1.m
%%%%%%%%%%%%%%%%%%%%%%%%%%%%%%%%%%%%%%%%%%
%
% Discipline-1
%
function y1 = discipline1(x)
y1 = x(1)+x(2)+x(3)^2-0.2*x(5);
%
%%%%%%%%%%%%%%%%%%%%%%%%%%%%%%%%%%%%%%%%%%


%%%%%%%%%%%%%%%%%%%%%%%%%%%%%%%%%%%%%%%%%%
% MATLAB code discipline2.m
%%%%%%%%%%%%%%%%%%%%%%%%%%%%%%%%%%%%%%%%%%
%
% Discipline-2
%
function y2 = discipline2(x)
y2 = x(3)+x(2)+sqrt(x(4));
%
%%%%%%%%%%%%%%%%%%%%%%%%%%%%%%%%%%%%%%%%%%


%%%%%%%%%%%%%%%%%%%%%%%%%%%%%%%%%%%%%%%%%%
% MATLAB code eqconstr_val.m
%%%%%%%%%%%%%%%%%%%%%%%%%%%%%%%%%%%%%%%%%%
% computes value of equality constraint
function h = eqconstr_val(x)
y1 = discipline1(x);
y2 = discipline2(x);
h(1) = y1-x(4);
h(2) = y2-x(5);
%%%%%%%%%%%%%%%%%%%%%%%%%%%%%%%%%%%%%%%%%%


%%%%%%%%%%%%%%%%%%%%%%%%%%%%%%%%%%%%%%%%%%
% MATLAB code ineqconstr_val.m
%%%%%%%%%%%%%%%%%%%%%%%%%%%%%%%%%%%%%%%%%%
% computes value of inequality constraint
function g = ineqconstr_val(x)
```

```
y1 = discipline1(x);
y2 = discipline2(x);
g(1) = 1-y1/3.16;
g(2) = y2/24-1;
g(3) = -x(1);
g(4) = -x(2);
%%%%%%%%%%%%%%%%%%%%%%%%%%%%%%%%%%%%%%%%%


%%%%%%%%%%%%%%%%%%%%%%%%%%%%%%%%%%%%%%%%%
% MATLAB code func_val1.m
%%%%%%%%%%%%%%%%%%%%%%%%%%%%%%%%%%%%%%%%%
% computes function value
%
function y = func_val1(x)
y = x(1)^2 + x(2) + x(4) +exp(-x(5));
%%%%%%%%%%%%%%%%%%%%%%%%%%%%%%%%%%%%%%%%%


%%%%%%%%%%%%%%%%%%%%%%%%%%%%%%%%%%%%%%%%%
% MATLAB code func_val.m
%%%%%%%%%%%%%%%%%%%%%%%%%%%%%%%%%%%%%%%%%
% computes augmented Lagrangian function value
function y = func_val(x)
global LAMBDA RK BETA EQCONSTR ICONSTR FVALUE
y = x(1)^2 + x(2) + x(4) +exp(-x(5));
g = ineqconstr_val(x);
h = eqconstr_val(x);
EQCONSTR = h;
ICONSTR = g;
FVALUE = y;
y = y + LAMBDA*EQCONSTR' + RK*EQCONSTR*EQCONSTR' + sum(BETA.*
    max([ICONSTR; -BETA./(2*RK)])) + sum(RK*(max([ICONSTR;
    -BETA./(2*RK)])).^2);
%%%%%%%%%%%%%%%%%%%%%%%%%%%%%%%%%%%%%%%%


%%%%%%%%%%%%%%%%%%%%%%%%%%%%%%%%%%%%%%%%%
% MATLAB code grad_vec_eqcon.m
%%%%%%%%%%%%%%%%%%%%%%%%%%%%%%%%%%%%%%%%%
function deriv = grad_vec_eqcon(x,delx,n_of_eqcons)
xvec = x;
xvec1 = x;
for j = 1:n_of_eqcons
for i = 1:length(x)
    xvec = x;
    xvec1 = x;
    xvec(i) = x(i) + delx;
    xvec1(i) = x(i) - delx;
    h = eqconstr_val(xvec);
    h1 = eqconstr_val(xvec1);
```

Appendix B

```matlab
    deriv(j,i) = (h(j) - h1(j))/(2*delx);
end
end
%%%%%%%%%%%%%%%%%%%%%%%%%%%%%%%%%%%%%


%%%%%%%%%%%%%%%%%%%%%%%%%%%%%%%%%%%%%
% MATLAB code grad_vec_ineqcon.m
%%%%%%%%%%%%%%%%%%%%%%%%%%%%%%%%%%%%%
function deriv = grad_vec_ineqcon(x,delx,n_of_iqcons)
xvec = x;
xvec1 = x;
for j = 1:n_of_iqcons
for i = 1:length(x)
    xvec = x;
    xvec1 = x;
    xvec(i) = x(i) + delx;
    xvec1(i) = x(i) - delx;
    g = ineqconstr_val(xvec);
    g1 = ineqconstr_val(xvec1);
    deriv(j,i) = (g(j) - g1(j))/(2*delx);
end
end
%%%%%%%%%%%%%%%%%%%%%%%%%%%%%%%%%%%%%
```

Chapter 10

Code Name	Details
Gomory's Method (All-Integer Problem)	
simplex.m	Simplex method for solving linear programming problem (LPP)
dual_step1.m	Solves step 1 of problem
dual_step2.m	Solves step 2 of problem
Gomory's Method (Mixed-Integer Problem)	
simplex.m	Simplex method for solving LPP
dual_step.m	Dual simplex method
Branch and Bound Method	
simplex.m	Simplex method for solving LPP
subproblem1.m	Simplex method for subproblem 1
subproblem2.m	Simplex method for subproblem 2
node2_subproblem1.m	Simplex method for subproblem 1 (Node 2)
node2_subproblem2.m	Simplex method for subproblem 2 (Node 2)

Particle Swarm Optimization
pso.m   Main program
func1.m   Objective function
constr.m   Constraints

```
%%%%%%%%%%%%%%%%%%%%%%%%%%%%%%%%%%%%%%%%%%%%%%%%%%%%%%
% MATLAB code simplex.m
% Gomory's method (All-integer problem)
%%%%%%%%%%%%%%%%%%%%%%%%%%%%%%%%%%%%%%%%%%%%%%%%%%%%%%
%
% The matrix A and b corresponds to equation Ax=b
% c -> vector of cost coefficients
% basic_set -> set of basic variables
% nonbasic_set -> setof nonbasic variables
% B -> matrix containing basic variable columns of A
% N -> matrix containing nonbasic variable columns of A
% xb -> basic variables
% y -> simplex multipliers
% cb -> cost coefficients of basic variables
% cn -> cost coefficients of nonbasic variables
%
clear all
clc
format rational
format compact
A = [1 -1 1 0;
     4 7 0 1];
b = [5;50];
c = [-3;-2;0;0];
 basic_set = [3 4];
 nonbasic_set = [1 2];
for i = 1:length(basic_set)
    B(:,i) = A(:,basic_set(i));
    cb(i) = c(basic_set(i));
end
for i = 1:length(nonbasic_set)
    N(:,i) = A(:,nonbasic_set(i));
    cn(i) = c(nonbasic_set(i));
end
cn_cap = cn;
cb_ini = cb;
b_cap = b;
zz1 = 0;
% Rest of the code remains same as in simplex.m (Chapter 4)
%%%%%%%%%%%%%%%%%%%%%%%%%%%%%%%%%%%%%%%%%%%%%%%%%%%%%%
```

Appendix B

```matlab
%%%%%%%%%%%%%%%%%%%%%%%%%%%%%%%%%%%%%%%%%%%%%%%%%%%%
% MATLAB code dual_step1.m (Solves step 1 of the problem)
% Gomory's method (All-integer problem)
%%%%%%%%%%%%%%%%%%%%%%%%%%%%%%%%%%%%%%%%%%%%%%%%%%%%
% The matrix A and b corresponds to equation Ax=b
% c -> vector of cost coefficients
% basic_set -> set of basic variables
% nonbasic_set -> set of nonbasic variables
% B -> matrix containing basic variable columns of A
% N -> matrix containing nonbasic variable columns of A
% xb -> basic variables
% y -> simplex multipliers
% cb -> cost coefficients of basic variables
% cn -> cost coefficients of nonbasic variables
%
clear all
clc
format rational
format compact
A = [1 0  7/11  1/11  0;
     0 1 -4/11  1/11  0;
     0 0 -7/11 -1/11  1];
b = [85/11;30/11;-8/11];
c = [0;0;13/11; 5/11;0];
basic_set = [1 2 5];
nonbasic_set = [3 4];
for i = 1:length(basic_set)
    B(:,i) = A(:,basic_set(i));
    cb(i) = c(basic_set(i));
end
for i = 1:length(nonbasic_set)
    N(:,i) = A(:,nonbasic_set(i));
    cn(i) = c(nonbasic_set(i));
end
cn_cap = cn;
cb_ini = cb;
b_cap = b;
zz = -315/11;
% Rest of the code remains same as in dual.m (Chapter 4)
%%%%%%%%%%%%%%%%%%%%%%%%%%%%%%%%%%%%%%%%%%%%%%%%%%


%%%%%%%%%%%%%%%%%%%%%%%%%%%%%%%%%%%%%%%%%%%%%%%%%%
% MATLAB code dual_step2.m (Solves step 2 of the problem)
% Gomory's method (All-integer problem)
%%%%%%%%%%%%%%%%%%%%%%%%%%%%%%%%%%%%%%%%%%%%%%%%%%
% The matrix A and b corresponds to equation Ax=b
% c -> vector of cost coefficients
% basic_set -> set of basic variables
% nonbasic_set -> set of nonbasic variables
```

```
% B -> matrix containing basic variable columns of A
% N -> matrix containing nonbasic variable columns of A
% xb -> basic variables
% y -> simplex multipliers
% cb -> cost coefficients of basic variables
% cn -> cost coefficients of nonbasic variables
%
clear all
clc
format rational
format compact
A =[1 0 0 0    1   0;
    0 1 0 1/7 -4/7 0;
    0 0 1 1/7 -11/7 0;
    0 0 0 -1/7 4/7 1];
b = [7;22/7;8/7;-1/7];
c = [0;0;0;2/7;13/7;0];
basic_set = [1 2 3 6];
nonbasic_set = [4 5];
for i = 1:length(basic_set)
    B(:,i) = A(:,basic_set(i));
    cb(i) = c(basic_set(i));
end
for i = 1:length(nonbasic_set)
    N(:,i) = A(:,nonbasic_set(i));
    cn(i) = c(nonbasic_set(i));
end
cn_cap = cn;
cb_ini = cb;
b_cap = b;
zz = -191/7;
% Rest of the code remains same as in dual.m (Chapter 4)
%%%%%%%%%%%%%%%%%%%%%%%%%%%%%%%%%%%%%%%%%%%%%%%%%%%


%%%%%%%%%%%%%%%%%%%%%%%%%%%%%%%%%%%%%%%%%%%%%%%%%%%
% MATLAB code simplex.m
% Gomory's method (Mixed-integer problem)
%%%%%%%%%%%%%%%%%%%%%%%%%%%%%%%%%%%%%%%%%%%%%%%%%%%
% The matrix A and b corresponds to equation Ax=b
% c -> vector of cost coefficients
% basic_set -> set of basic variables
% nonbasic_set -> setof nonbasic variables
% B -> matrix containing basic variable columns of A
% N -> matrix containing nonbasic variable columns of A
% xb -> basic variables
% y -> simplex multipliers
% cb -> cost coefficients of basic variables
% cn -> cost coefficients of nonbasic variables
%
```

Appendix B

```
clear all
clc
format rational
format compact
A = [1 1 1 0;
     5 2 0 1];
b = [6;20];
c = [-3;-2;0;0];
 basic_set = [3 4];
 nonbasic_set = [1 2];
for i = 1:length(basic_set)
    B(:,i) = A(:,basic_set(i));
    cb(i) = c(basic_set(i));
end
for i = 1:length(nonbasic_set)
    N(:,i) = A(:,nonbasic_set(i));
    cn(i) = c(nonbasic_set(i));
end
cn_cap = cn;
cb_ini = cb;
b_cap = b;
zz1 = 0;
% Rest of the code remains same as in simplex.m (Chapter 4)
%%%%%%%%%%%%%%%%%%%%%%%%%%%%%%%%%%%%%%%%%%%%%%%%%

%%%%%%%%%%%%%%%%%%%%%%%%%%%%%%%%%%%%%%%%%%%%%%%%%
% MATLAB code dual_step.m
% Gomory's method (Mixed-integer problem)
%%%%%%%%%%%%%%%%%%%%%%%%%%%%%%%%%%%%%%%%%%%%%%%%%
% The matrix A and b corresponds to equation Ax=b
% c -> vector of cost coefficients
% basic_set -> set of basic variables
% nonbasic_set -> set of nonbasic variables
% B -> matrix containing basic variable columns of A
% N -> matrix containing nonbasic variable columns of A
% xb -> basic variables
% y -> simplex multipliers
% cb -> cost coefficients of basic variables
% cn -> cost coefficients of nonbasic variables
%
clear all
clc
format rational
format compact
A = [1 0 -2/3  1/3  0;
     0 1  5/3 -1/3  0;
     0 0 -5/3  1/3  1];
b = [8/3;10/3;-1/3];
c = [0;0;4/3; 1/3;0];
basic_set = [1 2 5];
```

```
nonbasic_set = [3 4];
for i = 1:length(basic_set)
    B(:,i) = A(:,basic_set(i));
    cb(i) = c(basic_set(i));
end
for i = 1:length(nonbasic_set)
    N(:,i) = A(:,nonbasic_set(i));
    cn(i) = c(nonbasic_set(i));
end
cn_cap = cn;
cb_ini = cb;
b_cap = b;
zz =-44/3;
% Rest of the code remains same as in dual.m (Chapter 4)
%%%%%%%%%%%%%%%%%%%%%%%%%%%%%%%%%%%%%%%%%%%%%%%%%%%%%

%%%%%%%%%%%%%%%%%%%%%%%%%%%%%%%%%%%%%%%%%%%%%%%%%%%%%
% MATLAB code simplex.m
% Branch and Bound method
%%%%%%%%%%%%%%%%%%%%%%%%%%%%%%%%%%%%%%%%%%%%%%%%%%%%%
%
% The matrix A and b corresponds to equation Ax=b
% c -> vector of cost coefficients
% basic_set -> set of basic variables
% nonbasic_set -> setof nonbasic variables
% B -> matrix containing basic variable columns of A
% N -> matrix containing nonbasic variable columns of A
% xb -> basic variables
% y -> simplex multipliers
% cb -> cost coefficients of basic variables
% cn -> cost coefficients of nonbasic variables
%
clear all
clc
format rational
format compact
A = [2  5 1 0;
     2 -3 0 1];
b = [16;7];
c = [-4;-5;0;0];
basic_set = [3 4];
nonbasic_set = [1 2];
for i = 1:length(basic_set)
    B(:,i) = A(:,basic_set(i));
    cb(i) = c(basic_set(i));
end
for i = 1:length(nonbasic_set)
    N(:,i) = A(:,nonbasic_set(i));
    cn(i) = c(nonbasic_set(i));
end
```

Appendix B

```
cn_cap = cn;
cb_ini = cb;
b_cap = b;
zz1 = 0;
% Rest of the code remains same as in simplex.m (Chapter 4)
%%%%%%%%%%%%%%%%%%%%%%%%%%%%%%%%%%%%%%%%%%%%%%%%%


%%%%%%%%%%%%%%%%%%%%%%%%%%%%%%%%%%%%%%%%%%%%%%%%%
% MATLAB code subproblem1.m
% Subproblem-1
%%%%%%%%%%%%%%%%%%%%%%%%%%%%%%%%%%%%%%%%%%%%%%%%%
%
% The matrix A and b corresponds to equation Ax=b
% c -> vector of cost coefficients
% basic_set -> set of basic variables
% nonbasic_set -> setof nonbasic variables
% B -> matrix containing basic variable columns of A
% N -> matrix containing nonbasic variable columns of A
% xb -> basic variables
% y -> simplex multipliers
% cb -> cost coefficients of basic variables
% cn -> cost coefficients of nonbasic variables
%
clear all
clc
format rational
format compact
A = [2  5  1  0  0;
     2 -3  0  1  0;
     1  0  0  0  1];
b = [16;7;5];
c = [-4;-5;0;0;0];
basic_set = [3 4 5];
nonbasic_set = [1 2];
for i = 1:length(basic_set)
    B(:,i) = A(:,basic_set(i));
    cb(i) = c(basic_set(i));
end
for i = 1:length(nonbasic_set)
    N(:,i) = A(:,nonbasic_set(i));
    cn(i) = c(nonbasic_set(i));
end
cn_cap = cn;
cb_ini = cb;
b_cap = b;
zz1 = 0;
% Rest of the code remains same as in simplex.m (Chapter 4)
%%%%%%%%%%%%%%%%%%%%%%%%%%%%%%%%%%%%%%%%%%%%%%%%%
```

```
%%%%%%%%%%%%%%%%%%%%%%%%%%%%%%%%%%%%%%%%%%%%%%%%%%%%
% MATLAB code subproblem2.m
% Subproblem-2
%%%%%%%%%%%%%%%%%%%%%%%%%%%%%%%%%%%%%%%%%%%%%%%%%%%%
%
% The matrix A and b corresponds to equation Ax=b
% c -> vector of cost coefficients
% basic_set -> set of basic variables
% nonbasic_set -> setof nonbasic variables
% B -> matrix containing basic variable columns of A
% N -> matrix containing nonbasic variable columns of A
% xb -> basic variables
% y -> simplex multipliers
% cb -> cost coefficients of basic variables
% cn -> cost coefficients of nonbasic variables
%
clear all
clc
format rational
format compact
A = [2  5  1  0  0;
     2 -3  0  1  0;
    -1  0  0  0  1];
b = [16;7;-6];
c = [-4;-5;0;0;0];
basic_set = [3 4 5];
nonbasic_set = [1 2];
for i = 1:length(basic_set)
    B(:,i) = A(:,basic_set(i));
    cb(i) = c(basic_set(i));
end
for i = 1:length(nonbasic_set)
    N(:,i) = A(:,nonbasic_set(i));
    cn(i) = c(nonbasic_set(i));
end
cn_cap = cn;
cb_ini = cb;
b_cap = b;
zz1 = 0;
% Rest of the code remains same as in simplex.m (Chapter 4)
%%%%%%%%%%%%%%%%%%%%%%%%%%%%%%%%%%%%%%%%%%%%%%%%%%%%


%%%%%%%%%%%%%%%%%%%%%%%%%%%%%%%%%%%%%%%%%%%%%%%%%%%%
% MATLAB code node2_subproblem1.m
% Subproblem-1 (Node-2)
%%%%%%%%%%%%%%%%%%%%%%%%%%%%%%%%%%%%%%%%%%%%%%%%%%%%
%
% The matrix A and b corresponds to equation Ax=b
% c -> vector of cost coefficients
```

Appendix B

```matlab
% basic_set -> set of basic variables
% nonbasic_set -> setof nonbasic variables
% B -> matrix containing basic variable columns of A
% N -> matrix containing nonbasic variable columns of A
% xb -> basic variables
% y -> simplex multipliers
% cb -> cost coefficients of basic variables
% cn -> cost coefficients of nonbasic variables
%
clear all
clc
format rational
format compact
A = [2  5 1 0 0;
     2 -3 0 1 0;
     0  1 0 0 1];
b = [16;7;1];
c = [-4;-5;0;0;0];
basic_set = [3 4 5];
nonbasic_set = [1 2];
for i = 1:length(basic_set)
    B(:,i) = A(:,basic_set(i));
    cb(i) = c(basic_set(i));
end
for i = 1:length(nonbasic_set)
    N(:,i) = A(:,nonbasic_set(i));
    cn(i) = c(nonbasic_set(i));
end
cn_cap = cn;
cb_ini = cb;
b_cap = b;
zz1 = 0;
% Rest of the code remains same as in simplex.m (Chapter 4)
%%%%%%%%%%%%%%%%%%%%%%%%%%%%%%%%%%%%%%%%%%%%%%%


%%%%%%%%%%%%%%%%%%%%%%%%%%%%%%%%%%%%%%%%%%%%%%%
% MATLAB code node2_subproblem2.m
% Subproblem-2 (Node-2)
%%%%%%%%%%%%%%%%%%%%%%%%%%%%%%%%%%%%%%%%%%%%%%%
%
% The matrix A and b corresponds to equation Ax=b
% c -> vector of cost coefficients
% basic_set -> set of basic variables
% nonbasic_set -> setof nonbasic variables
% B -> matrix containing basic variable columns of A
% N -> matrix containing nonbasic variable columns of A
% xb -> basic variables
% y -> simplex multipliers
% cb -> cost coefficients of basic variables
```

```
% cn -> cost coefficients of nonbasic variables
%
clear all
clc
format rational
format compact
A = [2  5  1  0  0;
     2 -3  0  1  0;
     0 -1  0  0  1];
b = [16;7;-2];
c = [-4;-5;0;0;0];
basic_set = [3 4 5];
nonbasic_set = [1 2];
for i = 1:length(basic_set)
    B(:,i) = A(:,basic_set(i));
    cb(i) = c(basic_set(i));
end
for i = 1:length(nonbasic_set)
    N(:,i) = A(:,nonbasic_set(i));
    cn(i) = c(nonbasic_set(i));
end
cn_cap = cn;
cb_ini = cb;
b_cap = b;
zz1 = 0;
% Rest of the code remains same as in simplex.m (Chapter 4)
%%%%%%%%%%%%%%%%%%%%%%%%%%%%%%%%%%%%%%%%%%%%%%%%%%%

%%%%%%%%%%%%%%%%%%%%%%%%%%%%%%%%%%%%%%%%%%%%%
%    File name pso.m
%    Welded beam problem
%%%%%%%%%%%%%%%%%%%%%%%%%%%%%%%%%%%%%%%%%%%%%
%
% lb -> lower bound of variables
% ub -> upper bound of variables
% x -> position of individual
% v -> velocity of individual
% rand -> random number from 0 to 1
% fitness -> fitness of individual
% pbest -> best fitness achieved by individual
% gbest -> best fitness of group
%
clear all
clc
format long
pop = 50;
phi_1 = 1.05;
phi_2 = 1.1;
nmax = 1000;
scale_factor = 10000000;
weight = linspace(1,0.3,nmax);
fprintf('_____\n')
lb = [0.1 0.1 0.1 0.1];
```

Appendix B

```
ub = [2 10 10 2];
for i = 1:length(lb)
    for j = 1:pop
        x(i,j) = lb(i) + (ub(i)-lb(i))*rand;
        v(i,j) = 0;
    end
end
for i = 1:pop
    fitness(i) = func1(x(:,i),scale_factor);
    pbest(i) = fitness(i);
    px(i,:) = x(:,i);
end
[gbest, location] = min(fitness);
gx = x(:,location);
for i = 1:nmax
    for j = 1:pop
        v(:,j) = weight(i)*v(:,j) + phi_1*rand*(px(j,:)'-x(:,j)) +
                 phi_2*rand*(gx-x(:,j));
        x(:,j) = x(:,j) + v(:,j);
        for k = 1:length(x(:,j))
            if x(k,j) < lb(k) || x(k,j) > ub(k)
                x(k,j) = lb(k) + (ub(k)-lb(k))*rand;
            end
        end
        x(2,:) = round(x(2,:));
        x(4,:) = round(x(4,:));
        fitness(j) = func1(x(:,j),scale_factor);
        if fitness(j) < pbest(j)
            pbest(j) = fitness(j);
            px(j,:) = x(:,j);
        end
    end
    [gbest, location] = min(pbest);
    gx = x(:,location);
    [gx' gbest];
    fprintf('%3d %8.3f %8.3f %8.3f %8.3f % 8.3f    \n',i,gx,gbest)
end
fprintf('_____ \n')
%%%%%%%%%%%%%%%%%%%%%%%%%%%%%%%%%%%%%%%%%%%%%%%

%%%%%%%%%%%%%%%%%%%%%%%%%%%%%%%%%%%%%%%%%%%%%%
%   File name func1.m
%   Objective function for welded beam problem
%%%%%%%%%%%%%%%%%%%%%%%%%%%%%%%%%%%%%%%%%%%%%%
function y = func1(x,scale_factor)
y = 1.10471*x(1)*x(1)*x(2) + 0.04811*x(3)*x(4)*(14+x(2));
%y = (x(1)-1)^2 + (x(2)-5)^2;
penalty = 0.0;
[h,g] = constr(x);
for i = 1:length(h)
    if h(i)~=0
        penalty = penalty + h(i)^2;
    end
end
```

```
for i = 1:length(g)
    if g(i)>0
        penalty = penalty + g(i)^2;
    end
end
y = y+penalty*scale_factor;
%%%%%%%%%%%%%%%%%%%%%%%%%%%%%%%%%%%%%%%%%%%%%%%%%%%%


%%%%%%%%%%%%%%%%%%%%%%%%%%%%%%%%%%%%%%%%%%%%%%%%%%%%
%   File name constr.m
%   Constraint function for welded beam problem
%%%%%%%%%%%%%%%%%%%%%%%%%%%%%%%%%%%%%%%%%%%%%%%%%%%%
function [h,g] = constr(x)
h(1) = 0;
% g(1) = -x(1)^2 + x(2) -4;
% g(2) = -(x(1)-2)^2 + x(2) -3;
load = 6000;
length = 14;
modulusE = 30e6;
modulusG = 12e6;
tmax = 13600;
sigmamax = 30000;
delmax = 0.25;
tdash = load/(sqrt(2)*x(1)*x(2));
R = sqrt(x(2)*x(2)/4 + ((x(1)+x(3))/2)^2);
M = load*(length + x(2)/2);
J = 2* ((x(1)*x(2)/sqrt(2)) *(x(2)^2/12 + ((x(1)+x(3))/2)^2));
tdashdash = M*R/J;
tx = sqrt(tdash^2 + 2*tdash*tdashdash*x(2)/(2*R)+tdashdash^2);
sigmax = 6*load*length/(x(4)*x(3)^2);
delx = 4*load*length^3/(modulusE*x(4)*x(3)^3);
pcx = (4.013*sqrt(modulusE*modulusG*x(3)^2*x(4)^6/36)/(length^2)) *
      (1- (x(3)/(2*length))*sqrt(modulusE/(4*modulusG)));
g(1) = tx/tmax -1;
g(2) = sigmax/sigmamax -1;
g(3) = x(1) - x(4);
g(4) = (.10471*x(1)*x(1) + 0.04811*x(3)*x(4)*(14+x(2)))/5 -1;
g(5) = 0.125 - x(1);
g(6) = delx/delmax -1;
g(7) = load/pcx -1;
g(8) = x(1)/2 -1;
g(9) = x(4)/2 -1;
g(10) = -x(1) + 0.1;
g(11) = -x(4) + 0.1;
g(12) = x(2)/10 -1;
g(13) = x(3)/10 -1;
g(14) = -x(2) + 0.1;
g(15) = -x(3) + 0.1;
%%%%%%%%%%%%%%%%%%%%%%%%%%%%%%%%%%%%%%%%%%%%%%%%%%%%
```

Appendix C: Solutions to Chapter Problems

Chapter 1

1. Let x = number of times fare is reduced by Rs. 300

$$\text{Revenue} = R = \text{price} \times \text{quantity} = (15{,}000 - 300x) \times (130 + 4x)$$

For maximization, $\dfrac{dR}{dx} = 0$

$$\therefore x = 8.75$$

Now, $\dfrac{d^2R}{dx^2} = -2400$

$\therefore R$ has a local maximum at $x = 8.75$

$$\text{Best fare} = (15{,}000 - 300x) = \text{Rs. } 12{,}375$$

$$\text{Number of passengers} = (130 + 4x) = 165$$

$$\text{Revenue} = \text{Rs. } 2{,}041{,}875$$

2. Let x = number of additional trees that need to be planted

$$\text{Yield} = y = (50 + x) \times (300 - 3x)$$

For maximization, $\dfrac{dy}{dx} = 0$

$$\therefore x = 25$$

Now, $\dfrac{d^2y}{dx^2} = -6$

$\therefore R$ has a local maximum at $x = 25$

401

3. Let r = the radius of the circle; w and h are the width and height of the rectangle to be inscribed in the circle (see Figure C.1).

$$r^2 = \left(\frac{w}{2}\right)^2 + \left(\frac{h}{2}\right)^2$$

$$w = \sqrt{4r^2 - h^2} \text{ (Constraint)}$$

$A = wh$ (Objective function to be maximized)

$$\therefore A = \sqrt{4r^2 - h^2}\, h$$

Plotting (h, A) gives (see Figure C.2) optimal value of A as 50 cm² at $h = 7.07$ cm.

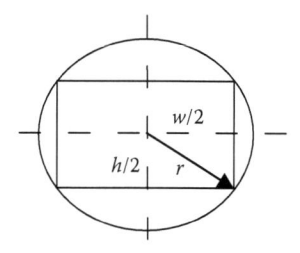

FIGURE C.1
Rectangle inscribed in a circle.

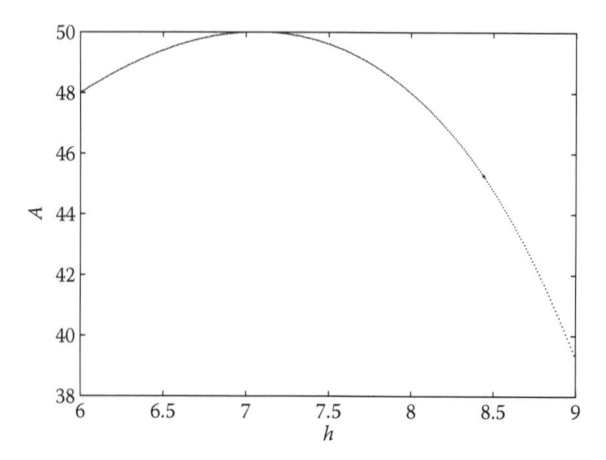

FIGURE C.2
h vs. A.

Appendix C 403

It is easy to show analytically $\left(\dfrac{dA}{dh}=0\right)$ that maximum area is given by $2r^2$ when $w=h=\sqrt{2}\,r$.

4. The fence needs to cover only three sides of the field because the river is flowing on one side. Thus the optimization problem can be stated as

Maximize
$$xy$$
subject to
$$2x + y = 300$$
where x and y are two adjacent sides of the rectangle.

5. Minimize
$$\sum_{i=1}^{n}\sum_{j=1}^{n} x_{ij} y_{ij} \quad (i \neq j)$$

subject to
$$\sum_{i=1}^{n} x_{ij} = 1$$

$$\sum_{j=1}^{n} x_{ij} = 1$$

where x_{ij} is an integer that takes a value 0 or 1.

6. Minimize
$$(r_r R) + r_s(W - R)$$

7. Minimize
$$(45 - m - c)^2 + (55 - 2m - c)^2 + (70 - 3m - c)^2 + (85 - 4m - c)^2 + (105 - 5m - c)^2$$
$$m^* = 15 \quad c^* = 27$$

8. $U = \dfrac{204{,}165.5}{330 - 2T} + \dfrac{10{,}400}{T - 20}$

$$40 \leq T \leq 90$$

9. $\nabla f = \begin{bmatrix} 5x_2 + \dfrac{4x_1}{2x_1^2 + 3x_2^2} \\ 5x_1 + \dfrac{6x_2}{2x_1^2 + 3x_2^2} \end{bmatrix}$ $\quad \nabla^2 f = \begin{bmatrix} \dfrac{12x_2^2 - 8x_1^2}{(2x_1^2 + 3x_2^2)^2} & 5 - \dfrac{24x_1 x_2}{(2x_1^2 + 3x_2^2)^2} \\ 5 - \dfrac{24x_1 x_2}{(2x_1^2 + 3x_2^2)^2} & \dfrac{12x_1^2 - 18x_2^2}{(2x_1^2 + 3x_2^2)^2} \end{bmatrix}$

10. Let x, y, and z denote the quantity of product A, B, and C respectively. The optimization problem can be stated as

Maximize

$$5x + 7y + 4z$$

subject to

$$12x + 25y + 7z \leq 28{,}000$$

$$11x + 6y + 20z \leq 35{,}000$$

$$15x + 6y + 5z \leq 32{,}000$$

11. Let

x_{11} = number of units transported from factory P to warehouse A
x_{12} = number of units transported from factory P to warehouse B
x_{13} = number of units transported from factory P to warehouse C
x_{14} = number of units transported from factory P to warehouse D
x_{15} = number of units transported from factory P to warehouse E
x_{21} = number of units transported from factory Q to warehouse A
x_{22} = number of units transported from factory Q to warehouse B
x_{23} = number of units transported from factory Q to warehouse C
x_{24} = number of units transported from factory Q to warehouse D
x_{25} = number of units transported from factory Q to warehouse E
x_{31} = number of units transported from factory R to warehouse A
x_{32} = number of units transported from factory R to warehouse B
x_{33} = number of units transported from factory R to warehouse C
x_{34} = number of units transported from factory R to warehouse D
x_{35} = number of units transported from factory R to warehouse E

Appendix C

Minimize

$$3x_{11} + 7x_{12} + 4x_{13} + 6x_{14} + 5x_{15} + 5x_{21} + 4x_{22} + 2x_{23} + 5x_{24} + x_{25} + 6x_{31} + 3x_{32} + 2x_{33} + 2x_{34} + 4x_{35}$$

subject to

$$x_{11} + x_{12} + x_{13} + x_{14} + x_{15} \leq 150$$

$$x_{21} + x_{22} + x_{23} + x_{24} + x_{25} \leq 110$$

$$x_{31} + x_{32} + x_{33} + x_{34} + x_{35} \leq 90$$

$$x_{11} + x_{21} + x_{31} \geq 50$$

$$x_{12} + x_{22} + x_{32} \geq 100$$

$$x_{13} + x_{23} + x_{33} \geq 70$$

$$x_{14} + x_{24} + x_{34} \geq 70$$

$$x_{15} + x_{25} + x_{35} \geq 60$$

12. The minimum value is −13.128 and occurs at $x = -0.47$.
    The maximum value (see Figure C.3) is 1.128 and occurs at $x = -3.53$.

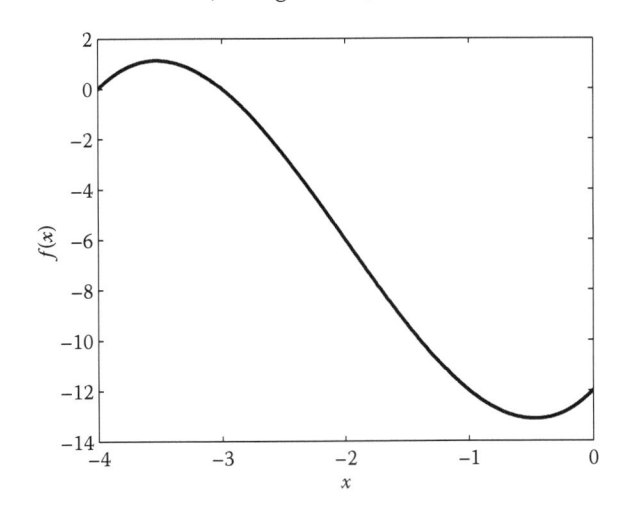

FIGURE C.3
Plot of the function.

13. Let x_{ij} = barrels of gasoline of type i used to make fuel of type j

F_j = barrels of fuel of type j

Profit = Revenue − Cost

The objective function and constraints can be written as
Minimize

$$90F_1 + 100F_2 - 60(x_{11} + x_{12}) - 65(x_{21} + x_{22}) \\ - 70(x_{31} + x_{32}) - 80(x_{41} + x_{42})$$

subject to

$$75x_{11} + 85x_{21} + 90x_{31} + 95x_{41} - 80F_1 \geq 0$$

$$75x_{12} + 85x_{22} + 90x_{32} + 95x_{42} - 90F_2 \geq 0$$

$$F_1 + F_2 \geq 6000$$

$$0 \leq (x_{11} + x_{12}) \leq 3000$$

$$0 \leq (x_{21} + x_{22}) \leq 4000$$

$$0 \leq (x_{31} + x_{32}) \leq 5000$$

$$0 \leq (x_{41} + x_{42}) \leq 4000$$

14. Functions at plots (a) and (d) are convex (see Figure C.4).
15. The Taylor series of a function $f(x)$ at $x = a$ is given by

$$f(a) + f'(a)(x-a) + \frac{f''(a)}{2!}(x-a)^2 + \cdots$$

The Taylor series for the function $\ln(x - 1)$ at $x = 3$ is given by

$$\ln 2 + \frac{1}{2}(x-3) - \frac{1}{8}(x-3)^2$$

Appendix C

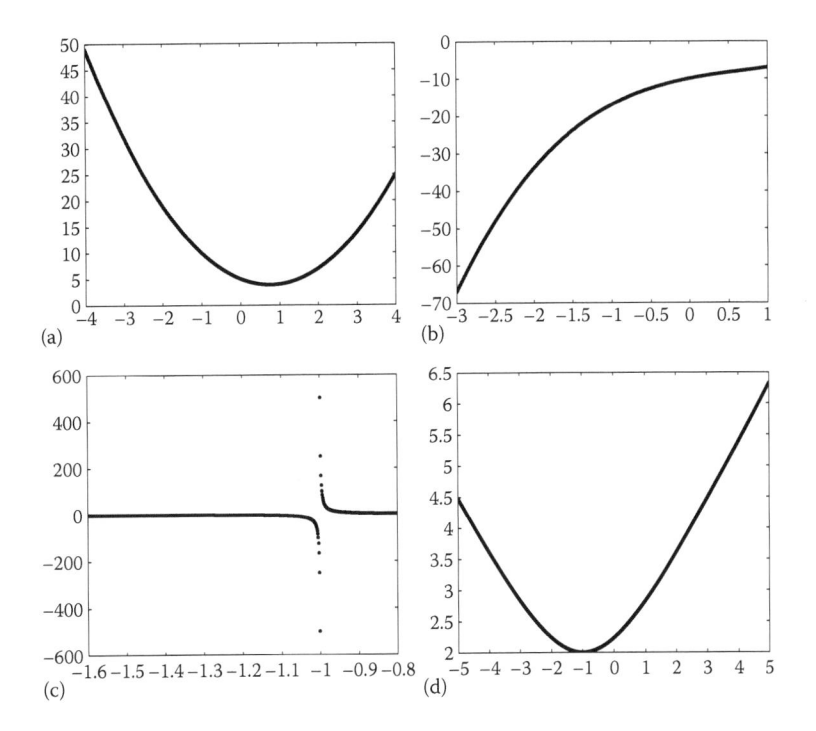

FIGURE C.4
Plot of four different functions.

16. The linear approximation of a function is given by the first two terms of the Taylor series. The linear expansion for the function $(1 + x)^{50} + (1 - 2x)^{60}$ at $x = 1$ is given by

$$L(x) = (1 + 2^{50}) + (120 + 50 \times 2^{49})(x - 1)$$

17. The Taylor series for the function e^x at $x = 3$ is given by

$$e^3 + e^3(x - 3) + \frac{e^3}{2}(x - 3)^2 + \frac{e^4}{6}(x - 3)^3$$

18. The Taylor series for the function $e^{\cos x}$ at $x = \pi$ is given by

$$\frac{1}{e} + \frac{1}{2e}(x - \pi)^2$$

19. The quadratic approximation of a function is given by the first three terms of the Taylor series. The quadratic expansion for the function $\ln(1 + \sin x)$ at $x = 0$ is given by

$$Q(x) = x - \frac{x^2}{2!}$$

20. The gradient of the function is given by

$$\nabla f = \begin{bmatrix} 2x_1 x_2 - x_2 x_3^2 \\ x_1^2 + 2x_2 x_3 - x_1 x_3^2 \\ x_2^2 - 2x_1 x_2 x_3 \end{bmatrix}$$

The gradient at $(1, 1, -1)$ is given by

$$\nabla f = \begin{bmatrix} 1 \\ -2 \\ 3 \end{bmatrix}$$

Now

$$\nabla f(x)^T u = \begin{bmatrix} 1 & -2 & 3 \end{bmatrix} \begin{bmatrix} 1/\sqrt{14} \\ 2/\sqrt{14} \\ 3/\sqrt{14} \end{bmatrix} = 6/\sqrt{14}$$

21. Both functions are convex (see Figure C.5).
22.
   i. The maximum value is 19,575/17 and occurs at $x = 99/17$ and $x = 48/17$ (see Figure C.6).
   ii. The maximum value is 120 and occurs at $x = 0$ and $x = 30$ (see Figure C.7).
23. The Jacobian is given by

$$J = \begin{bmatrix} 1 & 4x_2 & 9x_3 \\ 2x_1 x_2 x_3^2 & x_1^2 x_3^2 & 2x_3 x_2 x_1^2 \\ 3x_2 - 2x_3 & 3x_1 + 4x_3 & -2x_1 + 4x_2 \end{bmatrix}$$

Appendix C

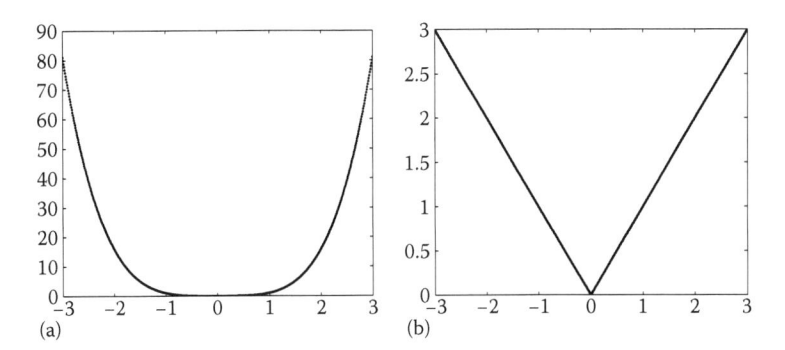

FIGURE C.5
Plot of two different functions.

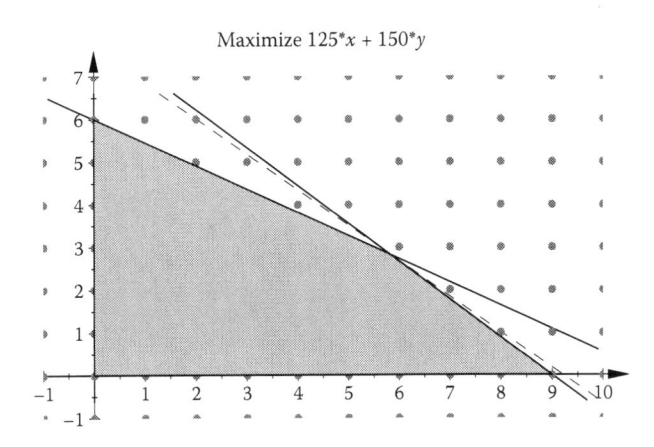

FIGURE C.6
Linear programming problem.

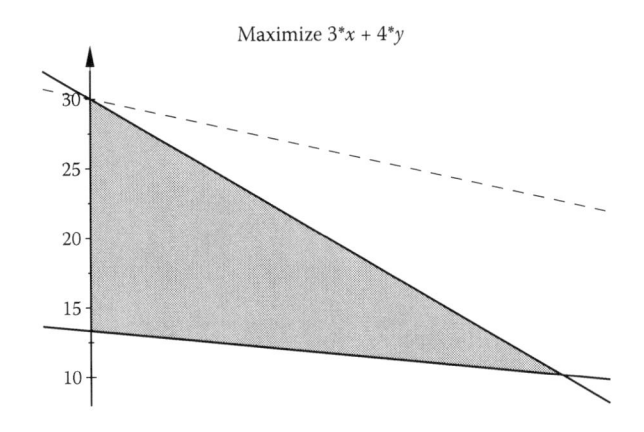

FIGURE C.7
Linear programming problem.

Chapter 2

1. $\left(\dfrac{L}{D}\right)_{max} = 1.71$ occurs at $\alpha^* = 20$ degrees

2.
   i. The minimum value of function is -3.517 and occurs at $x^* = -1.386$. The numbers of function evaluations are 16, 20, 40, and 58 by the golden section, cubic polynomial fit, bisection, and secant method respectively.
   ii. The minimum value of function is 4.369 and occurs at $x^* = 0.45$. The numbers of function evaluations are 15, 16, 36, and 44 by the golden section, cubic polynomial fit, bisection, and secant method respectively.
   iii. The minimum value of function is 0.691 and occurs at $x^* = 1.087$. The numbers of function evaluations are 15, 40, 32, and 128 by the golden section, cubic polynomial fit, bisection, and secant method respectively.
   iv. Minimum value of function is 11.052 and occurs at $x^* = 1.356$. The number of function evaluations is 16 by the golden section method. Other methods did not converge as the function is highly skewed.

3. The maximum value of function is 0.202 and occurs at $x^* = 3$.
4. The maximum value of function is 28.209 and occurs at $x^* = 3.577$.
5. The maximum value of function is 0.693 and occurs at $x^* = 0.0$. The minimum value of function is 0.526 and occurs at $x^* = 1.19$.
6. Let x and y be the length and depth of the beam and D be the diameter of the log. Then,

$$x^2 + y^2 = D^2$$

$$y^2 = 1 - x^2$$

Let S denote the strength of the beam. Then,

$$S = kxy^2$$

where k is a constant.

Now,

$$S = kx(1 - x^2)$$

For maximum,

$$\frac{dS}{dx} = 0$$

Therefore,

$$x = 0.5774 \text{ m} \quad \text{and} \quad y = 0.8165 \text{ m}$$

7. Total cost

$$C = 6\left(\frac{300}{x} + \frac{x}{3}\right) + 7 \times \frac{600}{x}$$

For maximum,

$$\frac{dC}{dx} = 0$$

Therefore,

$$x = 54.77 \text{ km/h}$$

8. Total time

$$t = \frac{7-x}{6} + \frac{\sqrt{25+x^2}}{2}$$

For minimum,

$$\frac{dt}{dx} = 0$$

Therefore,

$$x = 1.77 \text{ km}$$

9. $(T)_{min} = 41{,}375$ N occurs at $v^* = 149$ m/s

10. The global minimum value of function is −6.097 and occurs at $x^* = -1.18$. The local minimum value of function is −5.01 and occurs at $x^* = 0.43$. The regions ABC and CDE are convex. The region BCD is concave (see Figure C.8).
11. $f'(x) = 0$

$$\frac{k}{p_2} - 2x\frac{p_1}{p_2} = 0$$

Therefore,

$$x = \frac{k}{2p_1}$$

Hence,

$$f(x) = \frac{k^2}{4p_1 p_2} = 40.5$$

12. The minimum value of function is −4.899 and occurs at $x^* = 0.3$.
13. $r^* = 40.7$ mm and $h^* = 57.6$ mm.
14. $a^* = 2.24$ (see Figure C.9).

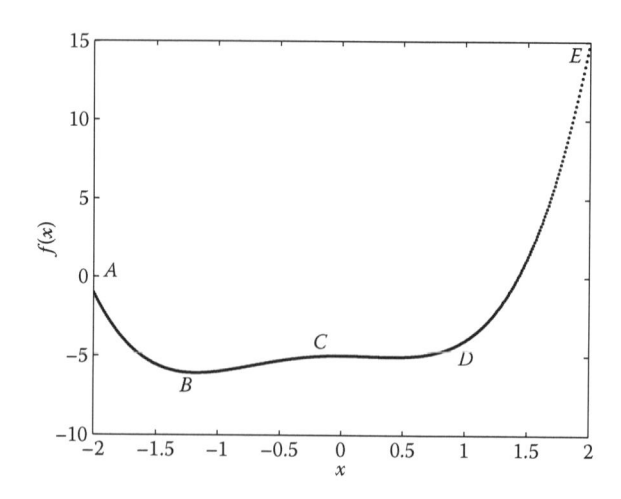

FIGURE C.8
Multimodal function.

Appendix C

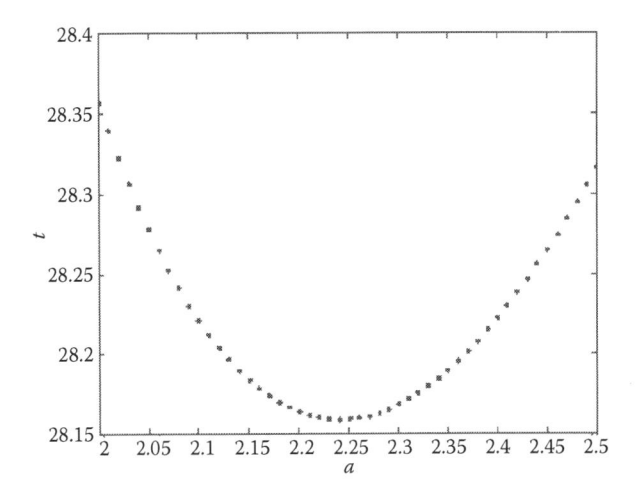

FIGURE C.9
Solution to bacteria problem.

Chapter 3

1. The gradient of the function is given by

$$\nabla f = \begin{bmatrix} 2x_1 + 3x_2 \\ 3x_1 + 4x_2 \end{bmatrix}$$

Therefore, search direction for the steepest descent method at (1, 2) is given by

$$S = -\nabla f = -\begin{bmatrix} 8 \\ 11 \end{bmatrix}$$

2. $x^* = (0, 0)$ with $f(x^*) = 0$
3. All the methods converge to the point $x^* = (3, 2)$ with $f(x^*) = 0$. The convergence history of different methods is given below:

i. Steepest descent method

```
Initial function value = 32.0000
No.        x-vector           f(x)      Deriv
```

No.	x-vector		f(x)	Deriv
1	2.777	1.706	4.239	46.648
2	-3.827	-2.255	34.894	23.591
3	-3.417	-2.938	8.305	63.644
4	3.276	1.088	7.130	31.920
5	2.843	1.804	2.045	10.994
6	3.036	1.921	0.095	16.863
7	2.993	1.992	0.004	2.150
8	3.001	1.997	0.000	0.792
9	3.000	2.000	0.000	0.089
10	3.000	2.000	0.000	0.018
11	3.000	1.999	0.000	0.007
12	3.000	1.999	0.000	0.032
13	3.000	2.000	0.000	0.024
14	3.000	2.000	0.000	0.010
15	3.000	1.999	0.000	0.005
16	3.000	1.999	0.000	0.017
17	3.000	2.000	0.000	0.029
18	3.000	2.000	0.000	0.013
19	3.000	1.999	0.000	0.007
20	3.000	2.000	0.000	0.022
21	3.000	2.000	0.000	0.012
22	3.000	1.999	0.000	0.007
23	3.000	2.000	0.000	0.019
24	3.000	2.000	0.000	0.011
25	3.000	1.999	0.000	0.007
26	3.000	2.000	0.000	0.018
27	3.000	2.000	0.000	0.011
28	3.000	1.999	0.000	0.006
29	3.000	2.000	0.000	0.017
30	3.000	2.000	0.000	0.010
31	3.000	1.999	0.000	0.006
32	3.000	2.000	0.000	0.017
33	3.000	2.000	0.000	0.010
34	3.000	1.999	0.000	0.006
35	3.000	2.000	0.000	0.016
36	3.000	2.000	0.000	0.010
37	3.000	1.999	0.000	0.006
38	3.000	2.000	0.000	0.016
39	3.000	2.000	0.000	0.009
40	3.000	1.999	0.000	0.006
41	3.000	2.000	0.000	0.015
42	3.000	2.000	0.000	0.009

Appendix C

ii. Newton's method

```
Initial function value = 32.0000
No.        x-vector            f(x)       Deriv

1        4.426    2.016       114.754     46.648
2        3.508    1.780         9.631    193.672
3        3.095    1.967         0.298     42.834
4        3.004    1.998         0.001      6.720
5        3.000    2.000         0.000      0.292
6        3.000    2.000         0.000      0.001
```

iii. Modified Newton's method

```
Initial function value = 32.0000
No.        x-vector            f(x)       Deriv

1        2.991    2.598         7.814     46.648
2        2.951    2.035         0.076     31.917
3        2.999    2.001         0.000      2.892
4        3.000    2.000         0.000      0.035
5        3.000    2.000         0.000      0.006
6        3.000    2.000         0.000      0.001
```

iv. Levenberg–Marquardt method

```
Initial function value = 32.0000
No.        x-vector            f(x)       Deriv

1        2.024    2.963        29.980     46.648
2        2.074    2.898        26.507     44.421
3        2.172    2.794        21.043     40.863
4        2.353    2.646        13.471     35.851
5        2.615    2.457         5.628     29.097
6        2.844    2.259         1.324     19.368
7        2.952    2.107         0.184      9.124
8        2.989    2.028         0.012      3.221
9        2.998    2.004         0.000      0.797
10       3.000    2.000         0.000      0.113
11       3.000    2.000         0.000      0.008
```

v. Conjugate gradient method

```
Initial function value = 32.0000
No.         x-vector          f(x)      Deriv
```

No.	x-vector		f(x)	Deriv
1	2.777	1.706	4.239	46.648
2	3.076	1.727	0.920	23.591
3	3.110	1.886	0.425	6.141
4	3.029	2.027	0.060	6.459
5	2.993	2.031	0.014	3.124
6	2.987	2.019	0.008	0.955
7	2.994	1.998	0.001	0.719
8	3.000	1.996	0.000	0.484
9	3.001	1.997	0.000	0.139
10	3.001	1.999	0.000	0.084
11	3.000	2.000	0.000	0.070

vi. DFP method

```
Initial function value = 32.0000
No.         x-vector          f(x)      Deriv
```

No.	x-vector		f(x)	Deriv
1	2.777	1.706	4.239	46.648
2	3.076	1.727	0.920	23.591
3	2.997	1.999	0.000	6.142
4	3.000	2.000	0.000	0.229
5	3.000	2.000	0.000	0.019
6	3.000	2.000	0.000	0.004
7	3.000	2.000	0.000	0.001

vii. BFGS method

```
Initial function value = 32.0000
No.         x-vector          f(x)      Deriv
```

No.	x-vector		f(x)	Deriv
1	2.777	1.706	4.239	46.648
2	3.076	1.727	0.920	23.591
3	2.997	2.000	0.000	6.141
4	3.000	2.000	0.000	0.234
5	3.000	2.000	0.000	0.020
6	3.000	2.000	0.000	0.004

Appendix C

viii. Powell method

```
Initial function value = 32.0000
No.          x-vector          f(x)

1       3.000      2.003      0.000
2       3.000      2.001      0.000
3       3.000      2.001      0.000
4       3.000      2.001      0.000
5       3.000      2.000      0.000
```

ix. Nelder–Meads method

Iteration	Deviation	f(x)
1	87.9962	58.789
2	29.8163	21.131
3	32.1499	10.859
4	13.5635	10.859
5	4.5179	6.998
6	5.4792	2.191
7	3.5857	0.483
8	1.3618	0.483
9	1.1302	0.483
10	0.3982	0.483
11	0.3662	0.116
12	0.2224	0.116
13	0.0865	0.072
14	0.0644	0.006
15	0.0417	0.006
16	0.0084	0.006
17	0.0089	0.002
18	0.0029	0.002
19	0.0023	0.001
20	0.0013	0.001
21	0.0011	0.000
22	0.0005	0.000
23	0.0002	0.000
24	0.0002	0.000

```
xc =
      2.9994    1.9987
```

4. The metric [A] approaches the inverse of the Hessian matrix in the DFP method.

```
Initial function value = 9.0000
No.         x-vector              f(x)       Deriv
─────────────────────────────────────────────────────
1        0.545    -0.183          0.165      13.928
2        0.001    -0.000          0.000       0.575
3       -0.000     0.000          0.000       0.001
4        0.000    -0.000          0.000       0.000
─────────────────────────────────────────────────────
```

```
>> A

A =

   0.909090909189563  -0.272727272447249
  -0.272727272447250   0.181818182613013

>> inv(hessian(x,delx,n_of_var))

ans =

   0.909090909090909  -0.272727272727273
  -0.272727272727273   0.181818181818182
```

5. The metric [A] approaches to the Hessian matrix in the BFGS method.

```
Initial function value = 67.0000
No.         x-vector              f(x)       Deriv
─────────────────────────────────────────────────────
1        0.818    -0.273          0.372      38.275
2        0.001    -0.000          0.000       0.862
3       -0.000     0.000          0.000       0.001
─────────────────────────────────────────────────────
```

```
>> A

A =

   1.999999985050531   2.999999955266635
   2.999999955266635   9.999999866143737

>> hessian(x,delx,n_of_var)

ans =

   2.000000000000000   3.000000000000000
   3.000000000000000  10.000000000000002
```

Appendix C

6. $x^* = (-0.656, -0.656)$ with $f(x^*) = -2.661$

```
Initial function value = 4.3891
No.         x-vector              f(x)         Deriv

1       0.098      0.184        -2.422       21.090
2      -0.764     -0.539        -2.601        0.987
3      -0.665     -0.599        -2.641        0.763
4      -0.655     -0.656        -2.661        0.845
5      -0.655     -0.656        -2.661        0.014
6      -0.656     -0.656        -2.661        0.010
7      -0.656     -0.656        -2.661        0.002
```

7. $x^* = (1, 0, 0)$ with $f(x^*) = 0$
8. Both the complex variable formula and the central difference formula give the same results.
9. The value of the analytical derivative at $x = 0.1$ is 10.995004165278026. The value of the derivative at $x = 0.1$ using the central difference formula is 10.995337352778689. The value of the derivative at $x = 0.1$ using the complex variable formula is 10.995004165278024.
10.

x_i	$f(x_i)$	S_i	α^*	$f(\alpha^*)$
(1, 1)	106	(2, 4)	0.431928	28.3361
(0, 0)	170	(1, 2)	1.41453	43.9167
(3, 2)	0	(1, 1)	–0.00366	0.00099

11. $x^* = (0, 0)$ with $f(x^*) = 0$
12.

$$x_1 = \frac{p-v}{v}$$

$$x_2 = \frac{p-w}{w}$$

Since the second derivative is negative $\left\{\dfrac{-p}{(1+x_1)^2}\right\}$, it corresponds to the maximum of the function

13. $x^* = (0.02, 1.6)$ with $f(x^*) = -25.632$

```
Initial function value = 0.0000
No.           x-vector              f(x)         Deriv

1          0.122      1.530       -24.751       32.102
2          0.020      1.593       -25.631       16.299
3          0.020      1.599       -25.632        0.167
4          0.020      1.599       -25.632        0.030
5          0.020      1.600       -25.632        0.022
6          0.020      1.600       -25.632        0.016
7          0.020      1.600       -25.632        0.011
8          0.020      1.600       -25.632        0.002
```

Chapter 4

1. Maximize

$$z = 5x + 7y$$

subject to

$$2x + 3y \le 42$$

$$3x + 4y \le 48$$

$$x, y \ge 0$$

The solution is $x = 0, y = 12, z = 84$

2.
   i. The solution is $x_1 = 0, x_2 = 5, z = -10$
   ii. The solution is $x_1 = 0, x_2 = 10, z = 50$
   iii. The solution is $x_1 = \dfrac{15}{7}, x_2 = \dfrac{110}{7}, z = \dfrac{610}{7}$
   iv. The solution is $x_1 = \dfrac{11}{5}, x_2 = \dfrac{6}{5}, z = \dfrac{1}{5}$

3. $\begin{bmatrix} x_1 \\ x_2 \\ x_3 \end{bmatrix} = \begin{bmatrix} 2/3 \\ -2 \\ 5/3 \end{bmatrix}$ (Infeasible)

Appendix C

$$\begin{bmatrix} x_1 \\ x_2 \\ x_4 \end{bmatrix} = \begin{bmatrix} 11/4 \\ 9/8 \\ 5/8 \end{bmatrix} \text{ (Feasible)}$$

$$\begin{bmatrix} x_1 \\ x_3 \\ x_4 \end{bmatrix} = \begin{bmatrix} 2 \\ 3/5 \\ 2/5 \end{bmatrix} \text{ (Feasible)}$$

$$\begin{bmatrix} x_2 \\ x_3 \\ x_4 \end{bmatrix} = \begin{bmatrix} -3 \\ 11/5 \\ -1/5 \end{bmatrix} \text{ (Infeasible)}$$

4. $k = 2$
5.
   i.

$$A = \begin{bmatrix} -1 & 2 & -3 & 1 & 0 & 0 \\ -2 & 1 & -4 & 0 & -1 & 0 \\ -3 & 2 & 5 & 0 & 0 & 1 \end{bmatrix}; c = \begin{bmatrix} 2 \\ 3 \\ -1 \\ 0 \\ 0 \\ 0 \end{bmatrix}; b = \begin{bmatrix} 5 \\ 5 \\ 7 \end{bmatrix}; x = \begin{bmatrix} x_1 \\ x_2 \\ x_3 \\ s_1 \\ e_2 \\ s_3 \end{bmatrix}$$

   ii.

$$A = \begin{bmatrix} 3 & -2 & -3 & 3 & -1 & 0 & 0 \\ -4 & -3 & 1 & -1 & 0 & -1 & 0 \\ 1 & 2 & 1 & -1 & 0 & 0 & 1 \\ 0 & 1 & 0 & 0 & 0 & 0 & 0 \end{bmatrix}; c = \begin{bmatrix} -2 \\ 3 \\ -4 \\ 4 \\ 0 \\ 0 \\ 0 \end{bmatrix}; b = \begin{bmatrix} 5 \\ 2 \\ 8 \\ 5 \end{bmatrix}; x = \begin{bmatrix} x_1' \\ x_2 \\ x_3' \\ x_3'' \\ e_1 \\ e_2 \\ s_3 \end{bmatrix}$$

6.

$$x_B = \begin{bmatrix} x_3 \\ x_4 \\ x_5 \end{bmatrix} = B^{-1}b = \begin{bmatrix} 1 & 0 & 0 \\ 0 & 1 & 0 \\ 0 & 0 & 1 \end{bmatrix}^{-1} \begin{bmatrix} 7 \\ 8 \\ 5 \end{bmatrix} = \begin{bmatrix} 7 \\ 8 \\ 5 \end{bmatrix}$$

7. The solution is $x_1 = 0$, $x_2 = 5$, $z = -10$

```
basic_set =
     3      4      5
nonbasic_set =
     1      2
Initial_Table =
     1      0      0      1      2      10
     0      1      0      2     -1       5
     0      0      1      4     -3       5
Cost =
     0      0      0      3     -2       0

basic_set =
     2      4      5
nonbasic_set =
     1      3
Table =
     1      0      0    1/2    1/2       5
     0      1      0    5/2    1/2      10
     0      0      1   11/2    3/2      20
Cost =
     0      0      0      4      1      10

------SOLUTION------
basic_set =
     2      4      5
xb =
     5
    10
    20
zz =
   -10
```

8. Because the cost coefficients of the nonbasic variables are not zeros, the LPP has a unique solution.

```
basic_set =
     3      4      5
nonbasic_set =
     1      2
```

Appendix C 423

```
Initial_Table =
         1          0          0          2         -4          2
         0          1          0         -1          1          3
         0          0          1          1          0          4
Cost =
         0          0          0          1         -2          0
```

```
basic_set =
         3          2          5
nonbasic_set =
         1          4
Table =
         1          0          0         -2          4         14
         0          1          0         -1          1          3
         0          0          1          1          0          4
Cost =
         0          0          0         -1          2          6
```

```
basic_set =
         3          2          1
nonbasic_set =
         4          5
Table =
         1          0          0          4          2         22
         0          1          0          1          1          7
         0          0          1          0          1          4
Cost =
         0          0          0          2          1         10
```

```
------SOLUTION------
basic_set =
         3          2          1
xb =
        22
         7
         4
zz =
       -10
```

9. $x_1 = 0, x_2 = 0, x_3 = 2$
10. The dual is

    Minimize

    $$z = 7x_1 + 6x_2$$

    subject to

    $$x_1 \geq 4$$

$$x_2 \geq 5$$
$$2x_1 + x_2 \geq 23$$
$$x_1 + 3x_2 \geq 24$$
$$x_1, x_2 \geq 0$$

The optimal solution for the dual problem is 93 at $x_1 = 9$, $x_2 = 5$.

The optimal solution for the primal problem is 93 at $y_1 = 0$, $y_2 = 0$, $y_3 = 3$, $y_4 = 1$.

11. $x_1 = 0.505$, $x_2 = 0.745$, $z = 1.25$

Chapter 5

1. String length = 17.
2. String length for each variable as 14, 17, and 12.
3. Rerun the codes by modifying the input parameters mentioned in the file in.m.
4. Roulette wheel slots can be constructed for each of the strings. For example, the first string will have slots from 0 to 0.065 (25/385). Other slots are made in a similar way. Ten uniformly distributed random numbers are generated between 0 and 1. The corresponding strings pointed out by the random numbers are then selected. The strings selected are S-1 (one copy), S-3 (two copies), S-4 (one copy), S-5 (two copies), S-6 (one copy), S-8 (two copies), and S-9 (one copy). No copies of the strings S-2, S-7, and S-10 are made (see Figure C.10).
5. A tour size of two is selected. Each string has to be paired randomly with any other string in the group using random number generation. The winner is decided by comparing the fitness values of the strings (see Table C.1).

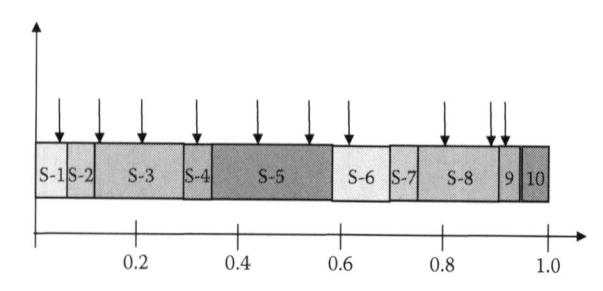

FIGURE C.10
Roulette wheel slots.

Appendix C

TABLE C.1

Tournament Selection

String	Competitor	Winner
S-1	S-5	S-5
S-2	S-10	S-10
S-3	S-10	S-3
S-4	S-2	S-2
S-5	S-10	S-5
S-6	S-5	S-5
S-7	S-8	S-8
S-8	S-1	S-8
S-9	S-4	S-9
S-10	S-9	S-9

6. Global minimum at $x^* = (0, 0)$ with $f(x^*) = 0$
7. Global minimum at $x^* = (0, 0)$ with $f(x^*) = 0$
8. The Himmelblau function has four distinct minima (see Figure C.11) as given below:

$$x_1 = 3, x_2 = 2, f(x^*) = 0$$

$$x_1 = 3.584, x_2 = -1.848, f(x^*) = 0$$

$$x_1 = -3.779, x_2 = -3.283, f(x^*) = 0$$

$$x_1 = -2.805, x_2 = 3.131, f(x^*) = 0$$

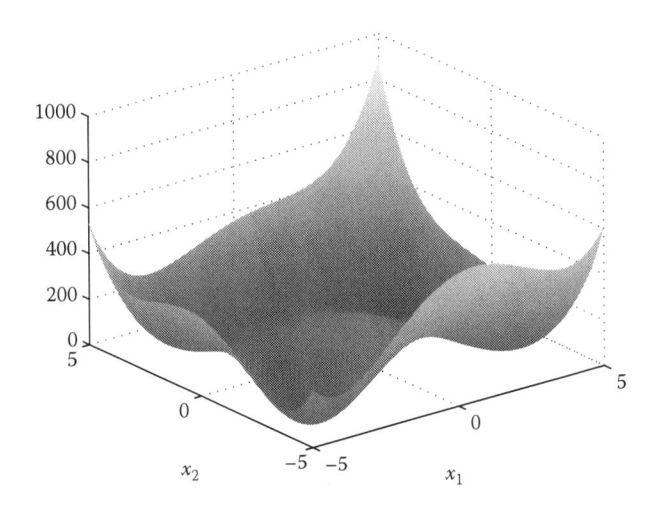

FIGURE C.11
Himmelblau function.

9. Global minimum $x^* = (-0.656, -0.656)$ with $f(x^*) = -2.661$
10. Global minimum $x^* = (1.139, 0.8996)$ with $f(x^*) = 1.9522$
11. Global minimum $x^* = (1, 3)$ with $f(x^*) = 0$
12. Global minimum $x^* = (0, 0)$ with $f(x^*) = 0$

Chapter 6

1.
   i. Infeasible
   ii. Feasible
   iii. Infeasible
   iv. Feasible
2. Only ii is active.
3. Substitute the value of $x_2 = 7 - x_1$ in the objective function

$$f(x) = (5x_1 - 14)^2 + (x_1 + 2)^2$$

Taking the first derivative as zero gives, $x_1 = \dfrac{34}{13}$

Therefore, $x_2 = \dfrac{57}{13}$ and $f(x^*) = \dfrac{288}{13}$

4. Writing the Lagrangian as

$$L(x,\lambda) = (3x_1 - 2x_2)^2 + (x_1 + 2)^2 + \lambda(x_1 + x_2 - 7)$$

The KKT conditions are given by the equations

$$20x_1 - 12x_2 + \lambda + 4 = 0$$

$$-12x_1 + 8x_2 + \lambda = 0$$

$$x_1 + x_2 - 7 = 0$$

Solving these equations gives the solution as $x_1 = 34/13$ and $x_2 = 57/13$, which is the optimum point with $\lambda = -48/13$. The minimum value of the function is $288/13$. Also,

$$\nabla^2 L = \begin{bmatrix} 20 & -12 \\ -12 & 8 \end{bmatrix} > 0$$

Appendix C

The Lagrange multipliers provide information on the sensitivity of objective function with respect to the right-hand side of the constraint equation (say, b). Then,

$$\Delta f = \mu \Delta b = -\frac{48}{13} \Delta b$$

Therefore,

$$f \approx \frac{288}{13} - \frac{48}{13} \Delta b$$

If the right-hand side of the constraint is changed by 1 unit, then the new value of the function minimum is 18.461 (approximately). The true minimum of the problem with the revised constraint is 18.615.

5.
Iteration 1

$$f(x) = 6; \; \nabla f(x) = \begin{bmatrix} 3 \\ -4 \end{bmatrix}; \; \nabla h = \begin{bmatrix} 1 \\ 1 \end{bmatrix}; \; \nabla g = \begin{bmatrix} -1 \\ -1 \end{bmatrix};$$

$$\nabla^2 L = \begin{bmatrix} 8 & -7 \\ -7 & 12 \end{bmatrix}$$

The quadratic problem is

Minimize

$$Q = \Delta x^T \begin{bmatrix} 3 \\ -4 \end{bmatrix} + \frac{1}{2} \Delta x^T \begin{bmatrix} 8 & -7 \\ -7 & 12 \end{bmatrix} \Delta x$$

subject to

$$-1 + [1 \;\; 1] \Delta x = 0$$

$$-1 + [-1 \;\; -1] \Delta x = 0$$

The solution of the quadratic problem is

$$\Delta x = \begin{bmatrix} 0.3529 \\ 0.6471 \end{bmatrix}$$

Now x is updated as

$$x = x + \Delta x = \begin{bmatrix} 1 \\ 1 \end{bmatrix} + \begin{bmatrix} 0.3529 \\ 0.6471 \end{bmatrix} = \begin{bmatrix} 1.3529 \\ 1.6471 \end{bmatrix}$$

Iteration 2

$$f(x) = 5.007; \; \nabla f(x) = \begin{bmatrix} 1.7056 \\ -1.9473 \end{bmatrix}; \; \nabla h = \begin{bmatrix} 1.64712 \\ 1.3529 \end{bmatrix}; \; \nabla g = \begin{bmatrix} -1 \\ -1 \end{bmatrix};$$

$$\nabla^2 L = \begin{bmatrix} 8.3751 & 2.7195 \\ 2.7195 & 6.6889 \end{bmatrix}$$

The quadratic problem is
Minimize

$$Q = \Delta x^T \begin{bmatrix} 1.7056 \\ -1.9473 \end{bmatrix} + \frac{1}{2} \Delta x^T \begin{bmatrix} 8.3751 & 2.7195 \\ 2.7195 & 6.6889 \end{bmatrix} \Delta x$$

subject to

$$0.2284 + [1.64712 \quad 1.3529] \Delta x = 0$$

$$-2 + [-1 \quad -1] \Delta x = 0$$

The solution of the quadratic problem is

$$\Delta x = \begin{bmatrix} -0.4279 \\ 0.3521 \end{bmatrix}$$

Appendix C 429

Now x is updated as

$$x = x + \Delta x = \begin{bmatrix} 1.3529 \\ 1.6471 \end{bmatrix} + \begin{bmatrix} -0.4279 \\ 0.3521 \end{bmatrix} = \begin{bmatrix} 0.9251 \\ 1.9991 \end{bmatrix}$$

In a similar manner, other iterations can be written. The values at the termination of the algorithm $\left(x = \begin{bmatrix} 1.0371 \\ 1.9284 \end{bmatrix} \right)$ are

$$f(x) = 4.4819; \quad \nabla f(x) = \begin{bmatrix} -0.8643 \\ -0.4648 \end{bmatrix}; \quad \nabla h = \begin{bmatrix} 1.9284 \\ 1.0371 \end{bmatrix}; \quad \nabla g = \begin{bmatrix} -1 \\ -1 \end{bmatrix}$$

$$\nabla^2 L = \begin{bmatrix} 17.4375 & 1.0098 \\ 1.0098 & 3.5976 \end{bmatrix}$$

6. Identical results are obtained.
7. The number of iterations will vary with different start values of the design variables.
8. Identical results are obtained.
9. Copy the SQP folder (of some other problem) to the working directory and make changes in function and constraint subroutines as follows.

```
function y = func_val(x)
y = 0.0064*x(1)*(exp(-0.184*x(1)^0.3*x(2))-1);

function y = func_val1(x)
y = 0.0064*x(1)*(exp(-0.184*x(1)^0.3*x(2))-1);

function h = eqconstr_val(x)
h(1) = 0;

function g = ineqconstr_val(x)
g(1) = ((3000+x(1))*x(1)^2*x(2))/1.2e13 -1;
g(2) = ( exp(0.184*x(1)^0.3*x(2)) )/4.1 -1;
```

```
%%%%%%%%%%%%%%%%%%%%%%%%%%%%%%%%%%%%%%%%%
% MATLAB code sqp.m
%%%%%%%%%%%%%%%%%%%%%%%%%%%%%%%%%%%%%%%%%
n_of_var = 2;
n_of_eqcons = 1;
n_of_iqcons = 2;
X = [30000 0.5];
```

Execute the SQP code with these modifications and the converged solution is obtained in five iterations.

No.	x-vector		f(x)	|Cons.|
1	30402.6828	0.384516	-153.906889	0.16689252491
2	31592.6073	0.344868	-153.329290	0.00924809461
3	31764.8743	0.342079	-153.711944	0.00001910009
4	31765.5812	0.342072	-153.714422	0.00000000010
5	31765.5812	0.342072	-153.714422	0.00000000000

10. $x^* = (0.05179, 0.3591, 11.1527)$ with $f(x^*) = 0.01267$

Chapter 7

1. The Pareto front is given in Figure C.12.
2. The Pareto front is given in Figure C.13.

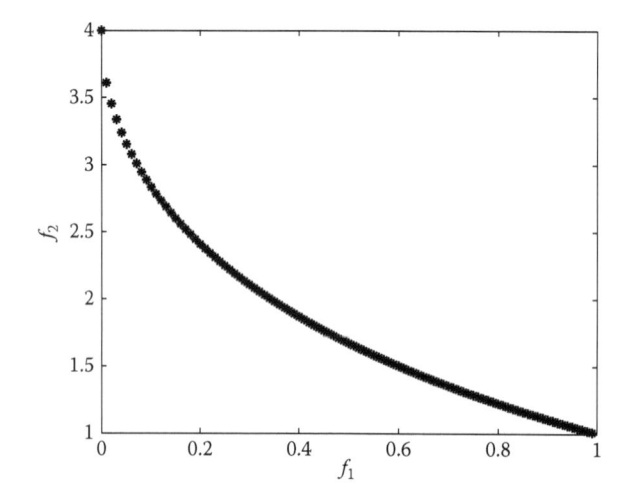

FIGURE C.12
Pareto front.

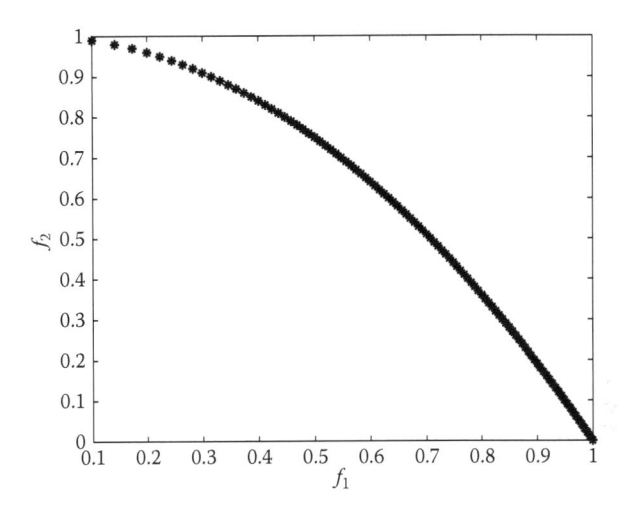

FIGURE C.13
Pareto front.

3. The Pareto front is given in Figure C.14.
4. The Pareto front is given in Figure C.15.
5. The Pareto front is given in Figure C.16.
6. The Pareto front is given in Figure C.17.

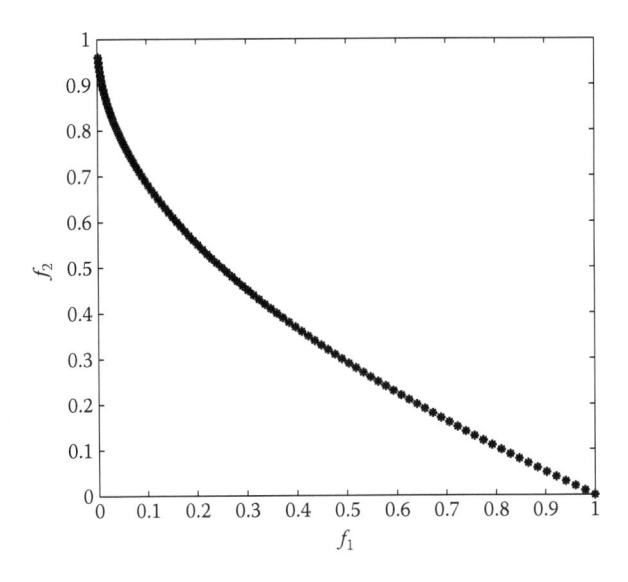

FIGURE C.14
Pareto front.

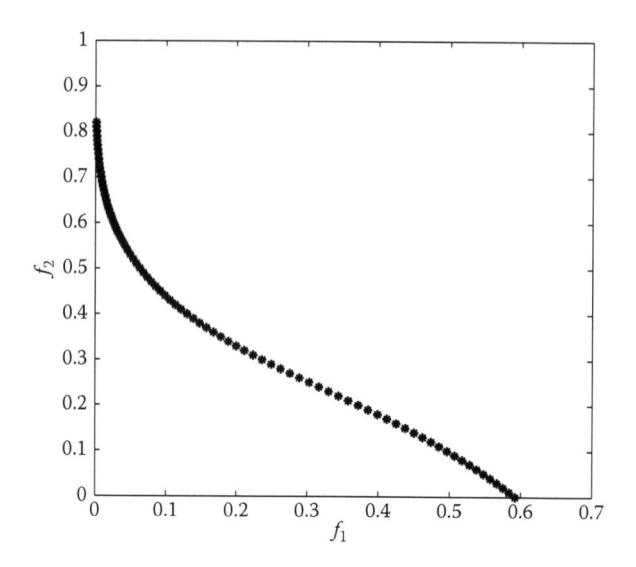

FIGURE C.15
Pareto front.

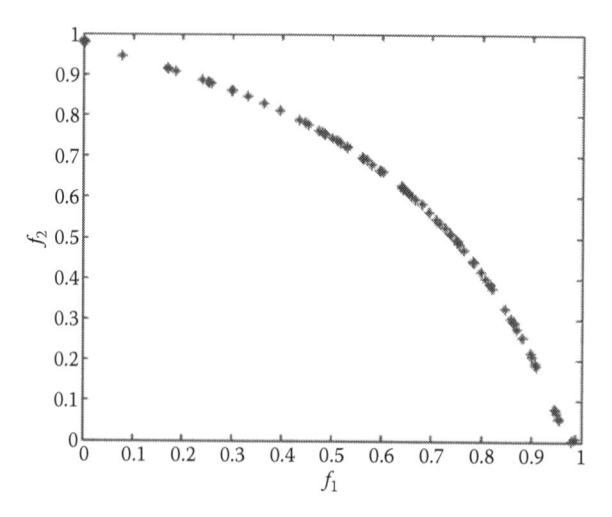

FIGURE C.16
Pareto front.

Appendix C 433

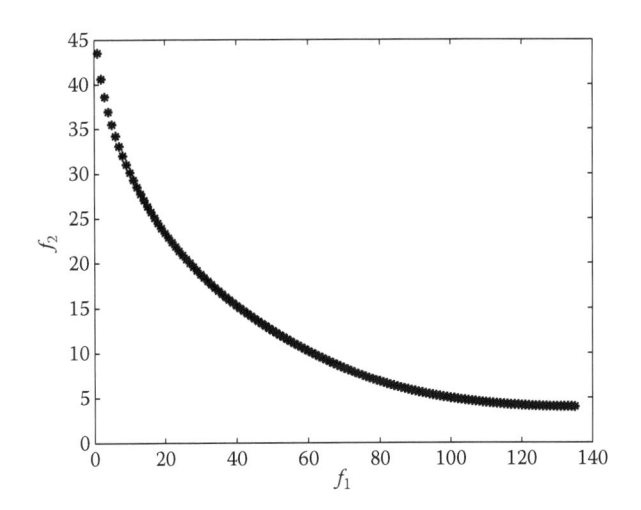

FIGURE C.17
Pareto front.

Chapter 8

1. $x^* = (1.305470, 1.390561, 0.4892672)$ with $f(x^*) = 125.9045$
2. $x^* = (0.3205667, 1.481980, 1.064722, 1.719745)$ with $f(x^*) = 47.47193$
3. $x^* = (0.5, 0.5)$ with $f(x^*) = 0.5$
4. $D^* = 0.922$ cm, $Q^* = 0.281$ m^3/s
5. $\Delta p^* = 400{,}000$ Pa, $Q^* = 7.5 \times 10^{-4}$ m^3/s, $C^* = \$477.19$
6. $\Delta t^* = 2.28°C$, $Q^* = 764.72$ m^3/m^2, $C^* = 1.163$ (\$/m^2)
7. $\omega^* = 469$ rad/s, $T^* = 262$ Nm

Chapter 9

1. Full factorial design

```
0.5     5       0.01
0.5     7.5     0.01
0.5     10      0.01
1.25    5       0.01
1.25    7.5     0.01
1.25    10      0.01
2       5       0.01
```

2	7.5	0.01
2	10	0.01
0.5	5	0.055
0.5	7.5	0.055
0.5	10	0.055
1.25	5	0.055
1.25	7.5	0.055
1.25	10	0.055
2	5	0.055
2	7.5	0.055
2	10	0.055
0.5	5	0.1
0.5	7.5	0.1
0.5	10	0.1
1.25	5	0.1
1.25	7.5	0.1
1.25	10	0.1
2	5	0.1
2	7.5	0.1
2	10	0.1

2. Central composite design

0.5	5
0.5	10
2	5
2	10
0.293	7.5
2.828	7.5
1.25	7.07
1.25	14.14
1.25	7.5

3. $a_0 = 0.16$, $a_1 = 0.572$

4. $y = 49.2682 + 0.02x_1 + 0.2745x_2 + 0.3084x_3 + 14.3068x_4$

5. $\left(z_1^*, z_2^*, x_1^*, x_2^*\right) = (0.7, 17, 7.3, 7.71532)$. The value of objective function is 2994.355.

Chapter 10

1. $x^* = (7, 1)$ with $f(x^*) = -23$
2. $x^* = (1, 0)$ with $f(x^*) = -1$
3. Let x_1 and x_2 be the number of chairs and tables to be produced. The integer programming problem is

Appendix C

Maximize

$$f(x) = 100x_1 + 160x_2$$

subject to

$$6x_1 + 14x_2 \leq 42$$

$$7x_1 + 7x_2 \leq 35$$

$$x_1, x_2 \geq 0$$

where x_1 and x_2 are integers.
The optimal solution is $x^* = (5, 0)$ with $f(x^*) = 500$.
4. $x^* = (1, 0, 1, 1, 0, 1, 0, 0, 1)$ with $f(x^*) = 72$
5. $x^* = (2, 2)$ with $f(x^*) = -16$
6. $x^* = (1, 5)$ with $f(x^*) = -39$
7. $x^* = (0, 5)$ with $f(x^*) = 10$
8. $x^* = (0, 1)$ with $f(x^*) = 2$

Chapter 11

1.
   i. $x^* = (0, 5)$ with $f(x^*) = -10$
   ii. $x^* = \left(\dfrac{1}{5}, \dfrac{52}{5}\right)$ with $f(x^*) = \dfrac{262}{5}$
   iii.
$$x^* = \left(\frac{15}{7}, \frac{110}{7}\right) \text{ with } f(x^*) = \frac{610}{7}$$

2. $x^* = (8, 0)$ with $f(x^*) = -24$
3. The optimal path is ADFGI and the minimum distance is 19.
4. Two numbers of component 1, two numbers of component 2, and one number of component 3, with the probability of the system = 0.9736.

Index

Page numbers followed by f and t indicate figures and tables, respectively.

A

Aerodynamic response surface models, 244
Aerospace applications
    weight minimization for, 3
Affine scaling method, for LPP, 125–126, 127f
All-integer programming problem, 263; *see also* Integer programming problem
Angle of attack (α), 244, 246, 255, 255f, 256
Annealing, 140, 154; *see also* Simulated annealing (SA)
Ant colony optimization (ACO) technique, 2, 160–163
    applications, 160
    background, 160
    formula, 165
Array operators, MATLAB®, 317
Arrays, MATLAB®, 309–312
Aspiration criteria, 163
Augmented Lagrange multipliers (ALM) method
    for constrained optimization problem, 175–176, 182–184
    formula, 198
    MATLAB® code, 183–184, 372–374

B

Backward difference formula, 17–18, 28
Backward difference method, 17
Balas' method, 264, 272–274, 286–287
Bank angle, 246
Barrier function methods, for LPP, 125

Basic feasible solution
    for LPP, 103
Basic solution, for LPP, 103–105
    feasible, 103
    optimal, 103
Bellman, Richard, 2, 289
BFGS method, *see* Broyden–Fletcher–Goldfarb–Shanno method
Bilevel integrated system synthesis (BLISS) architecture, of MDO, 252–253, 254f
Bisection method
    algorithm for, 39t
    comparison with other methods, 49–51, 50f, 51t
    for 1-D optimization problem, 38–40, 39f–40f
    MATLAB® code, 39, 328–329
Boyle's law, 5
Branch-and-bound method
    MATLAB® code, 279–281, 282–283, 394–398
    for nonlinear integer programming problems, 263–264, 278–284, 280f, 282f, 284f
Broyden–Fletcher–Goldfarb–Shanno (BFGS) method
    algorithm for, 73t
    MATLAB® code, 72–73, 345–347
    performance comparison with other methods
        for nonlinear function, 82t
        for quadratic function, 81t
        for Rosenbrock's Function, 79t
        for Wood's function, 83t
    search direction in (formula), 87
    for unconstrained optimization problems, 55, 72–73, 73t

437

C

Calculus
  history, 1
Calculus of variations, 1
  history, 1
Cantilever rod (example), 10–11, 11f
Cauchy, Augustin-Louis, 1
Central difference formula, 18, 28
  for second derivative, 28
Central difference method, 17
CFD analysis, *see* Computational fluid dynamics (CFD) analysis
Collaborative optimization (CO) architecture, of MDO, 251–252, 251f
  advantage, 252
  disadvantage, 252
Command window, MATLAB®, 309, 311f
Computational fluid dynamics (CFD) analysis, 190–191, 244, 254, 255
Computers
  development of, 2
Concave function, 15, 16f
Concurrent subspace optimization (CSSO) architecture, of MDO, 252, 253f
  formula, 259
Conjugate directions, 68, 68f
Conjugate gradient method
  algorithm for, 69t
  MATLAB® code, 69–70, 342–343
  performance comparison with other methods
    for nonlinear function, 82t
    for quadratic function, 81t
    for Rosenbrock's Function, 79t
    for Wood's function, 83t
  search direction in (formula), 87
  for unconstrained optimization problems, 68–70, 68f, 69t, 70f
  vs. steepest descent method, 69–70, 70f
Constrained optimization, 2
Constrained optimization problem, 169–196
  application to structural design, 195–196, 195f
  geometric programming, 231–235

optimality conditions, 171–174
  example, 173–174, 174f
  Karush–Kuhn–Tucker (KKT) conditions, 172–173
  Lagrange function, 171, 172
  Lagrange multipliers, 172, 173
  regular point, 172
overview, 169–171, 170f, 171f
solution techniques, 175–176
  augmented Lagrange multipliers (ALM) method, 175–176, 182–184
  feasible directions, method of, 176, 190–195
  penalty function method, 175, 176–182, 177f, 179f
  Rosen's gradient projection method, 176, 192–195, 193f
  sequential quadratic programming, 176, 184–190
  variable substitution method, 175
  Zoutendijk's method, 176, 191–192
vs. unconstrained problem, 169
Constraints, 4–5
  equality, 4–5, 170
  inequality, 4, 5, 170
Continuous data, 5
Contour plot, MATLAB®, 318f
Contraction operation, simplex, 75–76
Convergence method
  linear, 62
  quadratic, 62
  superlinear, 62
Convex function, 13–14, 14f, 16f
  examples, 14, 15f
Convexity, 13–16, 14f–16f
  MATLAB® code, 13, 322–323
Convex set, 13–14, 14f
Crossover operation, in GA, 147–148, 148t
CSSO (concurrent subspace optimization) architecture, of MDO, 252, 253f
  formula, 259
Cubic polynomial fit
  algorithm for, 45t
  comparison with other methods, 49–51, 50f, 51t
  for 1-D optimization problem, 44–45
  MATLAB® code, 45, 332–333

Index

Curse of dimensionality, in dynamic programming, 289
Cylindrical can manufacturing (example), 8–9, 8f

D

Dantzig, George, 2
Darwin' survival of the fittest principle, 140
Davidon–Fletcher–Powell (DFP) method
   algorithm for, 71t
   MATLAB® code, 71–72, 177, 344–345
   performance comparison with other methods
      for nonlinear function, 82t
      for quadratic function, 81t
      for Rosenbrock's function, 79t
      for Wood's function, 83t
   search direction in (formula), 87
   for unconstrained optimization problems, 55, 70–72
Decision variables, 3–4
de Fermat, Pierre, 1
Degeneracy
   simplex method for LPP, 114–116
Degree of difficulty, in geometric programming, 225–226
Demand–supply problem, 5–6
Dependent variable, 6
Derivative(s)
   concept of, 16–17, 17f
   directional, 16–22, 17f, 18f
   of function, 18–19, 18f
   inflection point (saddle point), 19, 19f
   MATLAB® code, 18, 323–324
Design of experiments (DoE), 256
Design variables, 3–4, 5
   for optimization problem, 3–4, 4t
Deterministic dynamic programming, 289–294
   concept of, 290–291, 290f
   example, 293–294
   stage 1, 291, 292t
   stage 2, 291, 292t
   stage 3, 291, 292f, 292t
   stage 4, 291, 291f, 291t
   structure of, 290f
Dichotomous search method
   for 1-D optimization problem, 38, 47–48, 48f
Diet problem, 1
   example, 6–8, 7t
Differential equation, 6
   solution for, 6
Directional derivative, 16–22, 17f, 18f
Direct search methods, 35, 38
   1-D optimization problem
      dichotomous search, 47–48, 48f
      Fibonacci method, 47, 49
      golden section method, 46–47, 47t
      interval halving method, 47, 48, 49f
   for unconstrained optimization problems
      Nelder–Mead algorithm, 55, 75–78, 75f, 76f, 77t
      Powell method, 55, 74, 74t
Discrete data, 5
Discrete programming problems, 263; *see also* Integer programming problem
Domination, principle of, 204
Dual problem
   geometric programming for, 229–231, 239
Dual simplex method, for LPP, 121–124
   algorithm for, 123t
   MATLAB® code, 121, 122, 356–358
   primal to dual conversion, transformation rules, 121–122, 122t
Dynamic programming, 289–296
   curse of dimensionality in, 289
   deterministic, 289–294
   limitations, 289
   for LPP (example), 293–294, 293f
   overview, 289
   principle of optimality in, 289
   probabilistic, 294–296, 295t–296t
   stages, 289
Dynamic programming problems
   history, 2

E

ε-constraints method
   concept of, 211, 211f

for multiobjective optimization
    problem, 210–212, 211f–212f
  nonconvex Pareto front, 211–212,
    212f
Elementary functions, in MATLAB®,
    313–314
End-effector, 83
Equality constraints, 4–5, 170
Euclid, 1
Euler, Leonhard, 1
Evolutionary methods, 139
  genetic algorithms, 140–142; see also
    Genetic algorithms (GAs)
    crossover and mutation, 147–148,
      148t
    fitness evaluation, 143, 144t
    initialize population, 142–143
    multimodal test functions,
      148–153, 149f, 151f, 152f
    reproduction, 143–147, 145f, 145t
    working principle, 141–142, 141f
  for nonlinear integer programming
    problems, 284–285, 285f, 285t
  PSO method, 284–285
  particle swarm optimization,
    157–158, 159f, 159t
Expansion operation, simplex, 76, 76f
Expressions, MATLAB®, 312–314
Exterior penalty function method,
    176–177, 177f

F

Feasible directions, method of, 176,
    190–195
  Rosen's gradient projection method,
    176, 192–195, 193f
  Zoutendijk's method, 176, 191–192
Feasible point, 5
Feasible solutions
  for LPP, 103
Fibonacci method
  for 1-D optimization problem, 38,
    47, 49
Finite element analysis, 254, 255
Fitness evaluation, in GA, 143, 144t
  modified, 145t
Fletcher–Reeves conjugate gradient
    method

algorithm for, 69t
MATLAB® code, 69–70, 343–344
performance comparison with other
  methods
  for nonlinear function, 82t
  for quadratic function, 81t
  for Rosenbrock's Function, 79t
  for Wood's function, 83t
  for unconstrained optimization
    problems, 68–70, 69t, 70f
  vs. steepest descent method, 69–70, 70f
Forward difference formula, 17, 28
Forward difference method, 17
Free (unrestricted) variable, 100
Full factorial design, 256, 256f
Function(s), 3; see also specific types
  derivative of, 18–19, 18f
  linear approximation, 23–25, 23f
  objective, 3
  quadratic approximation, 23–25, 23f,
    24f

G

Gauss, Carl Friedrich, 1
General solution, 6
Genetic algorithms (GAs), 2, 140–142, 245
  crossover and mutation, 147–148, 148t
  fitness evaluation, 143, 144t
    modified, 145t
  initialize population, 142–143
  MATLAB® code, 142, 148
  multimodal test functions, 148–153,
    149f, 151f, 152f
    Rastrigin's function, 149–151, 149f,
      151f
    Schwefel's function, 149, 151–153,
      152f
  reproduction, 143–147
    pie chart, 145–146, 146f
    Roulette wheel selection, 145–146,
      146f
    selection pressure, 145
    tournament selection, 145, 146–147,
      146t
  schema theorem, 147–148
  selection pressure, 145
  vs. gradient-based methods, 148–153
  working principle, 141–142, 141f

Index

Geometric programming, 223–238
  application (two-bar truss), 223, 235–238, 236f
  constrained optimization, 231–235
  degree of difficulty in, 225–226
  dual problem, 229–231
  objective function (posynomial form), 223–224
  overview, 223–224, 224f
  unconstrained problem, 224–229
Global optimum solutions, for nonconvex function, 13, 14f
Global variables
  MDO, 246
Goal programming method
  advantages, 213
  lexicographic, 214
  for multiobjective optimization problem, 212–214
    formula, 221
    Pareto front, 214
Golden section method
  advantages, 46
  algorithm for, 47t
  comparison with other methods, 49–51, 50f, 51t
  for 1-D optimization problem, 46–47
  MATLAB® code, 46, 58, 333–334
Gomory, Ralph, 2
Gomory constraint, 267, 268, 271
Gomory's cutting plane method
  for linear integer programming problems, 263, 265–272
  MATLAB® code, 266, 267–272, 390–394
Gradient-based algorithms, 14
Gradient-based 1-D optimization algorithms, 35, 38
  bisection method, 38–40, 39f–40f, 39t
  cubic polynomial fit, 44–45, 45t
  Newton–Raphson method, 40–42, 41f, 42t
  secant method, 42–43, 43f, 44t
Gradient-based search methods, 139
  for unconstrained optimization problems, 55, 60–62
    BFGS method, 55, 72–73, 73t
    DFP method, 55, 70–72, 71t

Fletcher–Reeves conjugate gradient method, 68–70, 68f, 69t, 70f
Levenberg–Marquardt method, 55, 66–67, 67t
modified Newton's method, 66, 66t
Newton's method, 55, 63–65, 65t
steepest descent method, 62–63, 63t
vs. GA, 149–153
Gradient(s)
  of function, 1, 16–22
  MATLAB® code, 20, 95, 324–325
Gradient vector, 16–22
  for objective function, 20–21, 20f
Graphical method, 11–13, 12f, 13f
  LPP solution with, 95–98
    feasible region, 95, 96f
    infeasible solution, 98
    infinite solutions, 96–97, 97f
    unbounded solution, 97–98, 98f
  MATLAB® code, 12, 95, 321–322
Guided random search methods, 139–164
  ant colony optimization, 160–163
  evolutionary methods, 139
  genetic algorithms, 140–142; *see also* Genetic algorithms (GAs)
    crossover and mutation, 147–148, 148t
    fitness evaluation, 143, 144t
    initialize population, 142–143
    multimodal test functions, 148–153, 149f, 151f, 152f
    reproduction, 143–147, 145f, 145t
    working principle, 141–142, 141f
  overview, 139–140, 140f
  particle swarm optimization, 157–158, 159f, 159t
  simulated annealing, 154–156, 155t, 156f–157f
  tabu search, 163–164, 163t

H

Hancock, Harris, 2
Hessian matrix (H), 16–22, 55, 63–64, 68
  example, 22
  inverse of, 70–71
  MATLAB® code, 21, 64, 340–341

positive definite, 21
for three-variable function, 29
Historical review, 1–2

I

Independent variable, 6
Individual discipline feasible (IDF)
  architecture, of MDO, 248, 248f
  advantage, 248
  formula, 248–249
Inequality constraints, 4, 5, 170
Infeasible solution
  for LPP, 98
Infinite solutions
  for LPP, 96–97, 97f
Inflection point, 19, 19f
Initialize population, in GA, 142–143
Integer programming problem, 263–285
  Balas algorithm, 264, 272–274
  development of, 2
  linear, 264–265, 265f
    Gomory's cutting plane method, 265–272
    zero-one problems, 272–277, 277f
  nonlinear, 277–278
    branch-and-bound method, 278–284, 280f, 282f, 284f
    evolutionary method, 284–285, 285f, 285t
  overview, 263–264, 264f
Interior penalty function method, 178–179, 179f
Interior-point methods, for LPP, 125–126, 125f, 126t
  affine scaling methods, 125–126, 127f
  algorithm for, 126t
  barrier function methods, 125
  MATLAB® code, 125, 358
  potential-reduction methods, 125
Interval halving method
  for 1-D optimization problem, 38, 47, 48, 49f

J

Jacobian (J) function, 20–21
  with three variables, 28
Job scheduling problem, 163

K

Kantorovich, Leonid, 2
Karmarkar, Narenndra, 125
Karush, William, 2
Karush–Kuhn–Tucker (KKT) conditions, 172–173
Kuhn, Harold, 2

L

Lagrange, Joseph-Louis, 1
Lagrange function
  for constrained optimization problem, 171, 172
  formula, 197
Lagrange multipliers, 172, 173, 182–183; *see also* Augmented Lagrange multipliers (ALM) method
Least squares method
  history, 1
Legendre, Adrien-Marie, 1
Leibniz, Gottfried Wilhelm, 1
Levenberg–Marquardt method
  algorithm for, 67t
  MATLAB® code, 67, 342–343
  performance comparison with other methods
    for nonlinear function, 82t
    for quadratic function, 81t
    for Rosenbrock's Function, 79t
    for Wood's function, 83t
  search direction in (formula), 87
  for unconstrained optimization problems, 55, 66–67
Lexicographic goal programming method, 214
Linear approximation, 23–25, 23f, 24f
  example, 24–25
Linear convergence method, 62
Linear function
  properties, 93
Linear integer programming problems, 264–265, 265f
  formula, 286–287
  Gomory's cutting plane method, 265–272
  zero-one problems, 272–277, 277f

Index

Linear programming (LP) model
  history, 2
Linear programming problem (LPP), 5, 93–131
  applications, 93
  basic feasible solution, 103
  basic solution, 103–105
  defined, 93
  dynamic programming for (example), 293–294, 293f
  feasible region for, 95, 96f
  feasible solution, 103
  graphical method, 95–98, 96f–98f
  infeasible solution for, 98
  infinite solutions for, 96–97, 97f
  interior-point method, 125–126, 125f, 126t, 127f
  optimal basic solution, 103
  overview, 93–94, 94f
  portfolio optimization, 127–131, 127t
  primal to dual conversion, transformation rules, 121–122, 122t
  simplex method, 105–120
    algorithm for, 109t
    degeneracy, 114–116
    dual, 121–124, 122t, 123t
    feasible region, 111–112, 111f
    multiple solutions, 112–114, 114f
    two-phase method, 116–120
  in standard form, 98–103
    formula, 133
  unbounded solution for, 97–98, 98f
Local minimum functions
  saddle point and, 19, 19f
Local optimum solutions, for nonconvex function, 13, 14f
Local variables
  MDO, 246
LPP, *see* Linear programming problem (LPP)

M

Machine allocation problem, 1
Mach number (M), 244, 255, 255f, 256
Mathematical models, 5
The MathWorks Inc., 309
MATLAB®, 12, 309–320

advantage, 309
array operators, 317
arrays, 309–312
command window, 309, 311f
elementary functions in, 313–314
expressions, 312–314
matrices, 309–312
matrix operations, 315–317
on Microsoft Windows, 309, 311f
operators, 312–313
overview, 309
plotting, 318, 318f–319f
programming, 319–320
MATLAB® code, 12, 18, 20, 321–400
  ALM method (*ALM.m*), 183–184, 372–374
  BFGS method (*BFGS.m*), 72–73, 345–347
  bisection method (*bisection.m*), 39, 328–329
  branch-and-bound method, 279–281, 282–283, 394–398
  convexity (*convexity.m*), 13, 322–323
  cubic polynomial fit (*cubic.m*), 45, 332–333
  derivative (*derivative.m*), 18, 323–324
  DFP method (*DFP.m*), 71–72, 177, 344–345
  dual simplex method, 121, 122, 356–358
  exhaustive.m, 37, 328
  Fletcher–Reeves conjugate gradient method (*conjugate.m*), 69–70, 343–344
  GA, 142, 148, 359–365
  golden section method (*golden.m*), 46, 58, 333–334, 335–336
  Gomory's cutting plane method, 266, 267–272, 390–394
  gradient (*grad.m*), 20, 324–325
  graphical method (*graph_exampl2.m*), 12, 95, 321–322
  Hessian matrix (*hessian.m*), 64, 340–341
  interior-point method (*interior.m*), 125, 358
  Levenberg–Marquardt method (*levenbergmarquardt.m*), 67, 342–343
  modified Newton's method (*modified_newton.m*), 66, 341–342

MuPad, 95, 96
Nelder–Mead algorithm
  (*neldermead.m*), 76–78, 348–350
Newton–Raphson method
  (*newtonraphson.m*), 41, 330
Newton's method (*newton.m*), 64–65, 339–340
positive definite matrix (*positive_definite.m*), 21, 325
Powell method (*powell.m*), 74, 347–348
PSO method (*pso.m*), 196, 209, 366–368, 369–371, 382–383, 398–400
quadratic approximation (*quadr.m*), 23, 326–327
Rastrigin's function, 149–150
robotics_nominal_traj.m, 84–85, 350–351
Rosenbrock function (*rosenbrock.m*), 59, 336–337
secant method (*secant.m*), 43, 330–331
simplex method for LPP (*simplex.m*), 109, 112–113, 118–120, 122, 124, 352–356
simulated annealing (*simann.m*), 155, 365–366
spring system (*springsystem.m*), 60, 337
SQP method (*sqp.m*), 187, 207, 250–251, 374–376, 378–380, 383–384, 386–389
steepest descent method (*steep_des.m*), 62, 63, 337–338
Matrices, MATLAB®, 309–312
MATrix LABoratory, *see* MATLAB®
Matrix operations, MATLAB®, 315–317
Microsoft Windows, MATLAB® on, 309, 311f
Mixed-integer programming problem, 263; *see also* Integer programming problem
Modeling, of optimization problem, 5–11
  cantilever rod (example), 10–11, 11f
  cylindrical can manufacturing (example), 8–9, 8f
  diet problem (example), 6–8, 7t
  reentry capsule (example), 9–10, 9f
Modified Newton's method
  algorithm for, 66t
  MATLAB® code, 66, 341–342
  performance comparison with other methods
    for nonlinear function, 82t
    for quadratic function, 81t
    for Rosenbrock's Function, 79t
    for Wood's function, 83t
    for unconstrained optimization problems, 66
Monotonic function, 35, 36f
Multidisciplinary design analysis (MDA), MDO, 245–246, 246f
  formula, 258
Multidisciplinary design feasible (MDF) architecture, of MDO, 247, 247f
  advantage, 247
  disadvantage, 247
  formula, 258
Multidisciplinary design optimization (MDO), 243–257
  advantages, 243–244
  for aerospace problems, 244
  architecture, 245–246
    BLISS architecture, 252–253, 254f
    CO architecture, 251–252, 251f
    CSSO architecture, 252, 253f
    example, 249–251
    IDF architecture, 248, 248f
    MDF analysis, 247, 247f
    multidisciplinary design analysis (MDA), 245–246, 246f
    SAND architecture, 249, 249f
  framework, 253–254
  global variables, 246
  local variables, 246
  overview, 243–245, 245f
  response surface methodology, 244, 254–257, 255f–257f, 256t
  single *vs.* two disciplines, 243, 244f
Multimodal functions, 14
Multimodal test functions, GA, 148–153
  Rastrigin's function, 149–151, 149f, 151f
  Schwefel's function, 149, 151–153, 152f
Multiobjective optimization problem, 203–219
  application (reentry bodies), 215–219, 215f, 217f, 219f
  ε-constraints method, 210–212, 211f–212f

Index

formula, 220
goal programming, 212–214
nondominated solutions, 204
objective functions, 204
overview, 203–205, 205f
Pareto optimal front, 204–205, 204f
principle of domination and, 204
utility function method, 214–215
weighted sum approach, 205–210, 207f–210f
Multiple plots, MATLAB®, 318f
Multiple solutions
  for LPP, 112–114, 114f
Multivariable function
  unidirectional search for, 58t
MuPad, 95, 96
Mutation operation, in GA, 147–148, 148t

N

Natural selection, 140
Nelder–Mead algorithm
  MATLAB® code, 76–78, 348–350
  for unconstrained optimization problems, 55, 75–78, 77t
Newton, Isaac, 1, 40
Newton–Raphson method
  algorithm for, 42t
  comparison with other methods, 49–51, 50f, 51t
  disadvantages, 42
  for 1-D optimization problem, 40–42, 41f, 45
  formula, 52
  MATLAB® code, 41, 330
Newton's law of cooling, 6
Newton's method, 23, 68
  algorithm for, 65t
  MATLAB® code, 64–65, 339–340
  modified, 66, 66t
  performance comparison with other methods
    for nonlinear function, 82t
    for quadratic function, 81t
    for Rosenbrock's function, 79t
    for Wood's function, 83t
  search direction in (formula), 86
  for unconstrained optimization problems, 55, 63–65

Nonconvex function, 14, 15f
  local and global optima for, 14, 15f
Nonconvex set, 13, 14f
Nondominated solutions
  for multiobjective optimization problem, 204; *see also* Multiobjective optimization problem
  ε-constraints method, 210–212, 211f–212f
  goal programming, 212–214
  utility function method, 214–215
  weighted sum approach, 205–210, 207f–210f
Non–gradient-based 1-D optimization algorithms, 35, 38
Non–gradient-based search methods; *see also* Direct search methods
  for unconstrained optimization problems, 55, 60
Nonlinear function
  contours of, 81f
  performance comparison of different methods for, 82t
  unconstrained optimization problems, 81–82
Nonlinear integer programming
  problems, 277–278
  branch-and-bound method, 278–284, 280f, 282f, 284f
  evolutionary method, 284–285, 285f, 285t

O

Objective function, 3
  geometric programming (posynomial form), 223–224
  multiobjective optimization problem
    cost minimization, 204
    efficiency maximization, 204
  for optimization problems, 3–4, 4t
  quadratic approximation of, 23, 24f
  tangent and gradient for, 20–21, 20f
  variables in, 3–4, 4t
Observations, defined, 5
One-dimensional (1-D) optimization algorithms, 35–51

gradient-based, 35, 38
monotonic function, 35, 36f
non–gradient-based, 35, 38
overview, 6f, 35–36
solution techniques, 38
    bisection method, 38–40, 39f–40f, 39t
    comparison of, 49–51, 50f, 51t
    cubic polynomial fit, 44–45, 45t
    dichotomous search method, 38, 47–48, 48f
    direct search methods, 35, 38
    Fibonacci method, 38, 47, 49
    golden section method, 46–47, 47t
    interval halving method, 38, 47, 48, 49f
    Newton–Raphson method, 40–42, 41f, 42t
    other methods, 47–49
    secant method, 42–43, 43f, 44t
    test problem (solar energy), 37, 37f
    unimodal function, 35, 36f
One-dimensional (1-D) optimization problem, 58
    defined, 35
    solution techniques, *see* One-dimensional (1-D) optimization algorithms
Operators, MATLAB®, 312–313
Optimal basic solution
    for LPP, 103
Optimality, principle of, 289
Optimality conditions, for constrained optimization problem, 171–174
    example, 173–174, 174f
    formula, 197
    Karush–Kuhn–Tucker (KKT) conditions, 172–173
    Lagrange function, 171, 172
    Lagrange multipliers, 172, 173
    regular point, 172
Optimization
    first textbook on, 2
    historical overview, 1–2
    meaning of, 1
    role of, 2
Optimization methods/techniques
    applications of, 2
    development of, 2
Optimization problem

constrained, *see* Constrained optimization problem
constraints, 4–5
convexity, 13–16, 14f–16f
1-D, 35; *see also* One-dimensional (1-D) optimization algorithms
described, 3–5
design variable for, 3–4, 4t
diet, 1
directional derivative, 16–22, 17f, 18f
function, 3
gradient vector, 16–22
graphical method, 11–13, 12f, 13f
Hessian matrix, 16–22
historical overview, 1–2
linear and quadratic approximations, 23–25, 23f, 24f
LPP, 5
machine allocation, 1
modeling of, 5–11
multiobjective, *see* Multiobjective optimization problem
objective function, 3–4, 4t
performance index, 3
present-day, 2
unconstrained, *see* Unconstrained optimization problem

P

Pareto optimal front
    multiobjective optimization problem, 204–205, 204f
    ε-constraints method, 211–212, 212f
    goal programming method, 214
    of reentry test body, 219, 219f
    weighted sum approach, 206–210, 207f–210f
Particle swarm optimization, 2
Particle swarm optimization (PSO) technique, 157–158, 205, 245
    algorithm for, 159t
    convergence, for Schwefel's function, 158, 159f
    formula, 165
    MATLAB® code, 196, 209, 366–368, 369–371, 382–383, 398–400
    nonconvex Pareto front generated with, 209, 209f

Index 447

for nonlinear integer programming problems, 284–285
Penalty function method, for constrained optimization problem, 175, 176–182
  advantages, 178
  disadvantages, 178
  exterior, 176–177, 177f
  formula, 198
  interior, 178–179, 179f
  welded beam (example), 179–182, 180f
Performance index, 3
Pheromone, 160, 161
Pie chart, reproduction in GA, 145–146, 146f
Plotting, MATLAB®, 318
  contour plot, 318f
  multiple plots, 318f
Poisson distribution, 147
Polynomial-time algorithm (Karmarkar), 125
Portfolio optimization problem, 93, 127–131, 127t
Positive definite Hessian matrix (H), 21
  MATLAB® code, 21, 325
Posynomials, in geometric programming techniques, 223–224, 225, 231, 232, 238
Potential-reduction methods, for LPP, 125
Powell method, 84
  algorithm for, 74t
  MATLAB® code, 74, 347–348
  performance comparison with other methods
    for nonlinear function, 82t
    for quadratic function, 81t
    for Rosenbrock's Function, 79t
    for Wood's function, 83t
    for unconstrained optimization problems, 55, 74, 74t
Primal problem, 229
Principle of optimality, in dynamic programming, 289
Probabilistic dynamic programming, 294–296, 295t–296t
  stage 1, 296t
  stage 2, 296t
  stage 3, 295t
Programming, MATLAB®, 319–320

Q

Quadratic approximation, 23–25, 23f, 29
  example, 24–25
  MATLAB® code, 23, 326–327
  of objective function, 23, 24f
Quadratic convergence method, 62
Quadratic function
  contours of, 80f
  performance comparison of different methods for, 81t
  unconstrained optimization problems, 79–81
Quadratic problem; see also Sequential quadratic programming (SQP) method
  formula, 198
Quasi-Newton method, 71; see also Davidon–Fletcher–Powell (DFP) method

R

Raphson, Joseph, 40
Rastrigin's function
  in GA, 149–151, 149f, 151f
  MATLAB® code, 149–150
  SA convergence for, 155–156, 156f
Ratio test, 108
Reentry bodies, multiobjective optimization problem application, 215–219
  design variables, 215–216, 215f
  MATLAB code, 219
  Pareto front of, 219, 219f
  response surface matrix, 217–218, 217t
Reentry capsule (example), 9–10, 9f
Reflection operation, simplex, 75–76, 75f
Reproduction, in GA, 143–147, 145f, 145t
  pie chart, 145–146, 146f
  Roulette wheel selection method, 145–146, 146f
  selection pressure, 145
  tournament selection method, 145, 146–147, 146t
Response surface methodology (RSM), 244, 252, 254–257
  central composite design, 257, 257t
  design matrix, 256t

full factorial design, 256, 256f
  of lift coefficient, 255, 255f
Response surface model, 217–218
Robotics
  MATLAB® code, 84–85, 350–351
  unconstrained optimization
    problem application to, 83–85,
    85f
Rosenbrock function, 59f
  contours of, 78, 79f
  MATLAB® code, 59, 336–337
  performance comparison of different
    methods for, 78, 79t
  steepest descent method on, 80f
  unconstrained optimization
    problems, 78, 79f, 79t
  unidirectional search on, 58–59, 58t
Rosen's gradient projection method, 176,
    192–195
  example, 193–195
  formula, 198
  with restoration move, 193f
Roulette wheel selection method, in GA,
    145–146, 146f

S

Saddle point, 19, 19f
  surface-contour plot of function
    with, 57, 58f
SAND (simultaneous analysis and
    design) architecture, of MDO,
    249, 249f
  formula, 259
Schema theorem, in GA, 147–148
  formula, 165
Schwefel's function
  in GA, 149, 151–153, 152f
  PSO convergence for, 158, 159f
  SA convergence for, 156f
Secant method
  algorithm for, 44t
  comparison with other methods,
    49–51, 50f, 51t
  for 1-D optimization problem, 42–43,
    43f
  formula, 52
  MATLAB® code, 43, 330–331
Selection pressure, in GA, 145

Sequential quadratic programming (SQP)
    method, 176, 184–190, 207, 245
  example
    cylindrical pressure vessel,
      188–189
    optimized production rate,
      189–190
    welded beam, 187–188
  MATLAB® code, 187, 207, 250–251,
    374–376, 378–380, 383–384,
    386–389
  trust region approach, 185
Simplex
  defined, 75
  operations to move
    contraction, 75–76
    expansion, 76, 76f
    reflection, 75–76, 75f
Simplex method, for LPP, 105–120
  algorithm for, 109t
  degeneracy, 114–116
  dual, 121–124, 122t, 123t
  feasible region, 111–112, 111f
  MATLAB® code, 109, 112–113,
    118–120, 122, 124, 352–356
  multiple solutions, 112–114, 114f
  two-phase method, 116–120
Simulated annealing (SA), 154–156
  algorithm for, 155t
  convergence of
    for Rastrigin function, 155–156,
      156f
    for Schwefel's function, 156f
    for spring system test problem,
      157f
  MATLAB® code, 155, 365–366
Simultaneous analysis and design
    (SAND) architecture, of MDO,
    249, 249f
  formula, 259
Slack variable, 100
Solar energy problem, 37; *see also* One-
    dimensional (1-D) optimization
    algorithms
  cost function for, 37, 37f
  MATLAB® code, 37
  solution techniques, 38
    bisection method, 38–40, 39f–40f,
      39t

Index

comparison of, 49–51, 50f, 51t
cubic polynomial fit, 44–45, 45t
dichotomous search method, 38, 47–48, 48f
Fibonacci method, 38, 47, 49
golden section method, 46–47, 47t
interval halving method, 38, 47, 48, 49f
Newton–Raphson method, 40–42, 41f, 42t
other methods, 47–49
secant method, 42–43, 43f, 44t
Spring system, 59–60, 60f; *see also* Unconstrained optimization problem
additional test functions
nonlinear function, 81–82, 81f, 82t
quadratic function, 79–81, 80f, 81t
Rosenbrock function, 78, 79f, 79t
Wood's function, 82–83, 82f, 83t
MATLAB® code, 60, 337
SA convergence for test problem of, 157f
solution techniques, 60–62
BFGS method, 72–73, 73t
criteria for, 61–62
DFP method, 70–72, 71t
Fletcher–Reeves conjugate gradient method, 68–70, 68f, 69t, 70f
gradient-based search methods, 55, 60–62
Levenberg–Marquardt method, 66–67, 67t
modified Newton's method, 66, 66t
Nelder–Mead algorithm, 75–78, 75f, 76f, 77t
Newton's method, 63–65, 65t
non–gradient-based search methods, 55, 60
Powell method, 74, 74t
steepest descent method, 62–63, 63t
test problem, 59–60
Standard form, of LPP, 98–103
Steepest descent direction, 62
Steepest descent method, 68, 78
advantage of, 66
algorithm for, 63t
behavior on Rosenbrock function, 80f
history, 1
MATLAB® code, 62, 63, 337–338
performance comparison with other methods
for nonlinear function, 82t
for quadratic function, 81t
for Rosenbrock's Function, 79t
for Wood's function, 83t
search direction in (formula), 86
for unconstrained optimization problems, 62–63, 64f
vs. conjugate gradient method, 69–70, 70f
Structural design
constrained optimization problem application to, 195–196, 195f
Superlinear convergence method, 62
Surface-contour plot of function, 57, 57f
with saddle point, 57, 58f
Surplus variable, 100
Survival of the fittest principle (Darwin), 140

T

Tabu search, 163–164
algorithm for, 163t
Taylor series approximation, 23
Taylor's series, 17, 41
Tournament selection method, in GA, 145, 146–147, 146t
Traveling salesman problem, 163
Trust region approach, 185
Tucker, Albert, 2
Two-phase method
for LPP, 116–120

U

Unbounded solution
for LPP, 97–98, 98f
Unconstrained optimization problem, 1, 5, 55–85
additional test functions
nonlinear function, 81–82, 81f, 82t
quadratic function, 79–81, 80f, 81t

Rosenbrock function, 78, 79f, 79t
Wood's function, 82–83, 82f, 83t
application to robotics, 83–85, 85f
geometric programming, 224–229, 239
overview, 55–57, 56f
solution techniques, 60–62
   BFGS method, 72–73, 73t
   criteria for, 61–62
   DFP method, 70–72, 71t
   Fletcher–Reeves conjugate gradient method, 68–70, 68f, 69t, 70f
   gradient-based search methods, 55, 60–62
   Levenberg–Marquardt method, 66–67, 67t
   modified Newton's method, 66, 66t
   Nelder–Mead algorithm, 75–78, 75f, 76f, 77t
   Newton's method, 63–65, 65t
   non–gradient-based search methods, 55, 60
   Powell method, 74, 74t
   steepest descent method, 62–63, 63t
surface-contour plot of function, 57, 57f
   with saddle point, 57, 58f
test problem, 59–60
unidirectional search, 57–59
vs. constrained problem, 169
Unidirectional search, 57–59
   formula, 86
   for multivariable function, 58t
   on Rosenbrock function, 58–59, 58t, 59f
Unimodal function, 35, 36f
Unrestricted (free) variable, 100
Utility function method
   formula, 221
   for multiobjective optimization problem, 214–215

V

Variable metric method, 70–71; *see also* Davidon–Fletcher–Powell (DFP) method
Variable(s); *see also* specific types
   decision, 3–4
   dependent, 6
   design, 3–4
   independent, 6
   in objective function, 3–4

W

Weighted sum approach
   for multiobjective optimization problem, 205–210, 207f–210f
      advantages, 208
      disadvantages, 209
      example, 210
      formula, 220–221
      incomplete Pareto front, 208, 208f
      nonconvex Pareto front generated with PSO, 209, 209f
      Pareto optimal front, 206–207, 207f
Weight minimization
   for aerospace applications, 3
Wood's function
   contours of, 82f
   performance comparison of different methods for, 83t
   unconstrained optimization problems, 82–83

Z

Zenedorous, 1
Zero-one programming problem, 263, 272–277, 277f
Zoutendijk's method, 176, 191–192